AAPS Advances in the Pharmaceutical Sciences Series

Volume 17

The AAPS Advances in the Pharmaceutical Sciences Series, published in partnership with the American Association of Pharmaceutical Scientists, is designed to deliver volumes authored by opinion leaders and authorities from around the globe, addressing innovations in drug research and development, and best practice for scientists and industry professionals in the pharma and biotech industries. Indexed in

Reaxys

SCOPUS Chemical Abstracts Service (CAS) SCImago

EMBASE

Jennica L. Zaro • Jeffrey Wang
Editors

Antibody-Drug Conjugates

The 21st Century Magic Bullets for Cancer

Second Edition

Springer

aaps®

Editors
Jennica L. Zaro
Department of Pharmacology
and Pharmaceutical Sciences
Alfred E. Mann School of Pharmacology
and Pharmaceutical Sciences
University of Southern California
Los Angeles, CA, USA

Jeffrey Wang
Department of Biotechnology
and Pharmaceutical Sciences
College of Pharmacy
Western University of Health Sciences
Pomona, CA, USA

ISSN 2210-7371 ISSN 2210-738X (electronic)
AAPS Advances in the Pharmaceutical Sciences Series
ISBN 978-3-031-83004-4 ISBN 978-3-031-83005-1 (eBook)
https://doi.org/10.1007/978-3-031-83005-1

This Springer imprint is published by the registered company Springer Nature Switzerland AG
The registered company address is: Gewerbestrasse 11, 6330 Cham, Switzerland

Preface

The field of antibody-drug conjugates (ADCs) has undergone remarkable advancements in recent years, marked by significant progress in both drug approvals and ongoing clinical development. Since the approval of the first ADC in 2010 (gemtuzumab ozogamicin, Mylotarg®), the landscape has expanded dramatically. Today, there are 11 FDA-approved ADCs, targeting a variety of cancers across multiple indications. The approved ADCs include a range of payloads, linkers, and antibodies, each optimized for a variety of specific therapeutic targets. The increasing diversity of ADCs reflects the growing potential of these innovative treatments to address a wide array of malignancies, from hematologic cancers to solid tumors.

This book aims to provide a comprehensive overview of the current state of the ADC field including the latest developments, challenges, and emerging trends, comprising expertise from a broad range of disciplines from basic research, industry, clinical practice, and regulatory affairs. We explore not only the scientific and technical aspects of ADC design—such as payloads, linkers, and antibody selection—but also the developmental hurdles and regulatory complexities that influence the success of ADCs in clinical practice. Real-world examples of ADCs that have made it from the lab to the clinic offer invaluable insights into the trials and triumphs that shape this dynamic field. It is our hope that this book will serve as both a valuable resource for experts in the field and an accessible introduction for those new to the exciting world of ADCs.

Los Angeles, CA, USA
Pomona, CA, USA

Jennica L. Zaro
Jeffrey Wang

Contents

About the Editors

Jennica L. Zaro, PhD Dr. Zaro earned her PhD in Pharmaceutical Sciences from the University of Southern California (USC). Her research focuses on the targeted delivery of peptide and protein drugs, as well as the development of recombinant bifunctional fusion proteins as therapeutics. Dr. Zaro is currently an Associate Professor at the USC School of Pharmacy and Pharmaceutical Sciences. She has held several academic administrative roles, including Core Director of the Translational Research Laboratory and Assistant Dean of Assessment for Graduate Education at USC, and Assistant Dean of Assessment at West Coast University. In addition to her academic roles, she has significant experience in the pharmaceutical industry, where she contributed to the development of formulations for aerosol products and designed and validated analytical methods for protein drugs.

Jeffrey Wang, PhD Dr. Wang received a BS degree from the School of Pharmacy, Shanghai Medical University, and a PhD degree in pharmaceutical sciences from the University of Southern California. Dr. Wang's research focuses on the application of principles of pharmacokinetics and drug metabolism in drug discovery and delivery, particularly in anticancer drug discovery and peptide drug delivery. He has published over 80 peer-reviewed articles and book chapters, and is the co-inventor of several US and international patents. Dr. Wang is currently a Professor of Biotechnology and Pharmaceutical Sciences and Associate Dean for Research and Global Advancement at the College of Pharmacy, Western University of Health Sciences.

Part I
Introduction

Chapter 1
Antibody–Drug Conjugates: A Historical Review

Wei-Chiang Shen

Abstract This chapter reviews the development of antibody–drug conjugates (ADCs) since the initiation of the concept of "magic bullet" by Paul Ehrlich more than 100 years ago. The main focus of the chapter is on the early exploratory approaches that were taken to prove the feasibility of antibody-mediated drug delivery. The progress of ADC development correlated closely with the advances of the knowledge and technology in immunology, conjugation chemistry, molecular biology, and cell biology. Results from these diverse studies in different scientific disciplines during the past half century have slowly established ADCs as a new category of biotherapeutics in cancer therapy. Currently, there are more than 10 ADCs that have been approved by the FDA for the treatment of various types of cancer.

Keywords History · ADC Development

The origin of the concept of antibody–drug conjugate (ADC) has been attributed mostly to the "magic bullet" idea conceived by Paul Ehrlich more than 100 years ago (Strebhardt and Ullrich 2008). Indeed, Ehrlich was able to demonstrate the selective absorption of dyes with different chemical structures by various tissues and thus the possibility of achieving a targeted delivery of drugs to the disease site. More importantly, Ehrlich was one of the scientists responsible for the discovery of antibodies and was the first one to describe the unique "receptors" on the target cells that could be recognized by antibodies (Strebhardt and Ullrich 2008). Ehrlich's early work on small-molecular dyes to target disease tissues, in combination with his later work on the specific recognition of target cells by antibodies, laid the foundation in the field of ADC for targeted drug delivery to treat human diseases.

W.-C. Shen (✉)
University of Southern California, Alfred E. Mann School of Pharmacy and Pharmaceutical Sciences, Department of Pharmacology and Pharmaceutical Sciences, Los Angeles, CA, USA
e-mail: weishen@usc.edu

J. L. Zaro, J. Wang (eds.), *Antibody-Drug Conjugates*, AAPS Advances in the Pharmaceutical Sciences Series 17, https://doi.org/10.1007/978-3-031-83005-1_1

However, the development of ADC, especially in cancer treatment, has made very little progress in the first half century since Ehrlich introduced the "magic bullet" concept in the early 1900s. The lack of any significant progress in ADC before 1970 is understandable. It was difficult to isolate and purify antibodies from animal or human serum. It was almost impossible to produce a large quantity of a specific antibody that could be used to prepare ADC and to test its efficacy in therapy. Nevertheless, several pioneers in the ADC field used partially purified immunoglobulin preparations, mostly from rabbits and goats, to make drug conjugates. They demonstrated the feasibility of using antibody as a targeted carrier for the treatment of different types of cancer. Since this book will focus on the recent developments of ADC with recombinant monoclonal antibodies, this chapter will cover mostly the early history of ADC development from the late 1960s to early 1980s, i.e., during pre- and early post-monoclonal antibody era.

Mathe in 1957 reported that cell-specific antiproliferation activity against L1210 leukemia cells could be achieved when methotrexate was conjugated via a diazo coupling reaction to anti-leukemia 1210 antigen immunoglobulins but not to normal gamma globulin (Mathé et al. 1958). Although Mathe subsequently made many important contributions in the field of cancer immunotherapy (Mathé 1969), he did not follow up his work on methotrexate–antibody conjugate after his historical report. After Mathe's publication, several other groups made more extensive investigations on ADC in the late 1960s and early 1970s. The most serious challenge at that time was how to translate the studies with animal immunoglobulins into clinical applications. The research was performed mostly in academic laboratories, with very little support from the pharmaceutical industry. Yet, with limitations of the quantity and purity of the immunoglobulins, early publications reported many observations that provided the foundation for further studies in the ADC field. For example, with alkylating chemotherapeutic agents, it was demonstrated in early 1970s by Ghose and collaborators at Dalhousie University in Canada (Ghose and Nigam 1972; Ghose et al. 1975) and Rowland and colleagues at Searle Research Laboratory in UK (Rowland et al. 1975) that a covalent conjugation between the immunoglobulin and the drug is essential to achieve the tumor targeting effect. Sela and colleagues at the Weismann Institute of Science in Israel reported in 1975 that daunomycin and adriamycin could be linked covalently to anti-BSA immunoglobulins with various covalent reactions, but the retention of both the drug and antibody activities was observed only with the periodate oxidation method (Hurwitz et al. 1975). This was the first report that indicated, with identical antibody and drug, that the activity of ADC was dependent on the conjugation method. This finding opened a new area in ADC, i.e., the linker chemistry, which played an important role in the later design of ADC (Blair and Ghose 1983). More impressively, with the use of isolated animal immunoglobulins, ADC was already tested in human patients in several studies in the mid-1970s and showed promising results (Ghose et al. 1977; Oon et al. 1974). Ghose and Blair published one of the first comprehensive reviews on ADC in 1978 which covered ADC studies with immunogobulins isolated from animal antiserum before 1977 (Ghose and Blair 1978). It is worthy to mention here that Moolten and Cooperband reported in 1970 on the preparation of a conjugate between anti-leukemia immunoglobulins and diphtheria toxin with a specific toxicity against the leukemia cells (Moolten and Cooperband 1970). Their finding initiated a new class of cancer therapeutics, i.e., immunotoxin (Vitetta and Uhr 1985), which focuses on the use of protein toxins rather than conventional drugs.

The field of ADC surged in 1980s for several reasons. First, and most importantly, was the successful production of monoclonal antibodies by Milstein and Kohler (Kohler and Milstein 1975). Monoclonal antibody technology has solved the issue in antibody production and purification. The first monoclonal antibody drug, Muromonab-CD3 (OKT3®), was approved by the FDA in 1986 as an immunosuppressive agent for kidney transplantation (Ortho Multicenter Transplant Study Group 1985). Subsequently, the application of recombinant technology to produce human-like antibody, first as chimeric antibody (Morrison et al. 1984) and later as humanized antibody (Jones et al. 1986), has greatly reduced the concern of immunogenicity of immunoglobulins as the classical murine monoclonal antibodies encountered. Second, many new biomarkers have been identified, such as HER2 (van de Vijver et al. 1988) and VEGF (Kim et al. 1993), which allowed immunologists focusing not only on the structure but also the function of antigens as targets in the design of anti-tumor monoclonal antibodies. Another reason that helped specifically to the progress of the ADC field in 1980s was the better understanding of protein uptake in mammalian cells. Scientists were able to understand the intracellular processing of ADC in antigen-bearing cells through the knowledge of endocytosis. It became clear that one of the limiting steps in the drug action of drug–macromolecular conjugates was the intracellular release of pharmacologically active drug (Shen and Ryser 1979). Thus, different types of linkages have been designed to facilitate the release of drug from the carrier macromolecules inside the cells based on the knowledge of cell biology such as endosomal/lysosomal proteolytic activity (Duncan et al. 1980; Monsigny et al. 1980; Trouet et al. 1982) and acidification (Shen and Ryser 1981). Many successful applications of linkage design in monoclonal antibody ADC preparation have been reported in 1980s (Gallego et al. 1984; Blattler et al. 1985; Dillman et al. 1988). To overcome the limitation of the amount of the active drug released inside the antigen-bearing cells, another approach was to increase the number of drugs that could be carried by an antibody molecule. Several approaches have been developed in 1980s to increase the drug loading per antibody, such as using dextran (Manabe et al. 1984), albumin (Garnett and Baldwin 1986), or liposomes (Allen et al. 1995) as intermediate carriers between the antibody and the drug, as well as using the IgM isotype that could accommodate more drugs per immunoglobulin molecule (Persiani et al. 1989).

Based on many successful feasibility studies, the field of ADC became a mature research area in immunotherapy in 1990s. Pharmaceutical and biotech companies began to invest into this new area of the 100-year-old Ehrlich's "magic bullet" concept. New technology in the production of human monoclonal antibodies such as the single-chain Fv polypeptides, scFv (Bird et al. 1988), and the phage-displaying method (McCafferty et al. 1990) has further promoted the idea of using monoclonal antibodies as drugs or drug carriers in general. In addition, the identification of highly cytotoxic agents and the design of reversible linkers between antibodies and cytotoxic agents has further contributed to the successful development of ADCs in cancer therapy (Riccardi et al. 2023). Consequently, the first ADC, gemtuzumab ozogamicin (Mylotarg®), was approved by the FDA in 2000 under an accelerated approval process for the treatment of elder patients with CD33-positive acute myeloid leukemia (Bross et al. 2001). Even though gemtuzumab ozogamicin did not receive a full approval from the FDA until 2017, due to the efficacy and toxicity issues in confirmatory clinical trials (Jen et al. 2018), its initial approval in 2000 is still considered as a milestone of ADC development in cancer therapies.

Following gemtuzumab ozogamicin, brentuximab vedotin (Adcetris®) was approved in 2011 for relapsed Hodgkin's lymphoma (van de Donk and Dhimolea 2012) and trastuzumab emtansine (Kadcyla®) in 2013 for metastatic breast cancer (Peddi and Hurvitz 2013). Since their approval more than 10 years ago, brentuximab vedotin and trastuzumab emtansine have continued to be the top best-selling ADCs on the market. Currently, there are 15 ADCs on the global market for the treatment of various types of cancer (Riccardi et al. 2023; Shen et al. 2024), with at least 11 of them having been approved by the FDA (Gogia et al. 2023). In addition, more than 20 ADCs for the treatment of non-oncologic deseases, including immunological and infectious diseases, are currently under different stages of development (Pal et al. 2023). At long last, ADCs have successfully been established as an important class of therapeutics for the targeted treatment of diffferent types of cancer, as well as non-oncologic diseases.

References

Allen TM, Brandeis E, Hansen CB, Kao GY, Zalipsky S (1995) A new strategy for attachment of antibodies to sterically stabilized liposomes resulting in efficient targeting to cancer cells. Biochim Biophys Acta 1237:99–108

Bird RE, Hardmann KD, Jacobson JW, Johnson S, Kaufman BM, Lee S, Lee T, Pope SH, Riordan GS, Whitlow M (1988) Single-chain antigen-binding proteins. Science 242:423–426

Blair AH, Ghose TI (1983) Linkage of cytotoxic agents to immunoglobulins. J Immunol Methods 59:129–143

Blattler WA, Kuenzi BS, Lambert JM, Senter PD (1985) A new heterobifunctional protein cross-linking reagent that forms an acid labile link. Biochemistry 24:1517–1524

Bross PF, Beitz J, Chen G, Chen XH, Duffy E, Kieffer L, Roy S, Sridhara R, Rahman A, Williams G, Pazdur R (2001) Approval summary: gemtuzumab ozogamicin in relapsed acute myeloid leukemia. Clin Cancer Res 7:1490–1496

Dillman RO, Johnson DE, Shawler DI, Koziol JA (1988) Superiority of an acid-labile daunorubucin-monoclonal antibody immunoconjugate compared to free drug. Cancer Res 48:6097–6102

Duncan R, Lloyd JB, Kopecek J (1980) Degradation of side chains of N-(2-hydroxypropyl) methacrylamide copolymers by lysosomal enzymes. Biochem Biophys Res Commun 94:284–290

Gallego J, Price MR, Baldwin RW (1984) Preparation of four daunomycin-monoclonal antibody 791T/36 conjugates with anti-tumor activities. Int J Cancer 33:737–744

Garnett MC, Baldwin RW (1986) An improved synthesis of a methotrexate-albumin-791T/36 monoclonal antibody conjugate cytotoxic to human osteogenic sarcoma cell lines. Cancer Res 46:2407–2412

Ghose T, Blair AH (1978) Antibody-linked cytotoxic agents in the treatment of cancer: current status and future prospects. J Natl Cancer Inst 61:657–676
Ghose T, Nigam SP (1972) Antibody as carrier of chlorambucil. Cancer 29:1398–1400
Ghose T, Guclu A, Tai J (1975) Suppression of an AKR lymphoma by antibody and chlorambucil. J Natl Cancer Inst 55:1353–1357
Ghose T, Norvell ST, Guclu A, Bodurtha A, Tai J, MacDonald AS (1977) Immunochemotherapy of malignant melanoma with chlorambucil-bound anti-melanoma globulins: preliminary results in patients with disseminated disease. J Natl Cancer Inst 58:845–852
Gogia P, Ashraf H, Bhasin S, Xu Y (2023) Antibody–drug conjugates: a review of approved drugs and their clinical level of evidence. Cancers 15:3886
Hurwitz E, Levy R, Maron R, Wilchek M, Arnon R, Sela M (1975) The covalent binding of daunomycin and adriamycin to antibodies, with retention of both drug and antibody activites. Cancer Res 35:1175–1181
Jen EY, Ko CW, Lee JE, Del Valle PL, Aydanian A, Jewell C, Norsworthy KJ, Przepiorka D, Nie L, Liu J, Sheth JM, Shapiro M, Farrell AT, Pazdur R (2018) FDA approval: gemtuzumab ozogamicin for the treatment of adults with newly diagnosed CD33-positive acute myeloid leukemia. Clin Cancer Res 24:3242–3246
Jones PT, Dear PH, Foote J, Neuberger MS, Winter G (1986) Replacing the complementarity determining regions in a human antibody with those from a mouse. Nature 321:522–525
Kim KJ, Li B, Winer J, Armanini M, Gillet N, Phillips HS, Ferrara N (1993) Inhibition of vascular endothelial growth factor-induced angiogenesis suppresses tumor growth in vivo. Nature 362:841–844
Kohler G, Milstein C (1975) Continuous cultures of fused cells secreting antibody of predefined specificity. Nature 256:495–497
Manabe Y, Tsubota T, Haruta Y, Kataoka K, Okazaki M, Haisa S, Nakamura K, Kimura I (1984) Production of a monoclonal antibody-methotrexate conjugate utilizing dextran T-40 and its biologic activity. J Lab Clin Med 104:445–454
Mathé G (1969) Approaches to the immunological treatment of cancer in man. Br Med J 4:7–10
Mathé G, Loc T, Bernard J (1958) Effet sur la leucemie 1210 de la souris d'un combinaison par-diazotation d'A-methopterin et de γ-globulines de hamsters porteurs de cette leucemie parheterograffe. C R Acad Sci 216:1626–1628
McCafferty J, Griffiths AD, Winter G, Chiswell DJ (1990) Phage antibodies: filamentous phage displaying antibody variable domains. Nature 348:552–554
Monsigny M, Kieda C, Roche A-C, Delmotte F (1980) Preparation and biological properties of a covalent antitumor drug-arm-carrier (DAC conjugate). FEBS Lett 119:181–186
Moolten FL, Cooperband SR (1970) Selective destruction of target cells by diphtheria toxin conjugated to antibody directed against antigens on the cells. Science 169:68–70
Morrison SL, Johnson MJ, Herzenberg LA, Oi VT (1984) Chimeric human antibody molecules: mouse antigen-binding domains with human constant region domains. Proc Natl Acad Sci USA 81:6851–6855
Oon CJ, Apsey M, Buckleton H, Cooke KB, Hanham I, Hazanika P, Hobbs JB, McLeod B (1974) Human immune γ-globulin treated with chlorambucil for cancer therapy. Behring Inst Mitt 56:228–255
Ortho Multicenter Transplant Study Group (1985) A randomized clinical trial of OKT3 monoclonal antibody for acute rejection of cadaveric renal transplants. NEJM 313:1–7
Pal LB, Bule P, Khan W, Chella N (2023) An overview of the development and preclinical evaluation of antibody–drug conjugates for non-oncological applications. Pharmaceutics 15:1807
Peddi PF, Hurvitz SA (2013) Trastuzumab emtansine: the first targeted chemotherapy for treratment of breast cancer. Future Oncol 9:319–326
Persiani S, Ballou B, Shen WC, Ryser HJ-P, Reiland JM, Hakala TR (1989) In vivo antitumor effect of methotrexate conjugated to anti-SSEA-1 monoclonal IgM antibody on MH-15 mouse teratocarcinoma. Cancer Immunol Immunother 29:167–170

Riccardi F, Dal Bo M, Macor P, Toffoli G (2023) A comprehensive overview on antibody-drug conjugates: from the conceptualization to cancer therapy. Front Pharmacol 14:1274088

Rowland GF, O'Neill GJ, Davies DAL (1975) Suppression of tumor growth in mice by a drug antibody conjugate using a novel approach to linkage. Nature 255:487–488

Shen WC, Ryser HJ-P (1979) Poly(L-lysine) and poly(D-lysine) conjugates of methotrexate: different inhibitory effect on drug resistant cells. Mol Pharmacol 16:614–622

Shen WC, Ryser HJ-P (1981) Cis-aconityl spacer between daunomycin and macromolecular carrier: a model of pH-sensitive linkage releasing drug from a lysosomotropic conjugate. Biochem Biophys Res Commun 102:1048–1054

Shen L, Sun X, Chen Z, Guo Y, Shen Z, Song Y, Xin W, Ding H, Ma X, Xu W, Zhou W, Che J, Tan L, Chen L, Chen S, Dong X, Fang L, Zhu F (2024) ADCdb: the database of antibody–drug conjugates. Nucleic Acids Res 52:D1097–D1109

Strebhardt K, Ullrich A (2008) Paul Ehrlich's magic bullet concept: 100 years of progress. Nat Rev Cancer 8:473–480

Trouet A, Masquelier M, Baurain R, Campeneere D-D (1982) A covalent linkage between daunomycin and protein that is stable in serum and reversible by lysosomal hydrolases, as required for a lysosomotropic drug-carrier conjugate: in vitro and in vivo studies. Proc Natl Acad Sci USA 79:626–629

van de Donk NWCJ, Dhimolea E (2012) Brentuximab vedotin. mAbs 4:458–465

van de Vijver MJ, Peterse JL, Mooi WJ, Wisman P, Lomans J, Dalesio O, Nusse R (1988) Neu-protein overexpression in breast cancer. Association with comedo-type ductal carcinoma in situ and limited prognostic value in stage II breast cancer. NEJM 319:1239–1245

Vitetta ES, Uhr JW (1985) Immunotoxins. Ann Rev Immunol 3:197–212

Part II
The ADC Construct

Chapter 2
Payloads of Antibody: Drug Conjugates

Chalet Tan

Abstract The payload of antibody–drug conjugates (ADCs) is the warhead of ADCs that kills the tumor cells. Five classes of cytotoxins including maytansinoids, auristatins, calicheamicins, pyrrolobenzodiazepine dimers, and camptothecins are employed as the payloads in a total of 12 ADCs currently approved by the FDA. The structures, mechanisms of action, and pharmacology of these small-molecule cytotoxins are reviewed and discussed. The payloads along with the approved ADCs and their therapeutic areas are summarized.

Keywords Antibody–drug conjugates · Payloads · Maytansinoids · Auristatins · Calicheamicins · Pyrrolobenzodiazepine dimers · Camptothecins

2.1 Introduction

Over the past several decades, four classes of exceedingly potent toxins have been discovered and evaluated for their antiproliferative activities, including maytansinoids, auristatins, calicheamicins, and pyrrolobenzodiazepine dimers (Cassady et al. 2004; Luesch et al. 2002; Thorson et al. 2000; Mantaj et al. 2017). However, these cytotoxins displayed serious toxicities in vivo at the dosing levels required for the anticancer efficacy, which precluded their further development as anticancer drugs. The therapeutic application of such cytotoxic molecules in cancer therapy is now revived via the conjugation to tumor-targeting monoclonal antibodies (mAbs) as the payloads (Drago et al. 2021; Dumontet et al. 2023). One additional class of

---

C. Tan (✉)
Department of Pharmaceutical Sciences, University of Tennessee Health Science Center, Memphis, TN, USA
e-mail: ctan5@uthsc.edu

J. L. Zaro, J. Wang (eds.), *Antibody-Drug Conjugates*, AAPS Advances in the Pharmaceutical Sciences Series 17, https://doi.org/10.1007/978-3-031-83005-1_2

cytotoxins, camptothecin analogs, although with less potency, has also been proven effective as ADC payloads (Govindan et al. 2019).

The molecular targets of these five classes of payloads being employed currently in 12 FDA-approved ADC drugs include microtubules (maytansinoids and auristatins) and DNA (calicheamicins, pyrrolobenzodiazepine dimers, and camptothecins). Because of the ubiquitous presence of such targets in all normal and tumor cells, it is of crucial importance to maximize the delivery of the toxins to the tumor while minimizing the exposure to normal cells. To identify the right combination of tumor target, internalizing mAb, payload, linker, drug/antibody ratio (DAR), and indication is essential for the development of a successful ADC therapeutic (Tarantino et al. 2023). Ideally, ADC can achieve targeted delivery of the payload to tumor cells with markedly improved therapeutic windows. Despite this prevailing dogma, mounting clinical evidence shows that the maximum tolerated dose (MTD) of FDA-approved ADCs is not significantly different from that of the corresponding small-molecule payloads (Colombo and Rich 2022). Nonetheless, ADCs can enhance the anticancer efficacy that demonstrates the promises and robustness of this therapeutic modality. Understanding the challenges and opportunities for the payload design and diversification will help advance the development of next-generation ADCs (Tsuchikama et al. 2024).

2.2 Maytansinoids

2.2.1 Chemistry and Anticancer Activity

Maytansine was initially isolated from an alcoholic extract of the bark of the African shrub *Maytenus serrate* and *Maytenus buchananii* (Kupchan et al. 1972, 1974). It was the first compound discovered in a class of ansa macrolide antibiotics named maytansinoids, which showed potent anticancer activity against human nasopharynx carcinoma KB cells (EC_{50} = 8 pM), murine lymphocytic leukemia P-388 cells (EC_{50} = 0.6 pM), and murine leukemia L1210 cells (EC_{50} = 2 pM) (Kupchan et al. 1978; Wolpert-Defilippes et al. 1975a). A series of maytansine analogs bearing a disulfide or thiol substituent were later synthesized to allow for the covalent linkage with mAbs (Widdison et al. 2006). Among these, DM1 and DM4 are currently being used as the ADC payloads.

The chemical structures of maytansine, DM1 and DM4, are presented in Fig. 2.1.

Studies on the structure–activity relationship of maytansinoids reveal that the C3 N-acyl-N-methyl-L-alanyl ester side chain, the C4–C5 epoxide moiety, the C9 carbinoyl functional group, and the conjugated C11 and C13 double bonds are all essential to the antitumor activity of maytansinoids (Kupchan et al. 1978; Widdison et al. 2006). By contrast, the phenyl ring and the N′-acyl group are modifiable. Importantly, the nature of the acyl group can be varied without a significant loss in the activity. DM1 and DM4 form stable covalent bonds with mAbs in an aqueous solution with a high efficiency and yield (Widdison et al. 2006). Sulfhydryl groups or their respective thiolated anions react readily with meleimido moieties in a

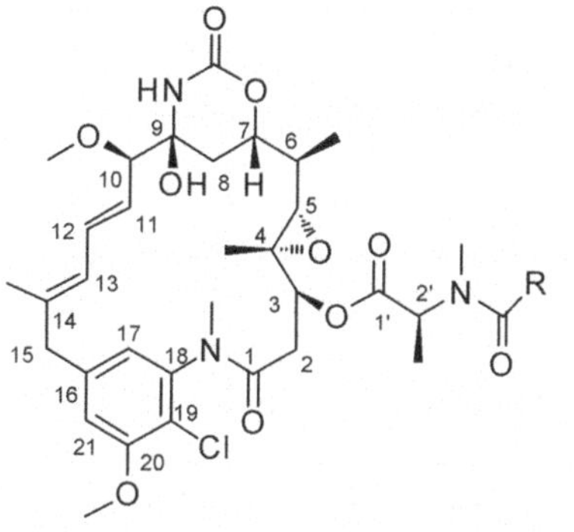

R= CH_3	maytansine
R= CH_2CH_2SH	DM1
R= $CH_2C(CH_3)_2SH$	DM4

Fig. 2.1 Structures of maytansine, DM1 and DM4

Michael-type addition reaction to form thioethers and with disulfide groups in a disulfide exchange reaction to form new disulfide bonds.

2.2.2 *Mechanism of Action*

Maytansine blocks the polymerization of tubulin, arrests the cell cycle at the G2/M phase, and inhibits mitosis (Remillard et al. 1975; Wolpert-DeFilippes et al. 1975b). Maytansine appears to share a common binding site on tubulin with vinca alkaloids (Mandelbaum-Shavit et al. 1976) but is about ten-fold more potent in inhibiting the binding of guanine nucleotides to tubulin at the exchangeable site (Huang et al. 1985). For DM1- or DM4-loaded ADCs to exert cytotoxicity, lysosomal processing is required, irrespective of the linker type (Erickson et al. 2006). In addition to the intracellular release of DM1 and DM4 from the ADCs, S-methyl-DM4 is also formed in the tumor cells, which is shown to be more potent than maytansine in suppressing the dynamic instability of the microtubules (Lopus et al. 2010).

2.2.3 *Early Preclinical Experiences*

Maytansine exerted potent growth inhibition against P-388 and L-1210 leukemia, Lewis lung carcinoma, and B-16 melanoma in mice at an intraperitoneal dose as low as 25 μg/kg/day for 9–10 consecutive days (reviewed in reference (Issell and Crooke 1978). In mice, maytansine was rapidly eliminated with a terminal half-life less than 20 min (Chari et al. 1992). Encouraged by its preclinical activity, maytansine was evaluated in several phase I clinical trials in patients with advanced solid tumors in the late 1970s (reviewed in reference (Wolpert-DeFilippes et al. 1975b). Maytansine was given intravenously, and the maximum tolerated dose was found to be 2–2.5 mg/m^2 every 3–4 weeks as a single dose or divided over three daily doses.

Table 2.1 FDA-approved ADCs that employ maytansinoids as payloads

ADC	Payload	DAR	Target	FDA-approved indications
Trastuzumab emtansine (T-DM1)	DM1	3.5:1	HER2	HER2$^+$ breast cancer (2013)
Mirvetuximab soravtansine	DM4	3.5:1	Folate receptor ⟨ (FRα)	FRα$^+$ ovarian, fallopian tube and peritoneal cancers (2022)

The dose-limiting toxicities of maytansine included nausea, vomiting, and neurotoxicity. Subsequently, the efficacy of maytansine was investigated in a number of phase II clinical trials in patients with ovarian, cervical, breast, head and neck, small-cell lung, or other advanced cancers, at a dose of 0.75–1.8 mg/m^2 divided over three daily doses every 1–2 weeks. These studies demonstrated that maytansine had little antitumor efficacy at its tolerated doses in cancer patients. Conjugation of DM1 with mAbs was shown to increase the MTD of maytansine by at least twofold, allowing the safe delivery of therapeutically effective levels of the payload to the tumor (Tolcher et al. 2003; Krop et al. 2012).

2.2.4 ADCs in the Clinic

As maytansinoids, DM1 and DM4 are employed as payloads in two ADCs that are approved by the FDA to treat solid malignancies. The ADCs, payloads, tumor targets, and therapeutic indications are summarized in Table 2.1.

2.3 Auristatins

2.3.1 Chemistry and Anticancer Activity

Dolastatin 10 was initially isolated from the sea hare *Dolabella auricularia* at a vanishingly low yield (~ 1 mg from each 100 kg) (Pettit et al. 1987). It is a unique linear pentapeptide comprising several unusual amino acids (Fig. 2.2a). Dolastatin 10 was found to possess a wide spectrum of anticancer activity against murine L1210 leukemia cells (IC_{50} = 0.01–1 nM) (Bai et al. 1990), lymphomas (IC_{50} = 0.1–1 pM) (Beckwith et al. 1993), ovarian (IC_{50} = 0.05–1.8 nM) (Aherne et al. 1996), colon (ovarian (IC_{50} = 0.02–0.2 nM) (Aherne et al. 1996), and lung cancer cells (IC_{50} = 0.03–0.18 nM) (Kalemkerian et al. 1999). Nevertheless, it was shown to be toxic toward hematopoietic progenitor cells (IC_{50} = 0.1–1 pM) (Jacobsen et al. 1991), which explains the potent myelosuppression associated with dolastatin 10 treatment in vivo. The C-terminal portion of dolastatin 10 is shown to be important

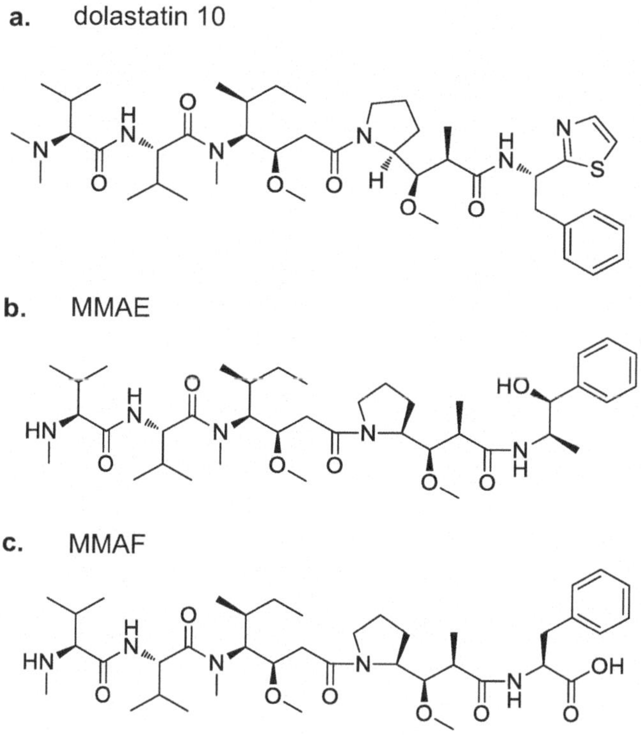

Fig. 2.2 Structures of dolastatin 10 (**a**), MMAE (**b**) and MMAF (**c**)

for its cytotoxicity, as the tripeptides missing dolaproine and dolaphenine moieties lack inhibitory activity against the cancer cells (Bai et al. 1993).

As analogs of dolastatin 10, monomethyl auristatin-E (MMAE, Fig. 2.2b) (Doronina et al. 2003) and monomethyl auristatin-F (MMAF, Fig. 2.2c) (Doronina et al. 2006) were designed as potent ADC payloads. Both MMAE and MMAF are peptide analogs, which have limited impact on the physicochemical properties of mAbs. MMAF differs from MMAE owing to a phenylalanine moiety at its C-terminus, contributing to its membrane impermeability. MMAE and MMAF are both highly stable molecules, showing no signs of degradation in plasma, human liver lysosomal extracts, or proteases such as cathepsin B. As free toxins, the cytotoxocity values of MMAE and MMAF are about 200-fold and 1000-fold less potent than those of dolastatin 10 in lymphoma cells, respectively. The conjugation of these toxins with cAC10, a monoclonal antibody specific for CD30, restored their cytotoxicity against $CD30^+$ lymphoma cells similar to the level of dolastatin 10 (Doronina et al. 2006; Francisco et al. 2003).

2.3.2 *Mechanism of Action*

Investigation into the mechanism of cell-cycle arrest by dolastatin 10 reveals that it impedes the polymerization of microtubules, suppresses tubulin-dependent GTP hydrolysis, and inhibits the binding of vinca alkaloids to tubulin (Bai et al. 1990). As a result, dolastatin 10 blocks the microtubule assembly and mitosis, arresting the cancer cells in the G2/M phase.

2.3.3 *Early Preclinical and Clinical Experiences*

In vivo, dolastatin 10 demonstrated exceptional anticancer efficacy against murine P388 leukemia at doses as low as 1 μg/kg (Pettit et al. 1987). A preclinical pharmacokinetic study demonstrated that dolastatin 10 was highly protein-bound in plasma (> 80%) and rapidly metabolized in the liver with an elimination half-life of 3.7 h in mice (Newman et al. 1994). The MTD in mice, rats, and dogs of dolastatin 10 was 1.35 μg/m^2 (0.45 μg/kg), 0.45 μg/m^2 (0.075 μg/kg), and no greater than 0.4 μg/m^2 (0.02 μg/kg), respectively (Mirsalis et al. 1999).

In patients with advanced solid tumors, the MTD of dolastatin 10 was reached at 0.3 μg/m^2, with granulocytopenia being the dose-limiting toxicity (Madden et al. 2000; Pitot et al. 1999). Peripheral neuropathy was also observed in some patients, a common toxicity for antimicrotubule agents. A three-compartment model adequately described the plasma concentration of dolastatin 10 vs. time profile showing a rapid distribution phase ($t_{1/2,\alpha}$ = 4–6 min). Metabolism turned out to be a minor elimination route in humans. Subsequently, dolastatin 10 was evaluated in phase II trials in patients with soft-tissue sacromas (Von Mehren et al. 2004), prostate (Vaishampayan et al. 2000), ovarian (Hoffman et al. 2003), and pancreatic/hepatobiliary carcinomas (Kindler et al. 2005). Although well tolerated at an i.v. dose of 0.4 μg/m^2 given every 3 weeks, dolastatin 10 failed to yield objective responses in cancer patients. These results clearly indicate the narrow therapeutic window of dolastatin 10. Conjugation of MMAE with anti-CD30 mAb allowed for the delivery of the payload in patients with CD30-positive hematological cancers at a dose over fivefold higher than the MTD of dolastatin 10, which was shown to be clinically efficacious (Katz et al. 2011).

2.3.4 *ADCs in the Clinic*

As auristatins, MMAE and MMAF are employed as the payloads in five ADCs approved by the FDA to treat hematological and solid tumors. Among these, belantamab mafoditin was withdrawn from the market in 2022 owing to the insufficient efficacy from a phase III trial. The ADCs, payloads, tumor targets, and therapeutic indications are summarized in Table 2.2.

Table 2.2 FDA-approved ADCs that employ auristatins as payloads

ADC	Payload	DAR	Target	FDA-approved indications
Brentuximab vedotin (SGN-35)	MMAE	4:1	CD30	Hodgkin's lymphoma, systemic anaplastic large-cell lymphoma, T-cell lymphoma (2011)
Polatuzumab vedotin	MMAE	3.5:1	CD79b	Diffuse large-cell lymphoma (2019)
Enfortumab vedotin	MMAE	3.8:1	Nectin-4	Urothelial cancer (2019)
Tisotumab vedotin	MMAE	4:1	Tissue factor	Cervical cancer (2021)
Belantamab mafoditin	MMAF	4:1	BCMA	Multiple myeloma (2020, withdrawn in 2022)

2.4 Calicheamicins

2.4.1 Chemistry and Anticancer Activity

Calicheamicin γ_1^I was initially isolated from a broth extract of a soil microorganism *Micromonospora echinospora calichensis*, which was found to possess the most potent antitumor activity among all members of the calicheamicins and was active against murine tumors at 0.5–1.5 μg/kg dose range (Lee et al. 1987a, b). It is an enediyne-containing anticancer antibiotic with unique structural features including a glycosylated hydroxyamino sugar and a labile methyltrisulfide group (Fig. 2.3a). A semisynthetic derivative of calicheamicin, N-acetyl-γ-calicheamicin 1, 2-dimethyl hydrazine (Fig. 2.3b), has been developed as a payload toxin for ADCs (Hinman et al. 1993). While γ_1^I hydrazide itself was ten-fold less cytotoxic than the parent γ_1^I, γ_1^I hydrazide-conjugated CT-M-01 antibody displayed the cytotoxicity equivalent to calicheamicin γ_1^I at the picomolar level (8 ng/ml) in human breast carcinoma MX-1cells.

2.4.2 Mechanism of Action

The potent cytotoxicity of calicheamicin γ_1^I is ascribed to its remarkable ability to cleave double-strand DNA at specific sites, resulting in cell death (Zein et al. 1988). Without exception, the preferred site of attack is at the following DNA sequence, with the cleavage positions indicated by arrows:

5' NNTCCT 3'
3' NNAGGA 5'

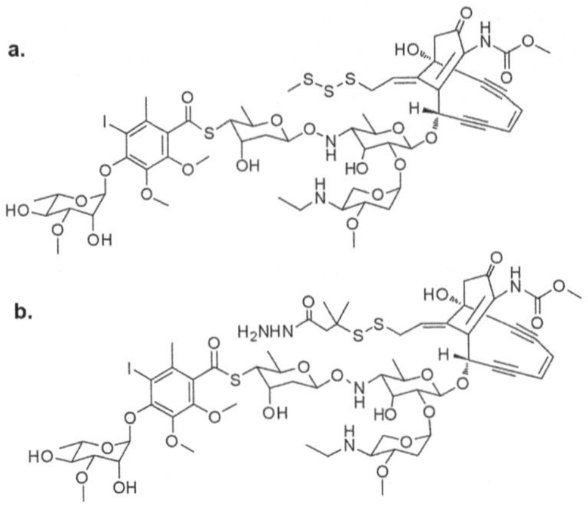

Fig. 2.3 Structures of calicheamicin γ_1^I (**a**) and its dimethyl hydrazide (**b**)

Calicheamicin γ_1^I binds to the double-helical DNA in the minor grove. The staggered nature of the cleavage sites on the double-stranded DNA and the remarkable efficiency/specificity of the scission reaction result from the unique fit between the compound and DNA (Zein et al. 1989; Ikemoto et al. 1995). For a double-strand scission to occur, the enediyne portion of the aglycone binds to the complementary cleavage site on the DNA. The presence of intracellular thiol cofactors initiates the aromatization of the diyne-ene moiety, followed by the generation of a nondiffusible 1, 4-dehydrobenzene-diradical that mediates oxidation strand scission by hydrogen abstraction on the deoxyribose ring (Zein et al. 1989). The association of the thiobenzoate-carbohydrate moiety with the minor grove not only enhances the DNA-binding affinity with the aglycone but also helps position the biradical for the optimal reactivity with the DNA (Drak et al. 1991).

2.4.3 Early Preclinical Experiences

In the P388 leukemia mouse model, the MTD and the therapeutic dose of calicheamicin γ_1^I following the intraperitoneal administration were 1.25 μg/kg and 5 μg/kg, respectively (Hinman et al. 1993). It became evident that although it had highly potent anticancer activity, calicheamicin γ_1^I caused lethal toxicities at a dose much lower than its tolerated dose. Calicheamicin γ_1^I had never been evaluated in the clinic due to its unfavorable therapeutic window. Conjugation of the dimethyl hydrazine derivative of calicheamicin γ_1^I with mAbs increased the MTD of calicheamicin γ_1^I more than ten-fold, allowing for the delivery of a therapeutically effective dose of the toxin to the tumor (Hinman et al. 1993). The dose-limiting toxicities of calicheamicin γ_1^I-loaded ADCs included thrombocytopenia and neutropenia.

Table 2.3 FDA-approved ADCs that employ calicheamicin as the payload

ADC	Payload	DAR	Target	FDA-approved indications
Gemtuzumab ozogamicin	calicheamicin	2–3:1	CD33	$CD33^+$ acute myeloid leukemia (2000, withdrawn 2011, re-approved 2017)
Inotuzumab ozogamicin	calicheamicin	5–7:1	CD22	$CD22^+$ acute lymphoblastic leukemia (2017)

2.4.4 ADCs in the Clinic

The dimethyl hydrazine derivative of calicheamicin γ_1^I has been employed in two FDA-approved ADCs (Table 2.3) to treat leukemia. In both cases, the calicheamicin derivative is covalently linked to mAbs via the acid-labile 4-(4′-acetylphenoxy) butanoic acid linker. The ADCs, tumor targets, and therapeutic indications are summarized in Table 2.3.

2.5 Pyrrolobenzodiazepine (PBD) Dimers

2.5.1 Chemistry and Anticancer Activity

Naturally occurring pyrrolobenzodiazepines (PBDs) are antitumor antibiotics isolated from various *Streptomyces* species, which include anthramycin, tomaymycin, sibiromycin, neothramycins, and DC-81 (Mantaj et al. 2017). These PBD monomers share a common tricyclic scaffold consisting of a substituted aromatic A-ring, a diazepine B-ring, and a pyrrolidine C-ring, with an (*S*)-configuration at the C11a-position between the B and C rings. When viewed from the C-ring toward the A-ring, the scaffold presents a right-handed helical twist, which enables the PBD molecules to mirror the curvature of B-form DNA and fit in the minor groove of DNA. PBD monomers are cytotoxins owing to the covalent binding of the N10–C11 imine moiety to the exocyclic C2–NH_2 group of guanine (G), which span three DNA base pairs and have a preference for binding to purine–G–purine triplets.

A large number of synthetic PBDs were generated and examined that culminated in a detailed understanding of structure–activity relationships around the PBD pharmacophore, which led to the development of SJG-136 (SG2000, Fig. 2.4), the first PBD dimer that reached the clinical testing stage (Hartley 2011). The rationale for developing PBD dimers was to generate cytotoxins that would retain the favorable inherent properties of PBD monomers but span greater lengths of DNA with an enhanced sequence selectivity. SJG-136 contains two PBD subunits that are linked head-to-head through the A-ring C8-positions (C8-C8’) via propylenedioxy tether, each with C2-*exo*-methylidene substitution on the C-ring. SJG-136 spans six base pairs in the DNA minor groove and covalently binds to spatially separated guanines on opposite strands, with a preference for binding to the 5′-purine-GATC-pyrimidine-3′ sequence (Martin et al. 2005). SJG-136 exhibits potent anticancer

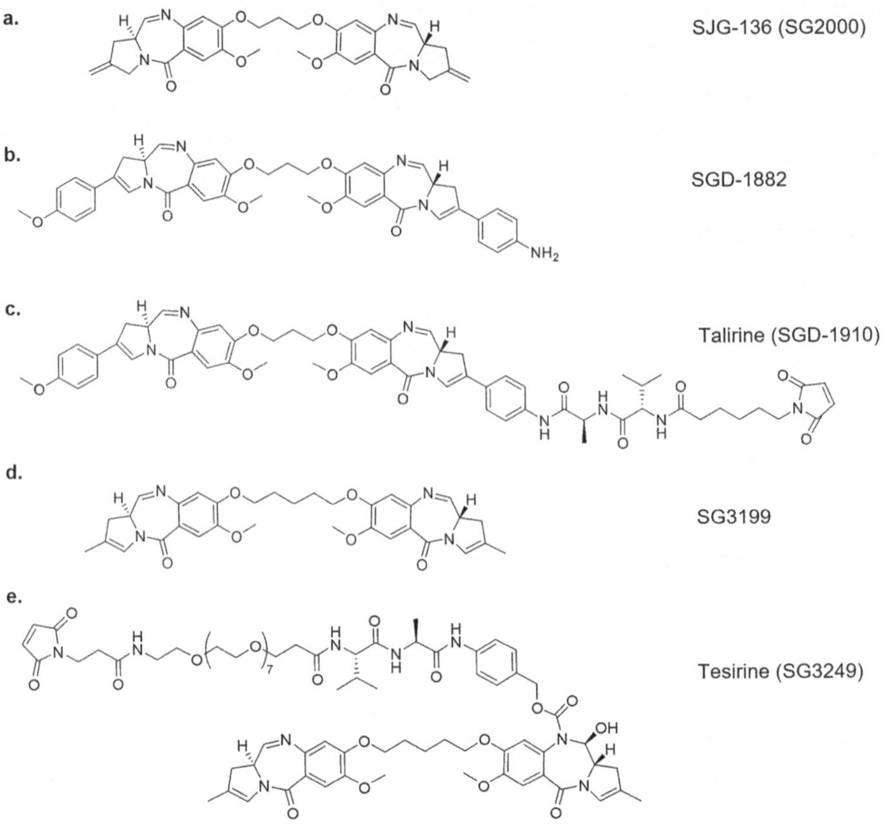

Fig. 2.4 Structures of PBD dimers SJG-136 (**a**), SGD-1882 (**b**) and the payload-linker talirine (**c**), SG-3199 and the payload-linker tesirine (**d**)

activity in the NCI 60-cell-line screen with a mean GI_{50} of 7.4 nM (range 0.14–320 nM) (Hartley et al. 2004).

The potency of PBD dimers was further improved by the introduction of a C2–C3-*endo*-unsaturated group with C2-aryl substitution that led to the development of SGD-1882 (Gregson et al. 2019; Hartley 2021). The C2 anilino group is coupled with the valine–alanine dipeptide trigger that is cleavable by cathepsin B, along with the maleimidocaproyl linker amenable for thiol conjugation. This complete payload-linker construct is named talirine (SGD-1910) (Gregson et al. 2019; Hartley 2021).

Another PBD dimer payload-linker tesirine (SG3249) has been rationally designed to reduce the hydrophobicity of talirine while maintaining the potency (Tiberghien et al. 2016). The warhead component of tesirine is SG3199. By replacing the C2-aryl with C2-methyl substitution that lowers the hydrophobicity, and by extending the C8–C8′ tether to five carbons that enhances the isohelicity with the DNA minor groove, SG-3199 is as potent as SGD-1882. One of the N10 positions of SG3199 is capped with a para-aminobenzyl (PAB) self-immolative unit, which is coupled with a valine–alanine dipeptidyl trigger, a PEG_8 spacer to further enhance aqueous solubility, and the maleimidocaproyl linker, resulting in the complete structure of tesirine. Results from a screen against a panel of 38 human solid tumors and hematological cell lines, SG3199 inhibits the proliferation of human cancer cells with a mean GI_{50} of 151.5 pM (range 0.79 pM–1.05 nM) (Hartley et al. 2018). Relative to the solid tumor cell lines, the hematological cell lines are in general more sensitive to SG3199.

2.5.2 *Mechanism of Action of PBD Dimers*

The anticancer activity of PBD dimers is accounted by their ability to efficiently form minor-groove DNA interstrand cross-links, which cause the cell cycle arrest at the G_2–M boundary, ultimately resulting in apoptosis at low nanomolar to picomolar concentrations (Hartley 2011). These cross-links form more rapidly in cells and persist longer than those produced in the major groove of DNA by conventional DNA cross-linking agents. The nondistorting nature of the cross-links in the DNA minor groove contributes to the persistence by evading excision repair mechanisms, conferring this ADC warhead the important ability to kill slowly proliferating cells (Hartley 2011).

2.5.3 *Preclinical and Clinical Experiences with PBD Dimers*

The efficacy of SJG-136 was extensively studied in athymic mice bearing human tumor xenografts. SJG-136 exhibited potent anticancer activity against a broad spectrum of cancer types (ovarian, breast, lung, glioma, colon, melanoma, and leukemia) as well as cisplatin-resistant ovarian cancer (Hartley et al. 2004; Alley et al. 2004). Of all of the dosing regimens tested, IV bolus administrations for 5 consecutive days conferred the greatest efficacy. The maximum tolerated dose of SJG-136 was found to be around 120 g/kg/day.

Based on the promising preclinical findings, SJG-136 entered phase I clinical trials in patients with advanced solid tumors in 2004. Three different intravenous dosing regimens were evaluated; the MTD of SJG-136 was found to be 45 g/m^2 administered every 21 days, 45 g/m^2/day administered on days 1, 8, and 15 of a 28-day cycle, or 30 g/m^2 administered daily (Hochhauser et al. 2009; Janjigian et al. 2010; Puzanov et al. 2011). The dose-limiting toxicities were vascular leak syndrome and hepatotoxicity. SJG-136 was advanced to phase II trials in patients with ovarian cancer and leukemias. It is worth noting that as a potential pharmacodynamic biomarker for DNA damage, γ-H2AX foci formation was observed in lymphocytes of patients treated with SJG-136 (). However, these trials were terminated due to the lack of clinical efficacy.

Table 2.4 ADCs that employ PBD dimers as payloads

ADC	Payload	DAR	Target	Indication / Status
Loncastuximab tesirine (ADCT-402)	SG3199	2.3:1	CD19	Large B-cell lymphoma (2021)
Camidanlumab tesirine (ADCT-301)	SG3199	2.3:1	CD25	Hodgkin's lymphoma and advanced solid tumors (phase II)
Serclutamab talirine (ABBV-321)	SGD-1882	2:1	EGFR	Advanced solid tumors associated with EGFR overexpression (phase I completed)

The pharmacokinetics of SG3199 was studied in rats following IV bolus administration of [^{3}H]-SG3199 at 0.1–1 g/kg (Hartley et al. 2018). Over 97% of the dose was eliminated in feces via biliary secretion, with a minor renal excretion (<4%). SG3199 was cleared quickly from the bloodstream with a $t_{1/2}$ of 8–42 min. The very short half-life of SG3199 is an important property in the context of an ADC warhead. ADCs with the enzyme-cleavable linker-containing tesirine have been shown to have a significant bystander effect. This is an important advantage when targeting tumors with a heterogeneous target antigen expression. The short half-life of the released SG3199 warhead ensures that the bystander effect is restricted and that the systemic accumulation of free drug, which could contribute to off-target toxicity, is limited.

2.5.4 ADCs in the Clinic

Loncastuximab tesirine is the only FDA-approved ADC that employs a PBD dimer as the payload. Two other tesirine- and talirine-containing ADCs are currently undergoing clinical trials. The ADCs, payloads, tumor targets, and therapeutic indications are summarized in Table 2.4.

2.6 Camptothecins

2.6.1 Chemistry and Anticancer Activity

Camptothecin is a pentacyclic, plant alkaloid with exceedingly low solubility originally isolated from the bark of Chinese tree *Camptotheca acuminata* (Pommier 2006; Garcia-Carbonero and Supko 2002). Two water-soluble derivatives of camptothecin—topotecan and irinotecan—are FDA-approved anticancer drugs for the treatment of solid tumors.

Irinotecan (CPT-11) has an ethyl substituent at C7 and a dipiperidyl carbamate at C10, which is metabolized to the active metabolite SN-38 in vivo by the action of carboxylesterases (Fig. 2.5a). SN-38 has an IC_{50} value of 1.0–6.0 nM against

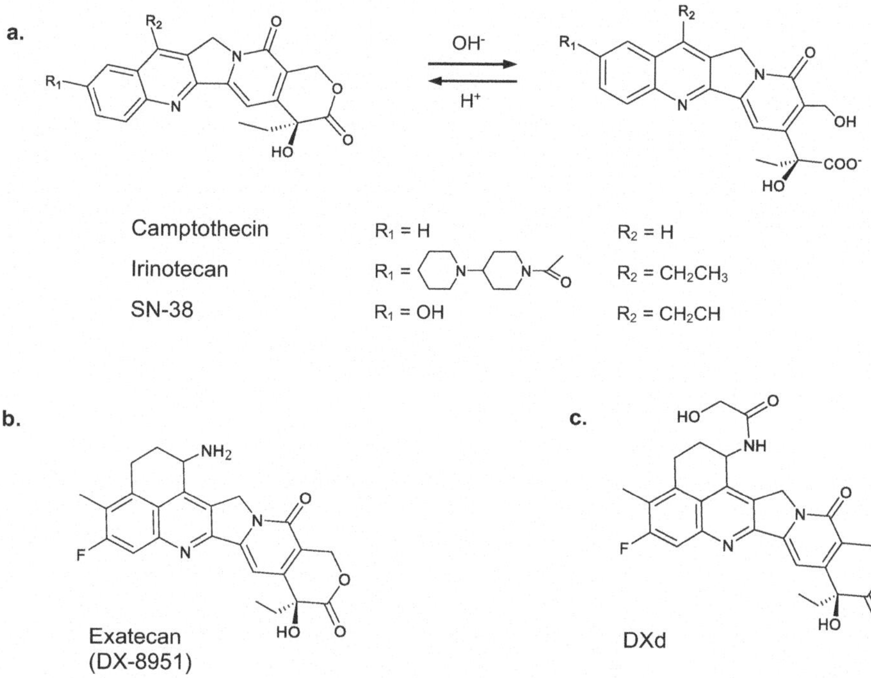

Fig. 2.5 Structures of camptothecin, irinotecan, and SN-38 (**a**), exatecan (**b**) and DXd (**c**)

various human solid tumor cell lines, which is 2 to 3 logs more potent than irinotecan (Goldenberg et al. 2015). SN-38 can undergo spontaneous opening of the lactone ring to the carboxylate form that diminishes its activity, as the lactone ring is essential for the stabilization of the DNA–topoisomerase I complex, and the cytotoxicity of the lactone form is significantly greater than the carboxylate form (Garcia-Carbonero and Supko 2002; Venditto and Simanek 2010).

Exatecan (DX-8951) is another water-soluble camptothecin analog with an amine-substituted cyclohexane ring (Mitsui et al. 1995). Exetecan has enhanced aqueous solubility and improved lactone stability owing to the cyclohexane ring connecting C7 and C9 with an exocyclic amine, as well as a methyl group at C10 and fluorine at C11 (Fig. 2.5b). The rigidity conferred by the cyclohexane ring is considered to favor the intact lactone ring in the lactone–carboxylate equilibrium. Without the need for metabolic activation, exetecan shows potent antiproliferative activity against a series of 32 human cancer lines with a mean GI_{50} of 3.9 nM (range 0.4–19.6 nM) (Mitsui et al. 1995). DXd is a 2-hydroxyacetic amide derivative of exatecan (Fig. 2.5c).

2.6.2 Mechanism of Action

Topoisomerases are nuclear enzymes that enable supercoiled DNA to relax for transcription and replication. Camptothecins bind and stabilize the transient single-strand DNA breaks generated by topoisomerase 1 (TOPO 1), which induce a double-strand cleavage of DNA as the cleavage complexes encounter the replication fork, resulting in apoptosis when the cells are in the S-phase of the cell cycle (Pommier 2006).

2.6.3 Preclinical and Clinical Experiences

It is the prodrug of SN-38, irinotecan, that is FDA-approved for the treatment of colorectal cancer, which also has activity in lymphoma, lung, ovarian, cervical, and gastric cancers. The dose-limiting toxicities of irinotecan are myelosuppression (neutropenia in particular) and delayed diarrhea, both elicited by SN-38 (Garcia-Carbonero and Supko 2002; Venditto and Simanek 2010). With irinotecan therapy, glucuronidated SN-38 (SN-38G) formed in the liver is biliary secreted into the intestines, which is subsequently deconjugated by β-glucuronidase (produced by the intestinal flora) and causes prolonged damages in the intestinal epithelium.

Exatecan mesylate (DX-8951f) displayed potent antitumor activity superior to irinotecan (CPT-11) in mice. DX-8951f at a total dose of 50–75 mg/kg (divided into three or four doses given every 4 days via IV bolus injection) greatly suppressed the tumor growth in subcutaneous xenograft models, including human gastric, colon, lung, breast, and renal cancers, more effectively than CPT-11 (260 mg/kg) (Mitsui et al. 1995; Kumazawa et al. 1998). Notably, DX-8951f was also efficacious against CPT-11-resistant human pancreatic SUIT-2 cancer cells in both subcutaneous and liver metastasis models (Takiguchi et al. 1997). In an orthotopic metastatic model using human pancreatic BxPC-3 cancer cells, DX-8951f not only retarded the growth of primary tumor in a dose–response manner, but it also demonstrated anti-metastatic activity in the late-stage model. It reduced the incidence of lymph node metastasis while eliminating lung metastasis, with significantly higher efficacy than gemcitabine, the front-line therapy of pancreatic cancer (Sun et al. 2003).

DX-8951f was advanced to clinical evaluation in late 1990s. DX-8951f was administered intravenously as 30-min IV infusion, and several dosing regimens had been evaluated in patients with solid malignancies during phase I trials. When DX-8951f was given as 30-min IV infusion daily for 5 consecutive days every 3 weeks, the MTD was found to be 0.3–0.5 mg/m^2/day (Rowinsky et al. 2020). When the drug was given once weekly for 3 consecutive weeks out of the 4-week cycle, the MTD was 2.10–2.75 mg/m^2/week (Braybrooke et al. 2003a). When patients received the treatment once every 3 weeks, the MTD was increased to 5–7 mg/m^2 (Boige et al. 2000; Minami et al. 2001). Weekly IV infusion and continuous 21-day infusion of DX-8951f were also tested in patients (Royce et al. 2001; Sharma et al. 2001; Garrison et al. 2003). The dose-limiting toxicity was myelosuppression (mostly neutropenia), while only mild diarrhea was observed infrequently

Table 2.5 FDA-approved ADCs that employ camptothecins as payloads

ADC	Payload	DAR	Target	FDA-approved indications
Sacituzumab govitecan (IMMU-132)	SN-38	7.7:1	TROP2	Triple-negative breast cancer (2020); urothelial carcinoma (2021)
Trastuzumab deruxtecan (T-DXd)	DXd	7.6:1	HER2	$HER2^+$ breast cancer (2019); $HER2^+$ gastric cancer (2021); $HER2^{low}$ breast cancer (2022)

in patients (Rowinsky et al. 2020; Braybrooke et al. 2003a; Boige et al. 2000; Minami et al. 2001; Royce et al. 2001; Sharma et al. 2001; Garrison et al. 2003). Subsequently, a series of phase II clinical trials were performed with a daily dose of 0.5 mg/m^2/day for 5 consecutive days every 3 weeks. DX-8951f only showed limited activity in patients with advanced cancer, including non-small-cell lung, breast, ovarian, colorectal, biliary tract, gastric cancer, and soft tissue sarcoma (Braybrooke et al. 2003b; Esteva et al. 2003; Verschraegen et al. 2004; Clamp et al. 2004; Royce et al. 2004; Abou-Alfa et al. 2005; Ajani et al. 2005; Reichardt et al. 2007). A phase III trial was conducted in patients with locally advanced or metastatic pancreatic adenocarcinoma to evaluate the combination therapy of exatecan mesylate (2 mg/m^2 on days 1 and 8, every 3 weeks) and gemcitabine. However, exatecan plus gemcitabine was not superior to gemcitabine alone with respect to the overall survival (Abou-Alfa et al. 2006).

2.6.4 ADCs in the Clinic

In contrast to the ultratoxic payloads (picomolar), SN-38 and deruxtecan are moderately cytotoxic (low nanomolar) that are linked to IgG at a DAR of around 8:1. Both camptothecins have been shown to elicit the bystander effect in the tumors. Currently, there are two FDA-approved ADCs that employ camptothecins as payloads to treat solid tumors (Table 2.5).

References

Abou-Alfa GK, Rowinsky EK, Patt YZ et al (2005) A Phase II study of intravenous exatecan administered daily for 5 days, every 3 weeks to patients with biliary tract cancers. Am J Clin Oncol 28:334–339

Abou-Alfa GK, Letourneau R, Harker G et al (2006) Randomized phase III study of exatecan and gemcitabine compared with gemcitabine alone in untreated advanced pancreatic cancer. J Clin Oncol 24:4441–4447

Aherne GW, Hardcastle A, Valenti M et al (1996) Antitumour evaluation of dolastatins 10 and 15 and their measurement in plasma by radioimmunoassay. Cancer Chemother Pharmacol 38:225–232

Ajani JA, Takimoto C, Becerra CR et al (2005) A phase II clinical and pharmacokinetic study of intravenous exatecan mesylate (DX-8951f) in patients with untreated metastatic gastric cancer. Investig New Drugs 23:479–484
Alley MC, Hollingshead MG, Pacula-Cox CM et al (2004) SJG-136 (NSC 694501), a novel rationally designed DNA minor groove interstrand cross-linking agent with potent and broad spectrum antitumor activity: part 2: efficacy evaluations. Cancer Res 64:6700–6706
Bai R, Petit GR, Hamel E (1990) Dolastatin 10, a powerful cytostatic peptide derived from a marine animal: inhibition of tubulin polymerization mediated through the vinca alkaloid binding domain. Biochem Pharmacol 39:1941–1949
Bai R, Roach MC, Jayaram SK et al (1993) Differential effects of active isomers, segments, and analogs of dolastatin 10 on ligand interactions with tubulin: correlation with cytotoxicity. Biochem Pharmacol 45:1503–1515
Beckwith M, Urba WJ, Longo DL (1993) Growth inhibition of human lymphoma cell lines by the marine products, dolastatins 10 and 15. J Nat Cancer Inst 85:483–488
Boige V, Raymond E, Faivre S et al (2000) Phase I and pharmacokinetic study of the camptothecin analog DX-8951f administered as a 30-minute infusion every 3 weeks in patients with advanced cancer. J Clin Oncol 18:3986–3992
Braybrooke JP, Boven E, Bates NP et al (2003a) Phase I and pharmacokinetic study of the topoisomerase I inhibitor, exatecan mesylate (DX-8951f), using a weekly 30-minute intravenous infusion, in patients with advanced solid malignancies. Ann Oncol 14:913–921
Braybrooke JP, Ranson M, Manegold C et al (2003b) Phase II study of exatecan mesylate (DX-8951f) as first line therapy for advanced non-small cell lung cancer. Lung Cancer 41:215–219
Cassady JM, Chan KK, Floss HG et al (2004) Recent developments in the maintained antitumor agents. Chem Pharm Bull 52:1–26
Chari RV, Martell BA, Gross JL et al (1992) Immunoconjugates containing novel maytansinoids: promising anticancer drugs. Cancer Res 52:127–131
Clamp A, Adams M, Atkinson R et al (2004) A phase IIA study of the topoisomerase I inhibitor, exatecan mesylate (DX-8951f), administered at two different dose schedules in patients with platinum- and taxane-resistant/refractory ovarian cancer. Gynecol Oncol 95:114–119
Colombo R, Rich JR (2022) The therapeutic window of antibody drug conjugates: a dogma in need of revision. Cancer Cell 40:1255–1263
Doronina SO, Toki BE, Torgov MY et al (2003) Development of potent monoclonal antibody auristatin conjugates for cancer therapy. Nat Biotechnol 21:778–784
Doronina SO, Mendelsohn BA, Bovee TD (2006) Enhanced activity of monomethylauristatin F through monoclonal antibody delivery: effects of linker technology on efficacy and toxicity. Bioconjug Chem 17:114–124
Drago JZ, Modi S, Chandarlapaty S (2021) Unlocking the potential of antibody-drug conjugates for cancer therapy. Nat Rev Clin Oncol 18:327–344
Drak J, Iwasawa N, Danishefsky S et al (1991) The carbohydrate domain of calicheamicin γ_1^I determines its sequence specificity for DNA cleavage. Proc Natl Acad Sci 88:7464–7468
Dumontet C, Reichert JM, Senter PD et al (2023) Antibody–drug conjugates come of age in oncology. Nat Rev Drug Discov 22:641–661
Erickson HK, Park PU, Widdison WC et al (2006) Antibody-maytansinoid conjugates are activated in targeted cancer cells by lysosomal degradation and linker-dependent intracellular processing. Cancer Res 66:4426–4433
Esteva FJ, Rivera E, Cristofanilli M et al (2003) A Phase II study of intravenous exatecan mesylate (DX-8951f) administered daily for 5 days every 3 weeks to patients with metastatic breast carcinoma. Cancer 98:900–907
Francisco JA, Cerveny CG, Meyer DL et al (2003) cAC10-vcMMAE, an anti-CD30–monomethyl auristatin E conjugate with potent and selective antitumor activity. Blood 102:1458–1465
Garcia-Carbonero R, Supko JG (2002) Current perspectives on the clinical experience, pharmacology, and continued development of the camptothecins. Clin Cancer Res 8:641–661

Garrison MA, Hammond LA, Geyer CE Jr et al (2003) A Phase I and pharmocokinetic study of exatecan mesylate administered as a protracted 21-day infusion in patients with advanced solid malignancies. Clin Cancer Res 9:2527–2537

Goldenberg DM, Cardillo TM, Govindan SV (2015) Trop-2 is a novel target for solid cancer therapy with sacituzumab govitecan (IMMU-132), an antibody-drug conjugate (ADC). Oncotarget 6:22496

Govindan SV, Cardillo TM, Goldenberg DM (2019) Topoisomerase inhibitors as antibody-drug conjugate (ADC) payloads. In: Thurston DE, Jackson PJM (eds) Cytotoxic payloads for antibody-drug conjugates. Royal Society of Chemistry, pp p166–p186

Gregson SJ, Tiberghien AC, Masterson LA, Howard PW (2019) Pyrrolobenzodiazepine dimers as antibody–drug conjugate (ADC) payloads. In: Thurston DE, Jackson PJM (eds) Cytotoxic payloads for antibody-drug conjugates. Royal Society of Chemistry, pp 296–331

Hartley JA (2011) The development of pyrrolobenzodiazepines as antitumour agents. Expert Opin Investig Drugs 20:733–744

Hartley JA (2021) Antibody-drug conjugates (ADCs) delivering pyrrolobenzodiazepine (PBD) dimers for cancer therapy. Expert Opin Biol Ther 21:931–943

Hartley JA, Spanswick VJ, Brooks N et al (2004) SJG-136 (NSC 694501), a novel rationally designed DNA minor groove interstrand cross-linking agent with potent and broad spectrum antitumor activity: part 1: cellular pharmacology, in vitro and initial in vivo antitumor activity. Cancer Res 64:6693–6699

Hartley JA, Flynn MJ, Bingham JP (2018) Pre-clinical pharmacology and mechanism of action of SG3199, the pyrrolobenzodiazepine (PBD) dimer warhead component of antibody-drug conjugate (ADC) payload tesirine. Sci Rep 8:10479

Hinman LM, Hamann PR, Wallace R et al (1993) Preparation and characterization of monoclonal antibody conjugates of the calicheamicins: a novel and potent family of antitumor antibiotics. Cancer Res 53:3336–3342

Hochhauser D, Meyer T, Spanswick VJ et al (2009) Phase I study of sequence-selective minor groove DNA binding agent SJG-136 in patients with advanced solid tumors. Clin Cancer Res 15:2140–2147

Hoffman MA, Blessing JA, Lentz SS (2003) A phase II trial of dolastatin-10 in recurrent platinum-sensitive ovarian carcinoma: a Gynecologic Oncology Group study. Gynecol Oncol 89:95–98

Huang AB, Lin CM, Hamel E (1985) Maytansine inhibits nucleotide binding at the exchangeable site of tubulin. Biochem Biophys Res Commun 128:1239–1246

Ikemoto N, Kumar RA, Ling TT et al (1995) Calicheamicin-DNA complexes: warhead alignment and saccharide recognition of the minor groove. Proc Natl Acad Sci 92:10506–10510

Issell BF, Crooke ST (1978) Maytansine. Cancer Treat Rev 5:199–207

Jacobsen SE, Ruscetti FW, Longo DL et al (1991) Antineoplastic dolastatins: potent inhibitors of hematopoietic progenitor cells. J Nat Cancer Inst 83:1672–1677

Janjigian YY, Lee W, Kris MG et al (2010) A phase I trial of SJG-136 (NSC# 694501) in advanced solid tumors. Cancer Chemother Pharmacol 65:833–838

Kalemkerian GP, Ou X, Adil MR, Rosati R, Khoulani MM, Madan SK, Pettit GR (1999) Activity of dolastatin 10 against small-cell lung cancer in vitro and in vivo: induction of apoptosis and bcl-2 modification. Cancer Chemother Pharmacol 43(6):507–515

Katz J, Janik JE, Younes A (2011) Brentuximab vedotin (SGN-35). Clin Cancer Res 17:6428–6436

Kindler HL, Tothy PK, Wolff R et al (2005) Phase II trials of dolastatin-10 in advanced pancreaticobiliary cancers. Investig New Drugs 23:489–493

Krop IE, LoRusso P, Miller KD et al (2012) A phase II study of trastuzumab emtansine in patients with human epidermal growth factor receptor 2-positive metastatic breast cancer who were previously treated with trastuzumab, lapatinib, an anthracycline, a taxane, and capecitabine. J Clin Oncol 30:3234–3241

Kumazawa E, Jimbo T, Ochi Y et al (1998) Potent and broad antitumor effects of DX-8951f, a water-soluble camptothecin derivative, against various human tumors xenografted in nude mice. Cancer Chemother Pharmacol 42:210–220

Kupchan SM, Komoda Y, Court WA et al (1972) Maytansine, a novel antileukemic ansa macrolide from Maytenus ovatus. J Am Chem Soc 94:1354–1356
Kupchan SM, Komoda Y, Branfman AR et al (1974) Novel maytansinoids: structural interrelations and requirements for antileukemic activity. J Am Chem Soc 96:3706–3708
Kupchan SM, Sneden AT, Branfman AR et al (1978) Structural requirements for antileukemic activity among the naturally occurring and semisynthetic maytansinoids. J Med Chem 21:31–37
Lee MD, Dunne TS, Chang CC et al (1987a) Calichemicins, a novel family of antitumor antibiotics. 2. Chemistry and structure of calichemicin γ_1^I. J Am Chem Soc 109:3466–3468
Lee MD, Dunne TS, Siegel MM et al (1987b) Calichemicins, a novel family of antitumor antibiotics. 1. Chemistry and partial structure of calichemicin γ_1^I. J Am Chem Soc 109:3464–3466
Lopus M, Oroudjev E, Wilson L et al (2010) Maytansine and cellular metabolites of antibody-maytansinoid conjugates strongly suppress microtubule dynamics by binding to microtubules. Mol Cancer Ther 9:2689–2699
Luesch H, Harrigan GG, Goetz G et al (2002) The cyanobacterial origin of potent anticancer agents originally isolated from sea hares. Curr Med Chem 9:1791–1806
Madden T, Tran HT, Beck D et al (2000) Novel marine-derived anticancer agents: a phase I clinical, pharmacological, and pharmacodynamic study of dolastatin 10 (NSC 376128) in patients with advanced solid tumors. Clin Cancer Res 6:1293–1301
Mandelbaum-Shavit F, Wolpert-DeFilippes MK, Johns DG (1976) Binding of maytansine to rat brain tubulin. Biochem Biophys Res Commun 72:47–54
Mantaj J, Jackson PJ, Rahman KM et al (2017) From anthramycin to pyrrolobenzodiazepine (PBD)-containing antibody-drug conjugates (ADCs). Angew Chem Int Ed 56:462–488
Martin C, Ellis T, McGurk CJ et al (2005) Sequence-selective interaction of the minor-groove interstrand cross-linking agent SJG-136 with naked and cellular DNA: footprinting and enzyme inhibition studies. Biochemistry 44:4135–4147
Minami H, Fujii H, Igarashi T et al (2001) Phase I and pharmacological study of a new camptothecin derivative, exatecan mesylate (DX-8951f), infused over 30 minutes every three weeks. Clin Cancer Res 7:3056–3064
Mirsalis JC, Schindler-Horvat J, Hill JR et al (1999) Toxicity of dolastatin 10 in mice, rats and dogs and its clinical relevance. Cancer Chemother Pharmacol 44:395–402
Mitsui I, Kumazawa E, Hirota Y et al (1995) A new water-soluble camptothecin derivative, DX-8951f, exhibits potent antitumor activity against human tumors in vitro and in vivo. Jpn J Cancer Res 86:776–782
Newman RA, Fuentes A, Covey JM et al (1994) Preclinical pharmacology of the natural marine product dolastatin 10 (NSC 376128). Drug Metab Dispos 22:428–432
Pettit GR, Kamano Y, Herald CL et al (1987) The isolation and structure of a remarkable marine animal antineoplastic constituent: dolastatin 10. J Am Chem Soc 109:6883–6885
Pitot HC, McElroy EA, Reid JM et al (1999) Phase I trial of dolastatin-10 (NSC 376128) in patients with advanced solid tumors. Clin Cancer Res 5:525–531
Pommier Y (2006) Topoisomerase I inhibitors: camptothecins and beyond. Nat Rev Cancer 6:789–802
Puzanov I, Lee W, Chen AP et al (2011) Phase I pharmacokinetic and pharmacodynamic study of SJG-136, a novel DNA sequence selective minor groove cross-linking agent, in advanced solid tumors. Clin Cancer Res 17:3794–3802
Reichardt P, Nielsen OS, Bauer S et al (2007) Exatecan in pretreated adult patients with advanced soft tissue sarcoma: results of a phase II--study of the EORTC Soft Tissue and Bone Sarcoma Group. Eur J Cancer 43:1017–1022
Remillard S, Rebhun LI, Howie GA et al (1975) Antimitotic activity of the potent tumor inhibitor maytansine. Science 189:1002–1005
Rowinsky EK, Johnson TR, Geyer CE Jr et al (2020) DX-8951f, a hexacyclic camptothecin analog, on a daily-times-five schedule: a phase I and pharmacokinetic study in patients with advanced solid malignancies. J Clin Oncol 18:3151–3163

Royce ME, Hoff PM, Dumas P et al (2001) Phase I and pharmacokinetic study of exatecan mesylate (DX-8951f): a novel camptothecin analog. J Clin Oncol 19:1493–1500
Royce ME, Rowinsky EK, Hoff PM et al (2004) A phase II study of intravenous exatecan mesylate (DX-8951f) administered daily for five days every three weeks to patients with metastatic adenocarcinoma of the colon or rectum. Investig New Drugs 22:53–61
Sharma S, Kemeny N, Schwartz GK et al (2001) Phase I study of topoisomerase I inhibitor exatecan mesylate (DX-8951f) given as weekly 24-hour infusions three of every four weeks. Clin Cancer Res 7(12):3963–3970
Sun FX, Tohgo A, Bouvet M et al (2003) Efficacy of camptothecin analog DX-8951f (exatecan mesylate) on human pancreatic cancer in an orthotopic metastatic model. Cancer Res 63:80–85
Takiguchi S, Kumazawa E, Shimazoe T et al (1997) Antitumor effect of DX-8951, a novel camptothecin analog, on human pancreatic tumor cells and their CPt-11-resistant variants cultured in vitro and xenografted into nude mice. Jpn J Cancer Res 88:760–769
Tarantino P, Ricciuti B, Pradhan SM et al (2023) Optimizing the safety of antibody–drug conjugates for patients with solid tumours. Nat Rev Clin Oncol 20:558–576
Thorson JS, Sievers EL, Ahlert J et al (2000) Understanding and exploiting nature's chemical arsenal: the past, present and future of calicheamicin research. Curr Pharm Des 6:1841–1879
Tiberghien AC, Levy JN, Masterson LA et al (2016) Design and synthesis of tesirine, a clinical antibody–drug conjugate pyrrolobenzodiazepine dimer payload. ACS Med Chem 7:983–987
Tolcher AW, Ochoa L, Hammond LA et al (2003) Cantuzumab mertansine, a maytansinoid immunoconjugate directed to the CanAg antigen: a phase I, pharmacokinetic, and biologic correlative study. J Clin Oncol 21:211–222
Tsuchikama K, Anami Y, Ha SY et al (2024) (2024) Exploring the next generation of antibody-drug conjugates. Nat Rev Clin Oncol 21:203–223
Vaishampayan U, Glode M, Du W et al (2000) Phase II study of dolastatin-10 in patients with hormone-refractory metastatic prostate adenocarcinoma. Clin Cancer Res 6:4205–4208
Venditto VJ, Simanek EE (2010) Cancer therapies utilizing the camptothecins: a review of the in vivo literature. Mol Pharm 7:307–349
Verschraegen CF, Kudelka AP, Hu W et al (2004) A phase II study of intravenous exatecan mesylate (DX-8951f) administered daily for 5 days every 3 weeks to patients with advanced ovarian, tubal or peritoneal cancer resistant to platinum, taxane and topotecan. Cancer Chemother Pharmacol 53:1–7
Von Mehren M, Balcerzak SP, Kraft AS et al (2004) Phase II trial of dolastatin-10, a novel anti-tubulin agent, in metastatic soft tissue sarcomas. Sarcoma 8:107–111
Widdison WC, Wilhelm SD, Cavanagh EE et al (2006) Semisynthetic maytansine analogues for the targeted treatment of cancer. J Med Chem 49:4392–4408
Wolpert-Defilippes MK, Adamson RH et al (1975a) Initial studies on the cytotoxic action of maytansine, a novel ansa macrolide. Biochem Pharmacol 24:751–754
Wolpert-DeFilippes MK, Bono VH Jr, Dion RL et al (1975b) Initial studies on maytansine-induced metaphase arrest in 11210 mürine leukemia cells. Biochem Pharmacol 24:1735–1738
Wu J, Clingen PH, Spanswick VJ et al (2013) γ-H2AX foci formation as a pharmacodynamic marker of DNA damage produced by DNA cross-linking agents: results from 2 phase I clinical trials of SJG-136 (SG2000). Clin Cancer Res 19:721–730
Zein N, Sinha AM, McGahren WJ et al (1988) Calicheamicin γ_1^I: an antitumor antibiotic that cleaves double-stranded DNA site specifically. Science 240:1198–1201
Zein N, Poncin M, Nilakantan R et al (1989) Calicheamicin γ_1^I and DNA: molecular recognition process responsible for site-specificity. Science 244:697–699

Chapter 3
Selecting an Optimal Antibody for Antibody–Drug Conjugate Therapy

Michael Ritchie, Laird Bloom, Gregory Carven, and Puja Sapra

Abstract Antibody–drug conjugates (ADCs) represent an increasing proportion of therapeutics in preclinical and clinical development for the treatment of oncology indications. ADC-based therapies are attractive due to their potential to increase efficacy and reduce the systemic toxicity produced by other conventional anticancer regimens. ADCs are complex molecules, however, and their clinical success depends both on the properties of the target in the context of the disease and on the properties of the ADC itself: the antibody, the toxin payload, and the linker, to which it is coupled, and the site and chemistry of linkage. In this chapter, we review some of the strategies that can be utilized to generate and characterize optimal ADC antibodies, including selection of appropriate targets, development of tools for antibody generation and screening, approaches to antibody isolation, advanced screening strategies for lead antibody selection, antibody engineering to increase selectivity and potency, and selection of appropriate combinations of linker, payload, and linker attachment strategy.

Keywords Antibody · Antibody–drug conjugate · Screening · Internalization · Toxin · Hybridoma · Phage display

M. Ritchie · P. Sapra (✉)
Bioconjugates Discovery and Development, Oncology Research Unit, Pfizer Worldwide Research and Development, Pearl River, NY, USA
e-mail: Puja.Sapra@pfizer.com

L. Bloom · G. Carven
Global Biotherapeutic Technologies, Pfizer, Cambridge, MA, USA

J. L. Zaro, J. Wang (eds.), *Antibody-Drug Conjugates*, AAPS Advances in the Pharmaceutical Sciences Series 17, https://doi.org/10.1007/978-3-031-83005-1_3

3.1 Target Selection for an ADC

3.1.1 *Characteristics of an Optimal ADC Target*

Selection of appropriate targets is likely the most important consideration for the success of an ADC development program. Target selection for ADCs can be classified into two principal approaches: target-first and antibody-first, or agnostic (Fig. 3.1). In the target-first approach, a target is chosen based on a set of factors including the expression pattern, abundance, and internalization properties, and an

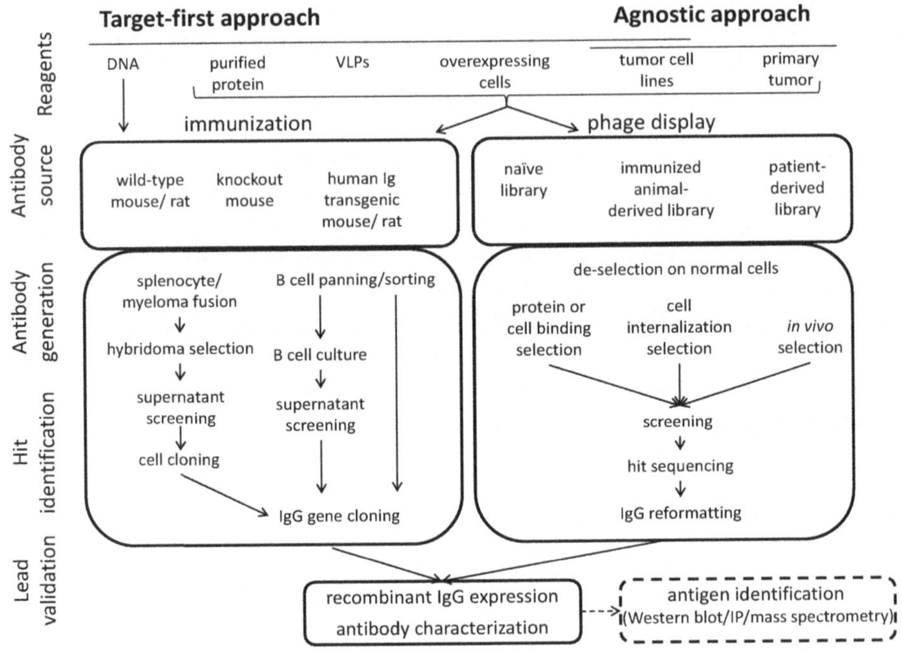

Fig. 3.1 Methods for antibody isolation and development via the agnostic and target-first approaches. Antibody isolation and validation can be performed with specific targets in mind ("target-first approach") or on tumor cells of interest without bias toward specific targets ("agnostic approach"). When the target is known, reagents for antibody generation can be as diverse as cDNA constructs for DNA immunization; purified target protein (full-length or domain); engineered cells or cell surrogates: virus-like particles (VLPs) or other target-enriched membranes; or tumor cells known to express the target. A target-agnostic approach will usually utilize primary tumor cells or tumor lines for antibody generation. The source for antibodies can be immunized animals, cancer patients, or phage display libraries with antibodies from naïve or tumor-exposed sources. A wide range of techniques, from conventional hybridoma generation or phage panning to B cell sorting and V gene isolation from immune sources or selection of phage binding to or internalizing in cells expressing the target. Once lead antibodies are cloned and selected, the antibodies can be recombinantly expressed and validated. If the target of an internalizing antibody is not known, target identification can often be done through mass spectrometry once a recombinant antibody is expressed

antibody generation campaign is focused on the isolation of antibodies against this target. In the agnostic approach, antibodies capable of binding to and internalizing in tumor cells are isolated, followed by retrospective identification of their targets. While the target-first approach may have the advantage of a higher degree confidence in the suitability and novelty of the target based on prior knowledge of target properties, it requires significant work before internalizing antibodies become available. By contrast, the agnostic approach leads quickly to reagents suitable for testing hypotheses directly, although this strategy tends to favor highly expressed surface antigens, which are likely to have been described previously.

3.1.2 Selection and Validation of Tumor-Associated Targets Using Omics Approaches

Most strategies used for the identification of targets amenable to the ADC approach resemble strategies used to identify tumor-associated cell surface antigens for other oncology biotherapeutic approaches, though with a few notable differences. The ADC concept of arming antibodies or antibody fragments with toxins or cytotoxic payloads relies on the selectivity of the antibody toward the target antigen to increase efficacy while reducing toxicity compared to nontargeted delivery. Thus, the ideal ADC target will be expressed highly in tumor cells and minimally across normal tissues. Multiple methods have been employed to identify differentially expressed proteins. Usually, this involves comparison of tumor samples (tissue biopsies, tumor cell lines, patient-derived xenografts, etc.) with closely related nontumor cells. Comparison can be done at the DNA, mRNA, or protein levels using genomic (e.g. CSH and FISH), transcriptomic (microarray), and/or proteomic techniques (2DE/MS, ICAT) reviewed elsewhere (Guo 2003; Shiio and Aebersold 2006; Yaziji and Gown 2004). The output of such omics studies is typically a list of genes for which expression is increased in tumor samples relative to normal tissues.

The actual number of tumor-associated targets that are identified using such techniques is dependent on the scale and the methodology used in the selection process. It is quite common for a single study to identify tens to hundreds of potential targets that meet the criteria for tumor association. However, since selectivity is only one of multiple critical factors involved in target selection for ADCs, such a list is often just a starting point for the validation of multiple targets in parallel.

Narrowing the list of potential targets can be achieved by additional diligence around each target to assess druggability, internalization, and other factors. For ADCs, the target molecule must reside on tumor cells in a location accessible to a systemically administered therapeutic antibody. Confirmation that a target is localized to the plasma membrane and contains extracellular regions that could be targeted with an antibody is a critical early validation step. For annotated proteins, assessment of topology may be performed using a protein knowledge base such as UniProt (2013) to ensure that the protein contains at least one predicted transmembrane region or is anchored to the cell surface through a glycosylphosphatidylinositol (GPI) link. If the target is novel, the EXPASY bioinformatic portal (www.expasy.org) (Gasteiger et al. 2003) contains numerous structure prediction tools that can be used to define features such as transmembrane helices, secondary structural elements such as intra–/extracellular domains, as well as sites for predicted protein modification.

For confirmation of cell surface localization, flow cytometry is commonly used if flow-compatible detection antibodies are available. Quantitative flow analysis can be useful to provide data on antigen density as the level of tumor (vs. normal) expression is an important component of target selection. For targets for which flow-compatible tools are unavailable, recombinant expression using genetic fusion to epitope tags (such as FLAG or myc) can be used. Alternatively, the target protein can be fused to a green fluorescent protein and subsequently probed by microscopy.

Determining whether (and to what extent) a given antigen can be shed from the cell membrane is another important consideration. Many cell surface receptors are shed, resulting in a portion (usually the extracellular domain(s)) being released into the circulation. Shed target protein at high levels may function as a sink, diverting the therapeutic antibody and its associated payload away from the tumor and thereby lowering the potency of the ADC and potentially causing increased liver clearance and associated systemic toxicity (Lin and Tibbitts 2012). It is known that a number of tumor-specific surface antigens, such as carcinoma antigen 125 (CA125), prostate-specific antigen (PSA), Human Epidermal Growth Factor receptor 2 (HER2), and others, are actively shed from cancer cells (Kulasingam and Diamandis 2008). The presence of circulating antigens can be expected to have a significant influence on the delivery of ADC payloads that use these antigens as the delivery target. While it is generally thought that targets that are shed from the cell surface should be avoided for the development of ADCs, the clinical efficacy of Kadcyla (anti-HER2-ADC)(Burris et al. 2011) indicates that the presence of shed extracellular domain need not rule out an otherwise-promising ADC target. Indeed, a recent study suggests that the presence of a soluble antigen may be beneficial for targeting solid tumors (Pak et al. 2012).

3.2 Antibody Characteristics

3.2.1 Optimal Antibody Characteristics

In addition to the careful selection of the tumor-specific antigen, characteristics of the antibody such as affinity and molecular size will have a significant impact on the efficacy of an ADC. Accordingly, there are a vast number of strategies available to characterize the antibody characteristics. While high-affinity antibodies often stand out in screening campaigns, it is worth noting that high affinity does not always correlate with high efficacy, as a high-affinity antibody binding to a rapidly internalizing target may be rapidly eliminated from circulation. High affinity may also reduce the ability of the antibody to penetrate far within the tumor due to tight binding to and/or rapid internalization into the first cells it encounters, a phenomenon termed the binding-site barrier (Rudnick et al. 2011; Adams et al. 2001). A recent study has examined how efficiently a panel of anti-HER2 antibodies with differing affinities can penetrate through a solid tumor. All of these antibodies bind to the same epitope of HER2, but do so with log-fold differences in affinity (Rudnick et al. 2011). The antibody with the lowest affinity (~200 nM K_D) penetrated through the tumor with the highest efficiency; however, this antibody possessed the lowest tumor retention due to its low affinity and the resultant inability to remain bound to the receptor (Rudnick et al. 2011). Conversely, antibodies with a higher affinity ($K_D < 1$ nM) for HER2 were largely confined to the perivascular space of the tumor (Rudnick et al. 2011). These high-affinity antibodies exhibited lower penetration because they are rapidly internalized and catabolized by the first cells encountered after extravasation. Conversely, the lower-affinity antibodies were not bound long enough to internalize and thus were able to continue moving through the tumor space (Rudnick et al. 2011). Accordingly, there exists a threshold whereby antibodies that possess a k_{off} value faster than the K_e value of receptor internalization can avoid internalization and catabolization by the cell, whereas antibodies with k_{off} slower than the K_e value of receptor internalization are internalized and catabolized more readily (Rudnick et al. 2011). A balance of the two rates, allowing effective delivery of toxin throughout the tumor, appears to be achieved by antibodies of moderate affinity (between 7 and 23 nM, in the case of the HER2 antibodies examined by Rudnick et al. (2011)).

Modeling and experimental analyses have shown that there is a fine balance between the molecular size and affinity of a therapeutic protein and the degree of tumor retention and penetration. Accordingly, a complex correlation exists in which modalities of intermediate size (~25 kDa) have the lowest tumor uptake, while therapeutic modalities of extremely small (<20 kDa) and large (>100 kDa) size exhibit a higher tumor uptake and retention (Schmidt and Wittrup 2009). This is due to the fact that small biotherapeutics can penetrate through a solid tumor rapidly without constraint, although a high affinity for the target antigen is required for the therapeutic to be retained (Schmidt and Wittrup 2009). However, while these smaller modalities can penetrate further through the tumor, they often suffer from faster serum clearance and shorter half-lives (Schmidt and Wittrup 2009). Alternatively, larger proteins usually exhibit reduced tumor penetration; however, they can achieve high accumulation due to their increased serum half-life (Schmidt and Wittrup 2009). Accordingly, intermediate-sized therapeutics (~25 kDa) appear to be the least optimal because they suffer from both a degree of restriction in tumor penetration and possess a reduced serum half-life.

Species cross-reactivity is also a practical consideration for many antibody discovery programs and is particularly important for ADC development due to the complex nature of toxicology studies. Alignment of the target amino acid sequence with that of its orthologs in other species can be useful to determine the identity or homology between human and another species such as nonhuman primate or rodent target. Sequence identity serves as a rough predictor of the likelihood for obtaining cross-reactive antibodies (i.e., there is a higher probability for an antibody to be cross-reactive to an ortholog when the antigens share high sequence identity). Similarly, it is difficult to raise cross-reactive antibodies when target identities are low (60% identical or lower). Thus, when faced with multiple potential targets coming from omics studies, priority may be given to follow-up on targets with high sequence homology. Linear sequence identity, however, does not always translate to the three-dimensional structure of a protein. For example, a ligand-binding site may be structurally well-conserved across species, despite lower overall homology in other areas of the protein. For this reason, it is common practice during an antibody generation campaign to screen for antibody candidates that bind to the human target and its nonhuman primate and/or rodent orthologs.

Finally, as the cellular uptake of an antibody is a critical component of its efficacy, assessment of a target's ability to internalize and subsequently shuttle its antibody and payload to the appropriate intracellular compartment (usually the lysosome) is of utmost importance. Strategies to measure internalization and trafficking of ADCs are presented in subsequent sections.

3.2.2 Engineering pH Dependency of Antibody Binding

Recently, the successful engineering of antibodies to have a dramatically lower affinity at pH 6.0 than at neutral pH has attracted interest as a method for the manipulation of the kinetics of antibody delivery to the lysosome Bouchard et al. 2009. pH dependence is achieved through the introduction of histidine residues at or near the antibody/antigen binding interface. These histidine residues are protonated < pH 6.0 and deprotonated >pH 6.0 and, if placed correctly, can have a profound effect on binding affinity in different pH environments. Antibodies against PCSK9 and IL-6R have been engineered to dissociate from their targets in the early endosome (pH ≤ 6.0) while retaining binding to the FcRn receptor, which returns them to the cell surface and prevents degradation in the lysosome (Chaparro-Riggers et al. 2012). A similar approach has been used to *increase* antibody trafficking to the lysosome in cells lacking endosomal FcRn expression Burris et al. 2011, which would be desirable for the delivery of ADCs requiring degradation for the release of the payload. However, the tumor microenvironment is often weakly acidic, which would reduce the cell-surface binding of pH-sensitive ADCs. Further investigation is required to understand the optimal degree of pH dependence for ADCs targeted to solid tumors, such that binding at the cell surface is not compromised.

3.2.3 Engineering Fc-mediated Interactions

Fc-mediated effector function accounts for most of the in vivo efficacy imparted by unconjugated mAbs in cancer therapy. However, the role of Fc effector function in the inhibition of tumor growth by an ADC in the clinic remains unclear. Several strategies have been employed to enhance immune-mediated effector functions of unconjugated mAbs, including afucosylation as well as mutations in the antibody Fc region, both of which can enhance FcγR binding and amplify antibody-dependent cell-mediated cytotoxicity (ADCC). While this could potentially increase the potency of an ADC, effector mechanisms may lead to undesirable side effects such as cross-linking-related agonistic effects, cytokine storm, or platelet aggregation (Chatenoud et al. 1990; Wing et al. 1996; Langer et al. 2005). Thus, effector functions may not be desirable when the targeted delivery of the payload is sufficient for antitumor activity.

Antibody isotypes IgG2 and IgG4 have inherently reduced binding to FcgRII. Accordingly, these isotypes have been utilized in ADCs where the effector function is not required for tumor killing (e.g., calicheamicin-based ADCs). However, given the high potency of these payloads, the concentration at which these ADCs are dosed in patients is low enough (1.8–9 mg/m^2) where eliciting effector function is likely to be minimal. However, ADCs possessing less potent payloads, such as microtubule inhibitors, may be dosed at higher levels (1–5 mg/kg) in the clinic, in which case effector functions may come into play.

While advances in antibody engineering have allowed for the enhancement and elimination of effector function-mediated cytotoxicity of a tumor, little has been done to dissect what role(s), if any, effector function has in ADC-mediated tumor killing. As a result, three isotypes (IgG1, IgG2, and IgG4) are being utilized for ADC development programs. One preclinical study of an anti-CD70 antibody conjugated to maleimidocaproyl monomethyl auristatin F (mcMMAF) compared the activity of this ADC with isotype variants (IgG1, IgG2, and IgG4) and mutations that eliminate FcγR binding (IgG1v1 and IgG4v3). Interestingly, IgG1v1-ADC lacking FcγIIIa binding was the most potent ADC in the preclinical models of renal cancer and glioblastoma. Compared to the parental IgG1 ADC, this mutant displayed enhanced tumor exposure of the drug and an improved therapeutic index in mice (McDonagh et al. 2008). Accordingly, the importance of eliciting effector function in ADC-mediated tumor killing remains an important, unanswered question.

3.3 Agnostic Approach to Antibody Production

Whole-cell immunization has been used for decades to identify antigens located on the surface of cells (Williams et al. 1977). Immunophenotyping of hematopoietic cells based on surface markers identified by antibodies raised against cells at different stages of growth became the basis for the "clusters of differentiation" (or CD antigen) classification system (Bernard and Boumsell 1984). Targets with selective expression on tumor cells can also be identified using hybridomas generated via immunization with tumor cells. Many well-known tumor antigens such as prostate-specific membrane antigen (PSMA) (Israeli et al. 1993) or Lewis^Y (Hellstrom et al. 1990) were identified in this manner. Typically, this includes a differential binding screen on tumor cells and a related nontumorigenic cell line. Screening can be performed using cell-based ELISA or FACS-based methods using intact cells.

The efficiency of identifying tumor antigens via whole-cell immunization approaches can be improved using techniques such as subtractive immunization (Bickel et al. 2000; Sleister and Rao 2002). Subtractive immunization is a technique that can be used to produce monoclonal antibodies specific for antigens (such as tumor antigens) that are present in low abundance in a protein mixture or are poorly immunogenic due to the presence of immunodominant epitopes present within the sample. Here, neonatal mice are tolerized by the injection of nontumorigenic tissue before thymic selection occurs, followed by a more traditional immunization with tumor tissues after the maturation of the mouse immune system.

Phage display has been used to identify panels of antibodies to surface antigens on a large variety of tumor cell lines, primary tumors, tumor tissue sections, cells captured by laser capture microdissection (Ruan et al. 2006), and even primary tumors in cancer patients, the latter injected with the phage several hours prior to tumor surgery (Shukla et al. 2013). In order to identify tumor-specific antibodies, libraries are typically exposed to non-target-bearing cells (e.g., noncancerous cell lines, normal cells from the same tissue, different cancer types from the same tissues, etc.) (Poul et al. 2000; Liu et al. 2004) in order to remove antibodies to common antigens, although this step is often applied following an initial round of selection on tumor cells to avoid the depletion of rare antibodies (Zhou et al. 2012). Methods for isolating tumor cell-binding phage include simple washing and elution with standard low- and high-pH buffers (Zhou and Marks 2012; Poul et al. 2000; Becerril et al. 1999), centrifugation through organic solvent to separate cells from an unbound phage (Akahori et al. 2009), and internalization (see below). Some extensive campaigns have examined over 50 tumor lines and over 10,000 resulting scFv (Bouchard et al. 2009; Kurosawa et al. 2009).

Selection for phage that internalizes into tumor cells can reduce the recovery of antibodies that bind non-tumor-specific or non-internalizing targets and has been used successfully on many tumor types (Liu et al., 2004); reviewed by Zhou et al., 2012; and also Zhou and Marks, 2012; Poul et al., 2009; Fransson and Borrebaeck, 2009). Selection for internalizing phage involves allowing the phage to bind to the cell surface at 4 °C to prevent internalization, transfer to 37 °C for a brief period (typically 15–60 min.) to allow internalization, followed by washing and stripping surface-bound phage with a low-pH buffer, and recovery of internalized phage by cell lysis and infection of *E. coli* with the cell lysate. Factors that can favor the recovery of internalized phage include a high abundance of the target on the cell surface, rapid constitutive target internalization, high affinity, accessible epitope, and multivalent phage (Zhou and Marks 2012; Zhou et al. 2012; Burris et al. 2011). While multivalent display (usually five copies per phage) allows greater recovery of lower-affinity clones than does monovalent display (Burris et al. 2011), and therefore is likely to lead to pools of higher diversity, numerous examples exist of monovalent phagemid systems giving rise to internalizing clones from naïve human scfv library selections (Osborn et al. 2013; Sapra et al. 2013; McDonagh et al. 2008).

Identification of antibody targets from whole-cell phage selection or whole-cell immunization can be challenging. Antigens are frequently identified by immunoprecipitation with the isolated cell-surface-binding antibodies, followed by proteolytic digestion and mass spectrometry to identify bound peptides (Kurosawa et al. 2008, 2009); see also (Poul et al. 2000; Goenaga et al. 2007). Overexpression of the candidate gene on a test cell or reduction of the candidate gene expression via siRNA can be used to confirm the identity of the candidate antigen. In one example, selection for phage internalizing in MDA-MB-231 breast cancer cells was followed by selection on yeast cells expressing either of the two candidate proteins, leading to the isolation of antibodies recognizing the two candidates (Zhou et al. 2010). The authors suggest that this method could be applied on a larger scale to isolate cell-binding antibodies to a large panel of candidate genes.

Large-scale tumor antibody screening has required a series of steps to narrow down candidates. Kurosawa and co-workers (Kurosawa et al. 2008, 2009, 2012) first used patterns of binding to multiple tumor cell lines or tissue sections to narrow down a panel of several hundred antibodies into clusters, followed by a combination of ELISA and western blotting against known antigens and immunoprecipitation/mass spectrometry to identify new antigens. Liu et al. (2004) selected and characterized a large panel of internalizing antibodies on prostate cancer cells, similarly using binding profiles to group cells into different epitope clusters.

Studies that have attempted to address the spectrum of tumor surface proteins from several approaches in parallel have suggested that certain common antigens are detected by different technologies (e.g., phage display, hybridoma, and proteomic approaches) but that the majority of antigens identified are different, pointing to the biases and technical limitations of relying on any single method (Rust et al. 2013).

3.4 Target-First Approach to Antibody Generation

3.4.1 Considerations and Preparation of Target for Antibody Screening

A panel of high-quality reagents is key to a successful campaign. Many of the tools used for target validation may also be useful during antibody generation and screening. Such tools typically include target DNA, expression plasmids, native and recombinant cell lines, purified proteins, and control or reference antibodies. When generating screening reagents, it is important to consider generating target orthologs in parallel which can be used to test for species' cross-reactivity of antibody hits.

The nature and inherent characteristics of target molecules selected for ADC intervention vary considerably. Nearly all are expressed on the cell surface, whether directly on tumor cells or the tumor vasculature. As integral membrane proteins, they are permanently attached to the cell membrane and typically contain one or more membrane-spanning alpha-helices. Such proteins can only be separated from the biological membranes using detergents, nonpolar solvents, or denaturing agents.

For this reason, the vast majority of protein antigens used for antibody generation and screening are broken into smaller soluble subdomains or extracellular domains of cell-surface receptors. These fragments can be designed to be recombinantly expressed for use as immunogens or selection reagents. The precise determination of the domain boundaries is often critical, and small changes in an expression construct may result in drastically different expression yields.

Recombinant production of target antigens can be done using either prokaryotic or eukaryotic hosts. The choice of the host likely depends on the nature of the target molecule. Most importantly, the yield should be a homogeneous, properly folded antigen such that the critical epitopes are accessible (Ebersbach and Geisse 2012). For rapid generation of material, transient expression in HEK293 or CHO cells is preferred. For many proteins, addition of tags, such as HA, FLAG, or His_6, will allow easy purification from complex cell culture media. Fc-based fusion proteins, composed of an immunoglobulin Fc domain genetically fused to the target antigen, may also be used. The Fc domain folds independently and can improve the solubility and stability of the partner molecule, both in vitro and in vivo, while also providing a convenient purification handle for protein-A/G affinity chromatography.

For oncology targets, transformed cell lines expressing the target are often readily available. These cell lines are potent immunogens for the generation of antibodies, resulting in a strong immune response in animals (Kung et al. 1979). Whole cells are complex antigens, made up of many cell-surface proteins, each a potential antigen. Thus, extensive screening is often required to generate antibodies specific for a singular antigen of interest. To reduce complexity during cell-based screening, recombinant overexpressing cell lines can be used. The full-length target antigen can be expressed in commonly used cell systems such as HEK293 or CHO cells. Nontransfected, parental control cells expressing low or no target molecules can be used for counter-screening.

To maximize the specific antibody response in the mouse upon immunization with whole cells, human target antigens can also be overexpressed on the surface of mouse cell lines derived from a strain syngenic to the immunization host (Ashley et al. 1997; Uchida et al. 2004). Unlike human cells, murine cells expressing recombinant antigen will typically exhibit a reduced background response as their entire cell surface components (apart from the transfected and expressed target molecule) are of mouse origin and therefore tolerized in the animal. An ideal antibody generation campaign may employ native expressing tumor cell lines and recombinant proteins as well as engineered cell lines on different backgrounds.

3.4.2 Approaches to Antibody Isolation

Many of the currently marketed therapeutic antibodies have been generated using traditional hybridoma technology developed by Kohler and Milstein (1975). Hybridoma cell lines are generated from the stable fusion of immortalized myeloma cells with immunoglobulin-producing B cells, which is inherently an inefficient process. Despite its limitations, hybridoma technology has been highly successful and has historically been the most widely used method for antibody generation. More recently, methods to identify antibodies in a more efficient manner have been employed. These include culture and expansion of antibody-producing B cells from immunized rodents using CD40/CD40L (Wen et al. 1987; Weber et al. 2003) as well as flow cytometry-based sorting of APCs using fluorescently labeled antigen, followed by RT-PCR from cells (Berry et al. 2009; Kurosawa et al. 2012).

Each of the above methods relies on the immunization of rodents or other animals as a source of antibody-producing immune cells. Mice are often used for immunization due to their well-defined genetics, short generation times, and low maintenance costs. The use of knockout mouse strains that have a targeted deletion of a particular gene of interest makes it easier to raise an immune response to some protein targets that are highly conserved among species. When knockout mice are not available, rats or other species can be used to facilitate the generation of human/mouse cross-reactive antibodies.

Human immunoglobulin transgenic rodents provide a method of obtaining fully human monoclonal antibodies using conventional hybridoma technology (Fishwild et al. 1996; Mendez et al. 1997; Osborn et al. 2013; Ma et al. 2013). Transgenic rodents offer a powerful way to generate antibodies to validated targets relatively rapidly without the need for humanization. The benefits of transgenic rodents must be balanced, however, against generally weaker immune responses, requiring more effort to generate equivalent panels of antibodies compared to wild-type rodents (Lonberg 2005). The relatively high technology access cost and limited availability of transgenic animals further contribute to the continued preference for nontransgenic animals among most researchers.

Immunization methods vary depending on the nature of the target and the reagents available during an antibody campaign. When the target is novel, there may be few available tools at the outset of a project. Requiring only plasmid DNA, genetic immunization can be initiated before other tools such as recombinant protein or engineered cell lines can be generated. DNA can be delivered by several means, including injection, a gene gun using DNA-coated gold beads, or intramuscular electroporation (Robinson and Pertmer 2001). The encoded antigen is taken up, translated, and presented by immune cells, such as dendritic cells, and elicits both a strong humoral and cellular immune response. Co-administration of DNA-encoding immune mediators such as GMCSF and Flt3L has been shown to increase the number of infiltrating dendritic cells at the immunization site, resulting in the overall increased antigen-specific responses.

Recent improvements to the technique have shown that DNA immunization is capable of generating immune responses against antigens such as GPCRs and ion channels, which are typically considered intractable to antibody generation (Hazen et al. 2014). Generally, immune responses from DNA immunization take longer to develop than those elicited by other methods, but the delayed response time is often offset by the time required to generate immunogens for more traditional approaches.

Rapidly generating a panel of antibodies against multiple targets can be very useful for ADC target validation efforts. Typical rodent immunization methods require two or more months of injections before immune responses are of sufficient titer to be used for antibody generation. Rapid immunization methods such as RIMMS (repetitive immunization, multiple sites) can be used to reduce the period of time between initiation and screening for a panel of hits, without sacrificing either the quantity or quality of hits in a given campaign (Kilpatrick et al. 1997). RIMMS capitalizes on rapid hypermutation and affinity maturation events that occur in secondary lymphatic tissue early in response to antigenic challenges. Small concentrations of antigen are injected at several subcutaneous sites three times per week over the course of 2 to 3 weeks. Following antigen challenge, draining lymph nodes from each of the injection sites are collected and used for fusion or B cell-based antibody screening.

Phage display provides the opportunity to guide antibody populations toward those with particular characteristics: for example, toward antibodies binding to a tumor cell type and not to a set of normal cells; toward antibodies binding to a known epitope on a known target; toward antibodies that are internalized rapidly; or a combination of these traits. Phage-derived antibodies generally have lower affinities than those derived from immunized animals (K_D ranging from 1–100 nM vs 0.1–10 nM), but due to the considerations of the binding site barrier described above and the likelihood that moderate-affinity antibodies will avoid internalization into low-expressing normal cells while still internalizing in high-expressing tumor cells, antibodies in the low double-digit nanomolar range are likely to be suitable for ADCs (Rudnick et al. 2011).

The most straightforward phage display approaches have used purified or recombinant antigens for phage selection in either solution or in solid phase, followed by screening for binding to the antigen expressed on the surface of engineered cells or tumor cells (Zhou et al. 2012). The advantage of this approach is that the target specificity of the antibodies is known, and binding to the more complex cell surfaces is only confirmatory. The chief disadvantages are that the approach requires the generation of high-quality protein reagents, and even following successful selections, antibodies selected for binding to purified antigen frequently fail to recognize the target in the context of the tumor cell. Selection of phage for binding to cells, including tumor cell lines, tumor tissue samples, or mammalian expression lines engineered to overexpress a specific target, reduces the risk that clones will fail to bind cell-surface target, but as cell surface complexity increases, the task of separating clones that bind to targets of interest from clones that do not becomes increasingly challenging.

Selection on cells overexpressing the surface protein of interest is also a path to isolating antibodies of known specificity. The advantage of this method is that it favors the isolation of antibodies that bind specifically to the cell-surface form of a known target (Hoogenboom et al. 1999). Generation of stable mammalian cell lines can be time-consuming, but often a transfected cell line and its untransfected parent can be used both for selection and screening of antibodies and for providing clear evidence for target specificity. Recombinant lines may overestimate the ability of low-affinity antibodies to bind cells expressing lower levels of a target and may have altered internalization mechanics, glycosylation, or association with other proteins, so validation of antibody properties on tumor cells is essential.

Often, a combination of methods is appropriate. If the target is known, phage selections using alternating rounds on purified proteins and cells can improve the likelihood that the resulting clones will bind specifically to the desired target as presented on tumor cells. Similarly, combining selection on purified protein and internalizing selections on tumor cells can focus the selected pool on clones of the desired phenotype. It is difficult to predict what characteristics observed for scFv of Fab displayed on the phage will translate to the characteristics in IgG format, and for this reason, it is often advantageous to convert as diverse a set of primary isolates to IgG as possible to confirm their binding and internalization profiles.

3.5 Screening for Internalization and Intracellular Trafficking

3.5.1 Screening by Microscopy and Flow Cytometry

Current linker chemistry has been designed such that cytotoxic payloads are cleaved from the antibody, via an assortment of mechanisms, within the intracellular space of the cell while remaining highly stable in the extracellular space. Thus, it is necessary to demonstrate efficient internalization of the selected antibodies. Depending on reagent availability, antibody binding and specificity can be determined in a relatively high throughput by ELISA or other protein-binding assays and in somewhat lower throughput by cell-based ELISA or flow cytometry on target-expressing engineered cell lines or tumor lines. Internalization is frequently tested in a subsequent step by either fluorescence-based internalization assays (Fig. 3.2a) or by testing the ability of the antibody to deliver toxins intracellularly (Fig. 3.2b).

When using fluorescence-based internalization assays, the screened antibodies can either be complexed with a fluorophore-conjugated modality (such as a Fab fragment or liposome) or directly conjugated with a fluorophore. The use of fluorophore-labeled secondary antibodies is the quickest way to tag an antibody or pool of antibodies. However, since the size and valency of a ligand can effect internalization, the presence of a secondary antibody may alter the internalization properties of the primary antibody (Nielsen et al. 2006). Alternatively, antibodies can be biotinylated with NHS-SS-biotin and subsequently complexed with streptavidin-linked fluorophore moieties (Nielsen et al. 2006). While this process is higher throughput than direct labeling, it also carries a risk of altering the binding affinity of the antibody to its antigen (Vira et al. 2010). The direct conjugation of a fluorophore to an antibody, while suitable only for a small number of antibodies in parallel, allows the user to monitor internalization and intracellular trafficking in real time and, importantly, also reduces the background noise that often arises from the use of secondary antibodies. Antibodies can be directly labeled with fluorophores via a number of

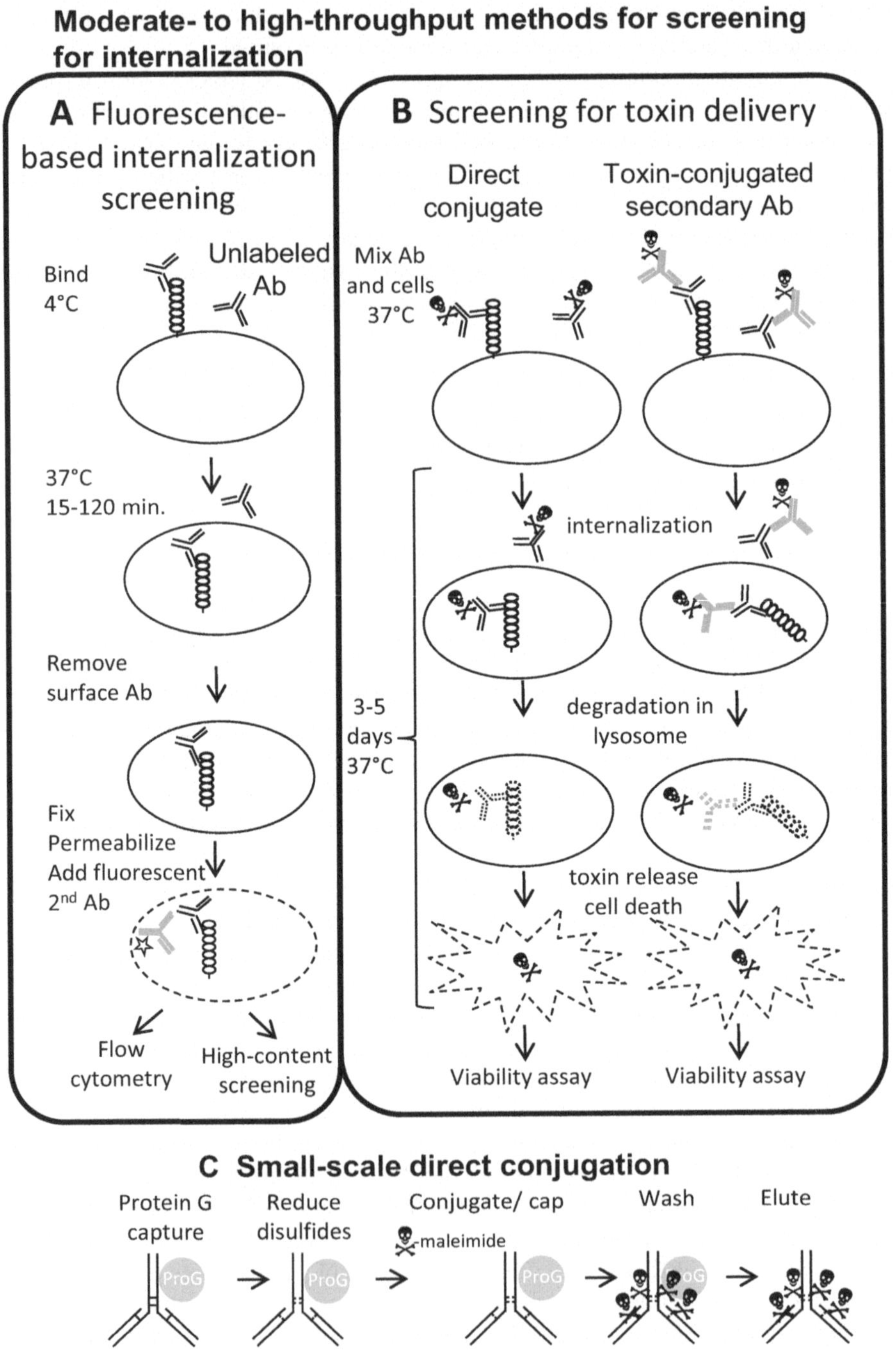

Fig. 3.2 Screening for internalizing antibodies. (**a**) Screening for internalization using fluorescence-based approaches. In fluorescence-based assays, target cells can be bound to antibodies at 4 °C, warmed to 37 °C to allow for internalization, washed to remove extracellular antibodies, and fixed and stained with fluorescent secondary reagents to visualize internalized antibodies.

commercially available conjugation kits that utilize random lysine or cysteine coupling of the fluorophore. Direct conjugation carries with it the risk of placing fluorophores at the antigen-binding interface, which may interfere with binding (Vira et al. 2010). Accordingly, after direct conjugation, it is important to verify that the binding affinity of the antibody has not been compromised (Vira et al. 2010).

To measure internalization, the fluorescent antibodies are added to cells seeded in a 96-well plate for a period of time at 4 °C to allow for binding without internalization (Nielsen et al. 2006; Harper et al. 2013). The cells are then moved to 37° C to allow for target-mediated internalization (Nielsen et al. 2006; Harper et al. 2013). After a selected period of time, non-internalized fluorescent antibodies are gently removed from the cell surface (e.g., by washing with a low pH buffer or, if NHS-SS-biotin was utilized to conjugate the fluorophore, by addition of a reducing agent), or the extracellular fluorescence is quenched by an antifluorophore antibody (Nielsen et al. 2006; Harper et al. 2013). Internalized fluorescence can then be quantified via a high-content fluorescence reader or flow cytometer, and the degree of internalization can be assessed by comparison with cells kept at 4 °C (Nielsen et al. 2006). High-resolution images can also be captured via confocal microscopy to verify the results (Fransson et al. 2004).

The engineering of second-generation ADCs that possess noncleavable, stable payload linkages to the antibody requires delivery to the lysosomal compartment of the cell for liberation of the payload from the antibody and subsequent delivery to the cell in an active form. Accordingly, screening for this intracellular delivery has become of paramount importance and can also be achieved with a moderately high throughput. This methodology requires the fluorescent labeling of the intracellular compartment of interest as well as the labeling of the internalizing antibodies with a different fluorophore by the methods described above. Organelle labeling can be achieved through various cell-permeable organelle fluorescent probes, through indirect immunostaining of the organelles, or through ectopic expression of a fluorophore-tagged organelle resident protein (Morelli et al. 2006; Starkuviene and Pepperkok 2007). Similar to the approach described above, cells are exposed to fluorescence-labeled antibodies at 4° C and are then moved to 37° C for a period of time before fixation. However, this method does not require quenching of non-internalized antibodies (Morelli et al. 2006; Starkuviene and Pepperkok 2007). The plates are read in a high-content fluorescence reader, and the magnitude of colocalization of the labeled antibody and the labeled organelle of interest are quantified (Starkuviene and Pepperkok 2007).

Fig. 3.2 (continued) Alternatively (not shown), antibodies directly conjugated to fluorophores can be used for internalization, and secondary antibodies omitted. Internalization can be measured qualitatively or quantitatively by microscopy or flow cytometry.(**b**) Screening for internalization by testing for intracellular toxin delivery, for target-mediated toxin delivery assays (right panel), antibodies can be directly conjugated to a toxin or can be complexed with a secondary antibody that is directly conjugated to a toxin. The conjugated antibody is then added to target cells for a period of 3–5 days, and cell viability is assessed. (**c**) Small-scale direct conjugation of antibodies. In a method described by Lyon et al. (2012), IgG are captured onto Protein G beads from small volumes of tissue culture medium and coupled to toxins via maleimide chemistry while remaining immobilized on the beads. A mixture of toxin/linker and a sulfhydryl capping reagent keeps the drug–antibody ratio within a controlled range

Conjugation to liposomes has also been used to measure antibody internalization (Nielsen et al. 2006; An et al. 2008). The antibodies can be coupled to fluorophore-labeled liposomes via either His_6/Ni-NTA or thiol coupling (Harper et al. 2013; An et al. 2008). In this method, conjugated liposomes are incubated with cells for a period of time, before non-internalized mAb/liposomes are gently removed and the cells are lysed (Harper et al. 2013). The remaining internalized fluorescence is measured via a microfluorimeter (Harper et al. 2013). Alternatively, liposomes can be loaded with a cytotoxin or immunotoxin in order to measure the ability of the antibody to confer target-mediated killing (Harper et al. 2013; An et al. 2008). The high valency of conjugated liposomes may result in internalization and trafficking properties that differ significantly from those of conventional IgG, however.

3.5.2 Screening for Intracellular Toxin Delivery

The ability of antibodies to internalize and deliver toxins to the cell type of interest can also be measured directly, and increasing numbers of methods are becoming available to allow rapid screening of hundreds to thousands of antibodies. Some investigators starting with antibodies derived from phage display have screened for internalization by generating genetic fusions of scFv to protein toxins such as *Pseudomonas* exotoxin and testing for cytotoxicity of the resulting immunotoxins (Yoshikawa et al. 2013; Cizeau et al. 2011). Expression in *E. coli* makes this method suitable for screening of large numbers of scFv, although Yoshikawa et al. (2013) showed that the cytotoxic potency of the selected fusion proteins was ~100-fold lower than that of the same molecules as neocarzinostatin-conjugated IgG, suggesting that only the most potent internalizing antibodies may be identified by this method.

The development of toxin-conjugated secondary reagents has facilitated screening of antibodies derived from both hybridoma and phage display. Conjugates of protein toxins to anti-IgG secondary reagents have been reported, either chemical conjugates to ricin A (Till et al. 1988) or pokeweed antiviral protein (Weltman et al. 1987) or genetic fusion of *Pseudomonas* exotoxin A to an anti-kappa single-domain antibody (Kellner et al. 2011) or to the Z domain of *Staphylococcus* Protein A (Mazor et al. 2007). Commercially available secondary antibodies conjugated to the ribosome-inactivating protein saporin have been widely used to confirm antibody internalization (Kohls and Lappi 2000; Nguyen et al. 2006; Fransson and Borrebaeck 2009; Sawada et al. 2011). Antibodies can be incubated with saporin-conjugated anti-mouse, anti-rat, or anti-human reagents and then incubated with tumor cells for 3–4 days. Uptake of the complexes and release of saporin from the endosome into the cytoplasm lead to cell death (Fig. 3.2b), which can be measured in a high-throughput format by viability reagents such as MTT or CellTiterGlo (Promega). The assay can be increased in sensitivity through the use of bivalent saporin-conjugated secondary IgG, which allows higher-order cross-linking of surface receptors and may enhance internalization, while Fab-saporin secondary conjugates maintain the level of cross-linking arising from the targeting antibody itself. For His-tagged scfv or Fabs, a saporin-conjugated anti-His_6 antibody provides a tool suitable for screening the internalization of His_6-tagged antibody fragments such as scFv. In addition to their use in antibody screening, saporin-conjugated secondary antibodies provide an opportunity to validate the internalization of newly identified cell surface proteins. For example, Fransson et al. (Fransson and Borrebaeck 2006) used a saporin conjugate to confirm the surface exposure and internalization of the nuclear protein Ku70/80.

The sensitivity of assays with secondary toxin conjugates can be further increased by using potent small-molecule toxins. Klussman et al. (2004) showed that the cytotoxicity of an anti-Lewis^Y antibody complexed to vcMMAE-conjugated secondary antibodies was 1–2 orders of magnitude greater than the cytotoxicity measured with the equivalent saporin-conjugated secondary antibodies. In addition, a vcMMAE-conjugated secondary antibody conferred nearly identical cytotoxicity to that of the direct vcMMAE conjugate, indicating that high-throughput screening with the secondary antibody may be predictive of the activity of the final conjugate. This high-sensitivity approach has recently been facilitated by the commercial availability of secondary reagents conjugated with representative payloads such as MMAE, MMAF, duocarmycin, and DM1 (Moradec, LLC, San Diego, CA).

Direct conjugation to a linker–payload is frequently a late step in screening for antibodies, carried out on only a small number of candidate antibodies. However, small-scale methods for the direct conjugation of toxin to antibodies are coming into increasingly widespread use, allowing early-stage screening of large panels of candidate antibodies. Lyon et al. (2012) describe a method for small-scale thiol-based conjugation of mcMMAF (Fig. 3.2c), capturing antibodies on Protein G resin and reducing interchain disulfides with TCEP to expose free thiols. The process uses less than 5 mL of conditioned medium in a 96-well block format, allowing for early-phase assessment of antibodies as ADC carriers prior to hybridoma subcloning, and by inclusion of both a maleimide linker–payload and a capping agent designed to minimize variations in drug–antibody ratios (DAR) despite the variable concentrations of antibody expected in crude supernatants. This critical step allows direct comparison of antibodies, since they carry similar levels of toxin. Lyon et al. describe preparation of 120 antibodies in parallel with a narrow range of drug–antibody ratios (DARs) slightly under 4. Other methods have been published for thiol-based conjugation at the 5 mg scale (Stefano et al. 2013). Site-specific conjugation via several alternative strategies (see below; reviewed in (Behrens and Liu 2013)) also holds the potential for application at early screening phases.

3.6 Payload Conjugation Strategies

The classes of cytotoxins currently being utilized to design ADCs fall into two categories: those that target microtubules and those that target DNA. The method by which a payload is conjugated to an antibody is of critical importance to the overall efficacy of an ADC. The chemistry involved and placement of the payload linkage on the antibody can have a profound effect on binding, tolerability, pharmacokinetic (PK) properties, and potency of an ADC.

Conventional conjugation strategies involve random conjugation to either lysine amines (as for Mylotarg (gemtuzumab ozogamicin), CMC-544, and Kadcyla(T-DM1)) or sulfhydryl groups on cysteines (for SGN-35 and CDX-011) of the antibody. Due to the high availability of multiple exposed lysines on the antibody surface (~70–90 per IgG1), lysine-based conjugation methods can produce a high degree of conjugation heterogeneity. For example, analysis of the recently approved ADC Kadcyla showed the presence of various drug-to-Ab ratio species ranging from 0 to 7 payloads, with an average of 3.5 payloads per antibody (Junutula et al. 2010), whereas Mylotarg possessed a large mixture of unconjugated species and conjugates with an average loading of 2.5 (Hamann et al. 2002). Alternatively, conventional cysteine-based random conjugation methods produce a more uniform degree of antibody loading. The presence of four reducible disulfide bonds within the IgG structure that covalently link the heavy and light chains together is an attractive feature that allows for the selective attachment of a drug to up to eight distinct sites using a thiol-reactive handle. The cysteine-conjugated Adcetris (SGN-35; brentuximab vedotin) contains a lower-level heterogeneous mixture of drug-loaded variants with an average of four payloads per antibody (Wakankar et al. 2011; Okeley et al. 2010). In theory, each of the loaded species in an ADC can represent a unique conjugate and thereby exhibit distinct PK and other biological properties. In fact, for anti-CD30-vcMMAE, decreasing drug loading from 8 to 4 or 2 drug molecules per Ab leads to slower ADC clearance and improvements in therapeutic index (Hamblett et al. 2004). In addition, batch-to-batch consistency in ADC production can be challenging and may require diligent manufacturing capabilities. Site-specific conjugation, in which a known number of linker–drugs are consistently conjugated to defined sites, is one way to overcome these challenges.

One of the first site-specific conjugation processes was established by Junutula et al. at Genentech who developed a cysteine-based site-specific conjugation approach called "'THIOMAB drug conjugates'" (TDCs) (Junutula et al. 2008). The authors engineered anti-MUC16 antibodies with additional cysteines on light and heavy chains that provide reactive thiol groups and did not perturb immunoglobulin folding and assembly, or alter antigen binding. The engineered Abs (ThiomAbs) were then reacted with maleimide-functionalized toxins, thereby yielding nearly homogeneous conjugates, TDCs (Junutula et al. 2008). TDCs possessed comparable antitumor activity as the conventionally conjugated ADCs, despite having a lower DAR (Junutula et al. 2008). Furthermore, the TDCs were tolerated at higher doses in Sprague–Dawley rats and cynomolgus monkeys and had an improved PK profile compared to conventional conjugates (Junutula et al. 2008). However, this method includes a reduction–reoxidation step that can potentially lead to unpaired, reactive sulfhydryl groups. Furthermore, depending on the location of the engineered cysteines, maleimide exchange processes could lead to drug loss and thus decrease the potency of the ADC.

Recently, the use of non-natural amino acids has also been employed for site-specific conjugation. In one study, a redundant stop codon was introduced at specific locations within the coding sequence of an anti-HER2 antibody, such that an amber suppressor, tRNA/aminoacyl-tRNA synthetase gene pair, can incorporate p-acetylphenylalanine into these locations. This allowed for the site-specific conjugation of an auristatin derivative through a stable oxime linkage. This ADC possessed excellent in vitro and in vivo potency with similar clearance rates as the parental IgG in rats (Axup et al. 2012). In a separate study, this technology was also used to produce an anti-5 T4 and an anti-HER2 site-specific ADC, containing two monomethyl auristatin D (MMAD) payloads per antibody (Sapra et al. 2013; Tian et al. 2014). These particular site-specific ADCs (NDCs) also possessed improved in vitro cytotoxicity, superior in vivo efficacy, and increased PK stability relative to the conventional conjugates and were well tolerated in preclinical toxicology studies (Sapra et al. 2013; Tian et al. 2014). Additionally, when NDCs were produced with protease-cleavable linkers, the site of conjugation on the antibody had a considerable effect on the stability of these rationally designed prodrug linkers, highlighting the importance of selecting suitable conjugation sites within the antibody (Tian et al. 2014).

In a different methodology to produce site-specific conjugates, the enzyme transglutaminase (mTG) from *Streptoverticillium mobaraense* was used to enzymatically conjugate payloads to antibodies in a site-specific manner and subsequently determine whether the site of conjugation affects the characteristics of the ADC (Strop et al. 2013). mTG does not recognize any of the naturally occurring glutamine residues in the Fc region of glycosylated antibodies but does recognize a "glutamine tag" that can be engineered into an antibody (Strop et al. 2013). The glutamine tag, LLQG, was engineered into different sites in the constant domain of an antibody targeting the epidermal growth factor receptor. mTG was then used to conjugate these sites with fluorophores or monomethyl dolastatin 10 (MMAD), and several sites were found to have good biophysical properties and a high degree of conjugation (Strop et al. 2013). Additionally, mTG was able to conjugate to glutamine tags present on anti-Her2 and anti-M1S1 antibodies (Strop et al. 2013). An anti-M1S1-vc-MMAD site-specific conjugate displayed strong in vitro and in vivo activities, suggesting that conjugation using this method does not alter antibody binding or affinity and demonstrates the utility of this approach in the site-specific conjugation of ADCs. The authors also showed that the site of conjugation has a significant impact on ADC stability and pharmacokinetics in a species-dependent manner (Strop et al. 2013). These differences were attributed to the position of the linkage rather than the chemical instability, as was observed with a maleimide linkage. The authors conclude that this method provides a conjugation strategy that produces homogeneous ADCs and allows for the adjustment of ADC properties in order to maximize the therapeutic window.

3.7 Conclusions and Perspectives

Current linker chemistry has enabled the development of ADCs that are highly stable in the circulation but are amenable to intracellular drug release after entry into the endosome or lysosomal compartment of the cell. While this approach may yield reduced toxicity, as the toxic drug is liberated from the antibody within the tumor cell, the advancement of linker chemistry may make it plausible to devise next-generation linkers that release the drug on the cell surface of cancer cells but not in plasma. The successful implementation of such linker chemistry would open new avenues in target identification and discovery of non-internalizing antigens that are uniquely present or highly overexpressed in tumors. The notion of non-internalizing antigens has been explored since the early period of radioimmuno-therapeutics development. Correspondingly, localized extracellular drug release by the ADCs is a direction being actively explored by several groups. However, it remains to be seen whether this approach will provide sufficient potency in preclinical settings and later in clinics.

Unique, tumor-activating antibody applications are also emerging as a popular strategy to enhance the potency and reduce the toxicity of monoclonal antibodies. In one such example, the authors developed a probody technology, in which antibody binding remains masked against antigen binding until becoming activated locally by proteases commonly active in the tumor microenvironment (Desnoyers et al. 2013). Hypothetically, this technology could help in developing ADCs againt targets which have high normal tissue expression. Furthermore, the technology could enable the development of safer ADCs and/or ADCs with an increased serum half-life. However, the utility of such technologies in the clinic remains to be evaluated.

In general, antibody-based therapies (i.e., ADCs, immunotoxins, or immunoliposomes) have performed better in hematological tumor settings compared to solid tumors, despite the fact that the overall expression level of target antigens is frequently expressed at lower levels in liquid tumors. However, in liquid tumors, antigens are expressed more homogeneously, and the entire tumor is more accessible for antibody-based therapy. Additionally, liquid tumors are usually more sensitive to cytotoxic compounds and have a more restricted antigen expression pattern on normal tissues. Indeed, a solid tumor environment poses unique challenges with regard to the number of antibody-based molecules that extravasate the tumor blood vessels and translocate through the tumor interstitium toward the surface of tumor cells (Rybak et al. 2007). Further, solid tumors have a heterogeneous blood supply and high interstitial pressures within tumor tissue, especially in necrotic zones, which may limit the diffusion of drugs or ADCs to poorly perfused areas (Stohrer et al. 2000). Additionally, the binding site barrier hypothesis suggests that antibodies bind to the first target cells they encounter, impairing their penetration through the entire tumor and limiting their therapeutic effects (Weinstein and van Osdol 1992; Rudnick et al. 2011). Therefore, the development of novel vehicles that allow increased penetration through the solid tumor mass may offer improved efficacy of therapeutic antibodies and ADCs.

The development of companion diagnostic tests will likely become an essential component of future ADC development programs. These diagnostic tests may help translate effective antibody-based therapeutics to successful, targeted treatments for populations that will best respond to the treatment, especially if the drug is being developed against an antigen that has heterogeneous distribution. Designing this companion diagnostics during the early stages of an ADC program may provide valuable information about the best indications to target but will also allow for the use of novel antibody tools being generated during the antibody isolation stage of the ADC program. It will be interesting to follow whether the field will continue to use the "gold standard" immunohistochemistry-based approaches to identify patient populations or move to less invasive procedures, such as circulating tumor cells or imaging modalities.

It remains to be determined whether amino acid alterations introduced to specific sites within an antibody will present any challenges for immunogenicity or stability of corresponding ADCs in the clinic. Although the site-specific ADCs appear to improve upon conventional conjugation strategies currently used in the clinic, these conclusions are based on fairly limited preclinical data and require a more rigorous analysis in the clinic. Further improvements in site-specific conjugation strategies are still possible, such as homogeneous conjugates with greater than two drugs per antibody. While higher drug loading has been correlated with an increased rate of clearance, those conclusions have been drawn from ADCs constructed using non-specific cysteine modification methods and may not be valid for site-specific ADCs. It will also be interesting to follow whether site-specific conjugates possess different off-target toxicity profiles compared to conventional conjugates. Furthermore, site-specific conjugates also promise increased drug exposure, and while higher exposure may be advantageous to achieve higher efficacy, it remains to be determined whether higher exposure results in more on-target toxicities. To our knowledge, none of the ADCs currently being evaluated in trials in humans used site-specific conjugation approaches; however, emerging preclinical data continue to build on the promise of site-specific conjugation technology to produce safer ADCs with controlled manufacturing and improved analytics.

ADCs have emerged as therapeutic modalities with great promise for the treatment of cancer. However, more may be needed to ensure the clinical success of this therapeutic class. Advances in antigen identification, antibody engineering, and linker and payload conjugation chemistry may help design more efficacious and safer ADCs.

Financial & Competing Interests Disclosure All authors are full-time employees of Pfizer Inc.

References

Adams GP, Schier R, McCall AM, Simmons HH, Horak EM, Alpaugh RK, Marks JD, Weiner LM (2001) High affinity restricts the localization and tumor penetration of single-chain fv antibody molecules. Cancer Res 61(12):4750–4755

Akahori Y, Kurosawa G, Sumitomo M, Morita M, Muramatsu C, Eguchi K, Tanaka M, Suzuki K, Sugiura M, Iba Y, Sugioka A, Kurosawa Y (2009) Isolation of antigen/antibody complexes through organic solvent (ICOS) method. Biochem Biophys Res Commun 378(4):832–835. https://doi.org/10.1016/j.bbrc.2008.11.129

An F, Drummond DC, Wilson S, Kirpotin DB, Nishimura SL, Broaddus VC, Liu B (2008) Targeted drug delivery to mesothelioma cells using functionally selected internalizing human single-chain antibodies. Mol Cancer Ther 7(3):569–578. https://doi.org/10.1158/1535-7163.MCT-07-2132

Ashley DM, Sampson JH, Archer GE, Batra SK, Bigner DD, Hale LP (1997) A genetically modified allogeneic cellular vaccine generates MHC class I-restricted cytotoxic responses against tumor-associated antigens and protects against CNS tumors in vivo. J Neuroimmunol 78(1–2):34–46

Axup JY, Bajjuri KM, Ritland M, Hutchins BM, Kim CH, Kazane SA, Halder R, Forsyth JS, Santidrian AF, Stafin K, Lu Y, Tran H, Seller AJ, Biroc SL, Szydlik A, Pinkstaff JK, Tian F, Sinha SC, Felding-Habermann B, Smider VV, Schultz PG (2012) Synthesis of site-specific antibody-drug conjugates using unnatural amino acids. Proc Natl Acad Sci USA 109(40):16101–16106. https://doi.org/10.1073/pnas.1211023109

Becerril B, Poul MA, Marks JD (1999) Toward selection of internalizing antibodies from phage libraries. Biochem Biophys Res Commun 255(2):386–393. https://doi.org/10.1006/bbrc.1999.0177

Behrens CR, Liu B (2013) Methods for site-specific drug conjugation to antibodies. MAbs 6(1):46

Bernard A, Boumsell L (1984) The clusters of differentiation (CD) defined by the first international workshop on human leucocyte differentiation antigens. Hum Immunol 11(1):1–10

Berry LM, Adams R, Airey M, Bracher MG, Bourne T, Carrington B, Cross AS, Davies GC, Finney HM, Foulkes R, Gozzard N, Griffin RA, Hailu H, Lamour SD, Lawson AD, Lightwood DJ, McKnight AJ, O'Dowd VL, Oxbrow AK, Popplewell AG, Shaw S, Stephens PE, Sweeney B, Tomlinson KL, Uhe C, Palframan RT (2009) In vitro and in vivo characterisation of anti-murine IL-13 antibodies recognising distinct functional epitopes. Int Immunopharmacol 9(2):201–206. https://doi.org/10.1016/j.intimp.2008.11.001

Bickel PE, Lodish HF, Scherer PE (2000) Use and applications of subtractive antibody screening. Biotechnol Genet Eng Rev 17:417–430

Bouchard L, Faucher G, Tchernof A, Deshaies Y, Lebel S, Hould FS, Marceau P, Vohl MC (2009) Comprehensive genetic analysis of the dipeptidyl peptidase-4 gene and cardiovascular disease risk factors in obese individuals. Acta Diabetol 46(1):13–21. https://doi.org/10.1007/s00592-008-0049-4

Burris HA 3rd, Rugo HS, Vukelja SJ, Vogel CL, Borson RA, Limentani S, Tan-Chiu E, Krop IE, Michaelson RA, Girish S, Amler L, Zheng M, Chu YW, Klencke B, O'Shaughnessy JA (2011) Phase II study of the antibody drug conjugate trastuzumab-DM1 for the treatment of human epidermal growth factor receptor 2 (HER2)-positive breast cancer after prior HER2-directed therapy. J Clin Oncol 29(4):398–405. https://doi.org/10.1200/JCO.2010.29.5865

Chaparro-Riggers J, Liang H, DeVay RM, Bai L, Sutton JE, Chen W, Geng T, Lindquist K, Casas MG, Boustany LM, Brown CL, Chabot J, Gomes B, Garzone P, Rossi A, Strop P, Shelton D, Pons J, Rajpal A (2012) Increasing serum half-life and extending cholesterol lowering in vivo by engineering antibody with pH-sensitive binding to PCSK9. J Biol Chem 287(14):11090–11097. https://doi.org/10.1074/jbc.M111.319764

Chatenoud L, Ferran C, Legendre C, Thouard I, Merite S, Reuter A, Gevaert Y, Kreis H, Franchimont P, Bach JF (1990) In vivo cell activation following OKT3 administration. Systemic cytokine release and modulation by corticosteroids. Transplantation 49(4):697–702

Cizeau J, Torres MG, Cowling SG, Stibbard S, Premsukh A, Entwistle J, MacDonald GC (2011) Fusogenics: a recombinant immunotoxin-based screening platform to select internalizing tumor-specific antibody fragments. J Biomol Screen 16(1):90–100. https://doi.org/10.1177/1087057110387425

Desnoyers LR, Vasiljeva O, Richardson JH, Yang A, Menendez EE, Liang TW, Wong C, Bessette PH, Kamath K, Moore SJ, Sagert JG, Hostetter DR, Han F, Gee J, Flandez J, Markham K, Nguyen M, Krimm M, Wong KR, Liu S, Daugherty PS, West JW, Lowman HB (2013) Tumor-specific activation of an EGFR-targeting probody enhances therapeutic index seek and destroy. Sci Transl Med 5(207). https://doi.org/10.1126/scitranslmed.3006682

Ebersbach H, Geisse S (2012) Antigen generation and display in therapeutic antibody drug discovery -- a neglected but critical player. Biotechnol J 7(12):1433–1443. https://doi.org/10.1002/biot.201200066

Fishwild DM, O'Donnell SL, Bengoechea T, Hudson DV, Harding F, Bernhard SL, Jones D, Kay RM, Higgins KM, Schramm SR, Lonberg N (1996) High-avidity human IgG kappa monoclonal antibodies from a novel strain of minilocus transgenic mice. Nat Biotechnol 14(7):845–851. https://doi.org/10.1038/nbt0796-845

Fransson J, Borrebaeck CA (2006) The nuclear DNA repair protein Ku70/80 is a tumor-associated antigen displaying rapid receptor mediated endocytosis. Int J Cancer 119(10):2492–2496. https://doi.org/10.1002/ijc.22212

Fransson J, Borrebaeck CA (2009) Selection and characterization of antibodies from phage display libraries against internalizing membrane antigens. Methods Mol Biol 480:113–127. https://doi.org/10.1007/978-1-59745-429-2_8

Fransson J, Ek S, Ellmark P, Soderlind E, Borrebaeck CA, Furebring C (2004) Profiling of internalizing tumor-associated antigens on breast and pancreatic cancer cells by reversed genomics. Cancer Lett 208(2):235–242. https://doi.org/10.1016/j.canlet.2003.11.036

Gasteiger E, Gattiker A, Hoogland C, Ivanyi I, Appel RD, Bairoch A (2003) ExPASy: the proteomics server for in-depth protein knowledge and analysis. Nucleic Acids Res 31(13):3784–3788

Goenaga AL, Zhou Y, Legay C, Bougherara H, Huang L, Liu B, Drummond DC, Kirpotin DB, Auclair C, Marks JD, Poul MA (2007) Identification and characterization of tumor antigens by using antibody phage display and intrabody strategies. Mol Immunol 44(15):3777–3788. https://doi.org/10.1016/j.molimm.2007.03.023

Guo QM (2003) DNA microarray and cancer. Curr Opin Oncol 15(1):36–43

Hamann PR, Hinman LM, Hollander I, Beyer CF, Lindh D, Holcomb R, Hallett W, Tsou HR, Upeslacis J, Shochat D, Mountain A, Flowers DA, Bernstein I (2002) Gemtuzumab ozogamicin, a potent and selective anti-CD33 antibody-calicheamicin conjugate for treatment of acute myeloid leukemia. Bioconjug Chem 13(1):47–58

Hamblett KJ, Senter PD, Chace DF, Sun MM, Lenox J, Cerveny CG, Kissler KM, Bernhardt SX, Kopcha AK, Zabinski RF, Meyer DL, Francisco JA (2004) Effects of drug loading on the antitumor activity of a monoclonal antibody drug conjugate. Clin Cancer Res 10(20):7063–7070. https://doi.org/10.1158/1078-0432.CCR-04-0789

Harper J, Mao S, Strout P, Kamal A (2013) Selecting an optimal antibody for antibody-drug conjugate therapy: internalization and intracellular localization. Methods Mol Biol 1045:41–49. https://doi.org/10.1007/978-1-62703-541-5_3

Hazen M, Bhakta S, Vij R, Randle S, Kallop D, Chiang V, Hötzel I, Jaiswal BS, Ervin KE, Li B, Weimer RM, Polakis P, Scheller RH, Junutula JR, Hongo J-AS (2014) An improved and robust DNA immunization method to develop antibodies against extra-cellular loops of multi-transmembrane proteins. MAbs 6(1):95. https://doi.org/10.4161/mabs.26761

Hellstrom I, Garrigues HJ, Garrigues U, Hellstrom KE (1990) Highly tumor-reactive, internalizing, mouse monoclonal antibodies to Le(y)-related cell surface antigens. Cancer Res 50(7):2183–2190

Hoogenboom HR, Lutgerink JT, Pelsers MM, Rousch MJ, Coote J, Van Neer N, De Bruine A, Van Nieuwenhoven FA, Glatz JF, Arends JW (1999) Selection-dominant and nonaccessible epitopes on cell-surface receptors revealed by cell-panning with a large phage antibody library. Eur J Biochem 260(3):774–784

Israeli RS, Powell CT, Fair WR, Heston WD (1993) Molecular cloning of a complementary DNA encoding a prostate-specific membrane antigen. Cancer Res 53(2):227–230

Junutula JR, Flagella KM, Graham RA, Parsons KL, Ha E, Raab H, Bhakta S, Nguyen T, Dugger DL, Li G, Mai E, Lewis Phillips GD, Hiraragi H, Fuji RN, Tibbitts J, Vandlen R, Spencer SD, Scheller RH, Polakis P, Sliwkowski MX (2010) Engineered thio-trastuzumab-DM1 conjugate with an improved therapeutic index to target human epidermal growth factor receptor 2-positive breast cancer. Clin Cancer Res 16(19):4769–4778. https://doi.org/10.1158/1078-0432.CCR-10-0987

Junutula JR, Raab H, Clark S, Bhakta S, Leipold DD, Weir S, Chen Y, Simpson M, Tsai SP, Dennis MS, Lu Y, Meng YG, Ng C, Yang J, Lee CC, Duenas E, Gorrell J, Katta V, Kim A, McDorman K, Flagella K, Venook R, Ross S, Spencer SD, Lee Wong W, Lowman HB, Vandlen R, Sliwkowski MX, Scheller RH, Polakis P, Mallet W (2008) Site-specific conjugation of a cytotoxic drug to an antibody improves the therapeutic index. Nat Biotechnol 26(8):925–932. https://doi.org/10.1038/nbt.1480

Kellner C, Bleeker WK, Lammerts van Bueren JJ, Staudinger M, Klausz K, Derer S, Glorius P, Muskulus A, de Goeij BE, van de Winkel JG, Parren PW, Valerius T, Gramatzki M, Peipp M (2011) Human kappa light chain targeted Pseudomonas exotoxin A--identifying human antibodies and Fab fragments with favorable characteristics for antibody-drug conjugate development. J Immunol Methods 371(1–2):122–133. https://doi.org/10.1016/j.jim.2011.06.023

Kilpatrick KE, Wring SA, Walker DH, Macklin MD, Payne JA, Su JL, Champion BR, Caterson B, McIntyre GD (1997) Rapid development of affinity matured monoclonal antibodies using RIMMS. Hybridoma 16(4):381–389

Klussman K, Mixan BJ, Cerveny CG, Meyer DL, Senter PD, Wahl AF (2004) Secondary mAb--vcMMAE conjugates are highly sensitive reporters of antibody internalization via the lysosome pathway. Bioconjug Chem 15(4):765–773. https://doi.org/10.1021/bc049969t

Kohler G, Milstein C (1975) Continuous cultures of fused cells secreting antibody of predefined specificity. Nature 256(5517):495–497

Kohls MD, Lappi DA (2000) Mab-ZAP: a tool for evaluating antibody efficacy for use in an immunotoxin. BioTechniques 28(1):162–165

Kulasingam V, Diamandis EP (2008) Strategies for discovering novel cancer biomarkers through utilization of emerging technologies. Nat Clin Pract Oncol 5(10):588–599. https://doi.org/10.1038/ncponc1187

Kung P, Goldstein G, Reinherz EL, Schlossman SF (1979) Monoclonal antibodies defining distinctive human T cell surface antigens. Science 206(4416):347–349

Kurosawa G, Akahori Y, Morita M, Sumitomo M, Sato N, Muramatsu C, Eguchi K, Matsuda K, Takasaki A, Tanaka M, Iba Y, Hamada-Tsutsumi S, Ukai Y, Shiraishi M, Suzuki K, Kurosawa M, Fujiyama S, Takahashi N, Kato R, Mizoguchi Y, Shamoto M, Tsuda H, Sugiura M, Hattori Y, Miyakawa S, Shiroki R, Hoshinaga K, Hayashi N, Sugioka A, Kurosawa Y (2008) Comprehensive screening for antigens overexpressed on carcinomas via isolation of human mAbs that may be therapeutic. Proc Natl Acad Sci USA 105(20):7287–7292. https://doi.org/10.1073/pnas.0712202105

Kurosawa G, Sumitomo M, Akahori Y, Matsuda K, Muramatsu C, Takasaki A, Iba Y, Eguchi K, Tanaka M, Suzuki K, Morita M, Sato N, Sugiura M, Sugioka A, Hayashi N, Kurosawa Y (2009) Methods for comprehensive identification of membrane proteins recognized by a large number of monoclonal antibodies. J Immunol Methods 351(1–2):1–12. https://doi.org/10.1016/j.jim.2009.09.003

Kurosawa N, Yoshioka M, Fujimoto R, Yamagishi F, Isobe M (2012) Rapid production of antigen-specific monoclonal antibodies from a variety of animals. BMC Biol 10:80. https://doi.org/10.1186/1741-7007-10-80

Langer F, Ingersoll SB, Amirkhosravi A, Meyer T, Siddiqui FA, Ahmad S, Walker JM, Amaya M, Desai H, Francis JL (2005) The role of CD40 in CD40L- and antibody-mediated platelet activation. Thromb Haemost 93(6):1137–1146. https://doi.org/10.1267/THRO05061137

Lin K, Tibbitts J (2012) Pharmacokinetic considerations for antibody drug conjugates. Pharm Res 29(9):2354–2366. https://doi.org/10.1007/s11095-012-0800-y

Liu B, Conrad F, Cooperberg MR, Kirpotin DB, Marks JD (2004) Mapping tumor epitope space by direct selection of single-chain Fv antibody libraries on prostate cancer cells. Cancer Res 64(2):704–710

Lonberg N (2005) Human antibodies from transgenic animals. Nat Biotechnol 23(9):1117–1125. https://doi.org/10.1038/nbt1135

Lyon RP, Meyer DL, Setter JR, Senter PD (2012) Conjugation of anticancer drugs through endogenous monoclonal antibody cysteine residues. Methods Enzymol 502:123–138. https://doi.org/10.1016/B978-0-12-416039-2.00006-9

Ma B, Osborn MJ, Avis S, Ouisse LH, Menoret S, Anegon I, Buelow R, Bruggemann M (2013) Human antibody expression in transgenic rats: comparison of chimeric IgH loci with human VH, D and JH but bearing different rat C-gene regions. J Immunol Methods 400-401:78–86. https://doi.org/10.1016/j.jim.2013.10.007

Mazor Y, Barnea I, Keydar I, Benhar I (2007) Antibody internalization studied using a novel IgG binding toxin fusion. J Immunol Methods 321(1–2):41–59. https://doi.org/10.1016/j.jim.2007.01.008

McDonagh CF, Kim KM, Turcott E, Brown LL, Westendorf L, Feist T, Sussman D, Stone I, Anderson M, Miyamoto J, Lyon R, Alley SC, Gerber HP, Carter PJ (2008) Engineered anti-CD70 antibody-drug conjugate with increased therapeutic index. Mol Cancer Ther 7(9):2913–2923. https://doi.org/10.1158/1535-7163.MCT-08-0295

Mendez MJ, Green LL, Corvalan JR, Jia XC, Maynard-Currie CE, Yang XD, Gallo ML, Louie DM, Lee DV, Erickson KL, Luna J, Roy CM, Abderrahim H, Kirschenbaum F, Noguchi M, Smith DH, Fukushima A, Hales JF, Klapholz S, Finer MH, Davis CG, Zsebo KM, Jakobovits A (1997) Functional transplant of megabase human immunoglobulin loci recapitulates human antibody response in mice. Nat Genet 15(2):146–156. https://doi.org/10.1038/ng0297-146

Morelli JK, Buehrle M, Pognan F, Barone LR, Fieles W, Ciaccio PJ (2006) Validation of an in vitro screen for phospholipidosis using a high-content biology platform. Cell Biol Toxicol 22(1):15–27. https://doi.org/10.1007/s10565-006-0176-z

Nguyen DH, Ball ED, Varki A (2006) Myeloid precursors and acute myeloid leukemia cells express multiple CD33-related Siglecs. Exp Hematol 34(6):728–735. https://doi.org/10.1016/j.exphem.2006.03.003

Nielsen UB, Kirpotin DB, Pickering EM, Drummond DC, Marks JD (2006) A novel assay for monitoring internalization of nanocarrier coupled antibodies. BMC Immunol 7:24. https://doi.org/10.1186/1471-2172-7-24

Okeley NM, Miyamoto JB, Zhang X, Sanderson RJ, Benjamin DR, Sievers EL, Senter PD, Alley SC (2010) Intracellular activation of SGN-35, a potent anti-CD30 antibody-drug conjugate. Clinical Cancer Res 16(3):888–897. https://doi.org/10.1158/1078-0432.CCR-09-2069

Osborn MJ, Ma B, Avis S, Binnie A, Dilley J, Yang X, Lindquist K, Menoret S, Iscache AL, Ouisse LH, Rajpal A, Anegon I, Neuberger MS, Buelow R, Bruggemann M (2013) High-affinity IgG antibodies develop naturally in Ig-knockout rats carrying germline human IgH/Igkappa/Iglambda loci bearing the rat CH region. J Immunol 190(4):1481–1490. https://doi.org/10.4049/jimmunol.1203041

Pak Y, Zhang Y, Pastan I, Lee B (2012) Antigen shedding may improve efficiencies for delivery of antibody-based anticancer agents in solid tumors. Cancer Res 72(13):3143–3152. https://doi.org/10.1158/0008-5472.CAN-11-3925

Poul M, Jarry G, Elhkim MO, Poul JM (2009) Lack of genotoxic effect of food dyes amaranth, sunset yellow and tartrazine and their metabolites in the gut micronucleus assay in mice. Food Chem Toxicol 47(2):443–448. https://doi.org/10.1016/j.fct.2008.11.034

Poul MA, Becerril B, Nielsen UB, Morisson P, Marks JD (2000) Selection of tumor-specific internalizing human antibodies from phage libraries. J Mol Biol 301(5):1149–1161. https://doi.org/10.1006/jmbi.2000.4026

Robinson HL, Pertmer TM (2001) Nucleic acid immunizations. Curr Protoc Immunol Chapter 2:Unit 2 14. https://doi.org/10.1002/0471142735.im0214s27

Ruan W, Sassoon A, An F, Simko JP, Liu B (2006) Identification of clinically significant tumor antigens by selecting phage antibody library on tumor cells in situ using laser capture microdissection. Mol Cell Proteomics 5(12):2364–2373. https://doi.org/10.1074/mcp.M600246-MCP200

Rudnick SI, Lou J, Shaller CC, Tang Y, Klein-Szanto AJ, Weiner LM, Marks JD, Adams GP (2011) Influence of affinity and antigen internalization on the uptake and penetration of Anti-HER2 antibodies in solid tumors. Cancer Res 71(6):2250–2259. https://doi.org/10.1158/0008-5472.CAN-10-2277

Rust S, Guillard S, Sachsenmeier K, Hay C, Davidson M, Karlsson A, Karlsson R, Brand E, Lowne D, Elvin J, Flynn M, Kurosawa G, Hollingsworth R, Jermutus L, Minter R (2013) Combining phenotypic and proteomic approaches to identify membrane targets in a 'triple negative' breast cancer cell type. Mol Cancer 12:11. https://doi.org/10.1186/1476-4598-12-11

Rybak JN, Roesli C, Kaspar M, Villa A, Neri D (2007) The extra-domain a of fibronectin is a vascular marker of solid tumors and metastases. Cancer Res 67(22):10948–10957. https://doi.org/10.1158/0008-5472.CAN-07-1436

Sapra P, Damelin M, Dijoseph J, Marquette K, Geles KG, Golas J, Dougher M, Narayanan B, Giannakou A, Khandke K, Dushin R, Ernstoff E, Lucas J, Leal M, Hu G, O'Donnell CJ, Tchistiakova L, Abraham RT, Gerber HP (2013) Long-term tumor regression induced by an antibody-drug conjugate that targets 5T4, an oncofetal antigen expressed on tumor-initiating cells. Mol Cancer Ther 12(1):38–47. https://doi.org/10.1158/1535-7163.MCT-12-0603

Sawada R, Sun SM, Wu X, Hong F, Ragupathi G, Livingston PO, Scholz WW (2011) Human monoclonal antibodies to sialyl-Lewis (CA19.9) with potent CDC, ADCC, and antitumor activity. Clin Cancer Res 17(5):1024–1032. https://doi.org/10.1158/1078-0432.CCR-10-2640

Schmidt MM, Wittrup KD (2009) A modeling analysis of the effects of molecular size and binding affinity on tumor targeting. Mol Cancer Ther 8(10):2861–2871. https://doi.org/10.1158/1535-7163.MCT-09-0195

Shiio Y, Aebersold R (2006) Quantitative proteome analysis using isotope-coded affinity tags and mass spectrometry. Nat Protoc 1(1):139–145. https://doi.org/10.1038/nprot.2006.22

Shukla GS, Krag DN, Peletskaya EN, Pero SC, Sun YJ, Carman CL, McCahill LE, Roland TA (2013) Intravenous infusion of phage-displayed antibody library in human cancer patients: enrichment and cancer-specificity of tumor-homing phage-antibodies. Cancer Immunol Immunother 62(8):1397–1410. https://doi.org/10.1007/s00262-013-1443-5

Sleister HM, Rao AG (2002) Subtractive immunization: a tool for the generation of discriminatory antibodies to proteins of similar sequence. J Immunol Methods 261(1–2):213–220

Starkuviene V, Pepperkok R (2007) The potential of high-content high-throughput microscopy in drug discovery. Br J Pharmacol 152(1):62–71. https://doi.org/10.1038/sj.bjp.0707346

Stefano JE, Busch M, Hou L, Park A, Gianolio DA (2013) Micro- and mid-scale maleimide-based conjugation of cytotoxic drugs to antibody hinge region thiols for tumor targeting. Methods Mol Biol 1045:145–171. https://doi.org/10.1007/978-1-62703-541-5_9

Stohrer M, Boucher Y, Stangassinger M, Jain RK (2000) Oncotic pressure in solid tumors is elevated. Cancer Res 60(15):4251–4255

Strop P, Liu SH, Dorywalska M, Delaria K, Dushin RG, Tran TT, Ho WH, Farias S, Casas MG, Abdiche Y, Zhou D, Chandrasekaran R, Samain C, Loo C, Rossi A, Rickert M, Krimm S, Wong T, Chin SM, Yu J, Dilley J, Chaparro-Riggers J, Filzen GF, O'Donnell CJ, Wang F, Myers JS, Pons J, Shelton DL, Rajpal A (2013) Location matters: site of conjugation modulates stability and pharmacokinetics of antibody drug conjugates. Chem Biol 20(2):161–167. https://doi.org/10.1016/j.chembiol.2013.01.010

Tian F, Lu Y, Manibusan A, Sellers A, Tran H, Sun Y, Phuong T, Barnett R, Hehli B, Song F, Deguzman MJ, Ensari S, Pinkstaff JK, Sullivan LM, Biroc SL, Cho H, Schultz PG, Dijoseph J, Dougher M, Ma D, Dushin R, Leal M, Tchistiakova L, Feyfant E, Gerber HP, Sapra P (2014) A general approach to site-specific antibody drug conjugates. Proc Natl Acad Sci USA 111(5):1766–1771. https://doi.org/10.1073/pnas.1321237111

Till M, May RD, Uhr JW, Thorpe PE, Vitetta ES (1988) An assay that predicts the ability of monoclonal antibodies to form potent ricin a chain-containing immunotoxins. Cancer Res 48(5):1119–1123

Uchida J, Lee Y, Hasegawa M, Liang Y, Bradney A, Oliver JA, Bowen K, Steeber DA, Haas KM, Poe JC, Tedder TF (2004) Mouse CD20 expression and function. Int Immunol 16(1):119–129

UniProt C (2013) Update on activities at the universal protein resource (UniProt) in 2013. Nucleic Acids Res 41(Database issue):D43–D47. https://doi.org/10.1093/nar/gks1068

Vira S, Mekhedov E, Humphrey G, Blank PS (2010) Fluorescent-labeled antibodies: balancing functionality and degree of labeling. Anal Biochem 402(2):146–150. https://doi.org/10.1016/j.ab.2010.03.036

Wakankar A, Chen Y, Gokarn Y, Jacobson FS (2011) Analytical methods for physicochemical characterization of antibody drug conjugates. MAbs 3(2):161–172

Weber M, Weiss E, Engel AM (2003) Combining EL4-B5-based B-cell stimulation and phage display technology for the successful isolation of human anti-Scl-70 autoantibody fragments. J Immunol Methods 278(1–2):249–259

Weinstein JN, van Osdol W (1992) Early intervention in cancer using monoclonal antibodies and other biological ligands: micropharmacology and the "binding site barrier". Cancer Res 52(9 Suppl):2747s–2751s

Weltman JK, Pedroso P, Johnson SA, Davignon D, Fast LD, Leone LA (1987) Rapid screening with indirect immunotoxin for monoclonal antibodies against human small cell lung cancer. Cancer Res 47(21):5552–5556

Wen L, Hanvanich M, Werner-Favre C, Brouwers N, Perrin LH, Zubler RH (1987) Limiting dilution assay for human B cells based on their activation by mutant EL4 thymoma cells: total and antimalaria responder B cell frequencies. Eur J Immunol 17(6):887–892. https://doi.org/10.1002/eji.1830170624

Williams AF, Galfre G, Milstein C (1977) Analysis of cell surfaces by xenogeneic myeloma-hybrid antibodies: differentiation antigens of rat lymphocytes. Cell 12(3):663–673

Wing MG, Moreau T, Greenwood J, Smith RM, Hale G, Isaacs J, Waldmann H, Lachmann PJ, Compston A (1996) Mechanism of first-dose cytokine-release syndrome by CAMPATH 1-H: involvement of CD16 (FcgammaRIII) and CD11a/CD18 (LFA-1) on NK cells. J Clin Invest 98(12):2819–2826. https://doi.org/10.1172/JCI119110

Yaziji H, Gown AM (2004) Testing for epidermal growth factor receptor in lung cancer: have we learned anything from HER-2 testing? J Clin Oncol 22(17):3646; author reply 3646–3648. https://doi.org/10.1200/JCO.2004.99.015

Yoshikawa M, Mukai Y, Okada Y, Tsumori Y, Tsunoda S, Tsutsumi Y, Aird WC, Yoshioka Y, Okada N, Doi T, Nakagawa S (2013) Robo4 is an effective tumor endothelial marker for antibody-drug conjugates based on the rapid isolation of the anti-Robo4 cell-internalizing antibody. Blood 121(14):2804–2813. https://doi.org/10.1182/blood-2012-12-468363

Zhou Y, Marks JD (2012) Discovery of internalizing antibodies to tumor antigens from phage libraries. Methods Enzymol 502:43–66. https://doi.org/10.1016/B978-0-12-416039-2.00003-3

Zhou Y, Zhao L, Marks JD (2012) Selection and characterization of cell binding and internalizing phage antibodies. Arch Biochem Biophys 526(2):107–113. https://doi.org/10.1016/j.abb.2012.05.007

Zhou Y, Zou H, Zhang S, Marks JD (2010) Internalizing cancer antibodies from phage libraries selected on tumor cells and yeast-displayed tumor antigens. J Mol Biol 404(1):88–99. https://doi.org/10.1016/j.jmb.2010.09.006

Chapter 4
Linker Design for Antibody–Drug Conjugates

E. Erica Hong and Ravi Chari

Abstract Antibody–drug conjugates (ADCs) seek to exploit the desirable features of their components, i.e., specificity in targeting tumor cells and long half-life in plasma conferred by the antibody, and the cytotoxic potency of the small effector molecule. A linker tethers the antibody to a cytotoxic agent and allows stable circulation of the cytotoxin in the form of pro-drug. Many linkers have been designed over the years and have showcased features that are more than a mere physical linkage between the two constituent parts. Linkers can confer several advantages to ADCs, such as modulating the rate of cytotoxin release in tumor cells, overcoming multidrug resistance, enhancing tumor penetration, and a better safety profile in humans. This review highlights the progress made in linker technology and how it has impacted the design of optimal ADCs for clinical evaluation.

Keywords Antibody–drug conjugate · Noncleavable linker · Cleavable linker · Reducible linker · Peptide linker · Acid-labile linker · Bystander activity · Lysosomal processing · MDR · Liver detoxification

4.1 Introduction

Antibody–drug conjugates are composed of a cytotoxic agent conjugated to a monoclonal antibody raised against an antigen that is preferentially expressed on cancer cells relative to normal tissues (Fig. 4.1) (Lambert 2013). A linker provides a covalent bridge between the cytotoxic agent and the antibody, which serves to deliver the cytotoxic agent specifically to cancer cells. The function of a linker is to keep the cytotoxic molecule stably attached to the antibody during the formulation and storage of the drug product and during the circulation in plasma following

E. E. Hong (✉) · R. Chari
ImmunoGen, Inc., Waltham, MA, USA
e-mail: erica.hong@immunogen.com; ravi.chari@immunogen.com

J. L. Zaro, J. Wang (eds.), *Antibody-Drug Conjugates*, AAPS Advances in the Pharmaceutical Sciences Series 17, https://doi.org/10.1007/978-3-031-83005-1_4

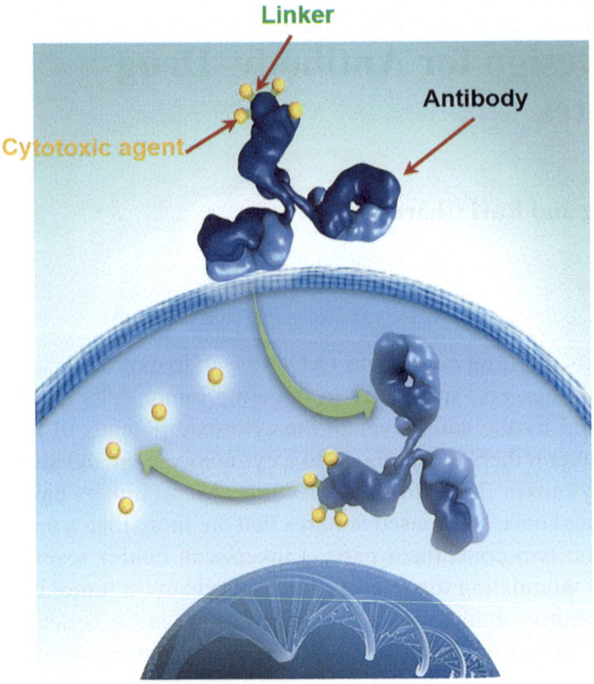

Fig. 4.1 Antibody–drug conjugate

administration. However, a linker must also allow the rapid and efficient release of the cytotoxic agent upon internalization of the ADC within cancer cells. This fine balance of extracellular stability and intracellular release is a necessary starting point for the design of a linker for ADCs.

Contemporary linker technologies have taken the linker beyond its basic role as a physical bridge between the antibody and a cytotoxic moiety. Linkers have been modified to improve the activation of ADCs, thereby allowing ADCs to release the cytotoxic agents inside the cells at a faster rate with higher efficiency utilizing different mechanisms for cytotoxin release. Modulation of the linker to effect a change in the polarity or charge of the final metabolite has also allowed improved activity toward multidrug-resistant (MDR) cells owing to better retention of the cytotoxic agent inside the cells. Some linker designs have facilitated the generation of catabolites that are capable of diffusing into proximal tumor cells, inducing bystander killing that results in greater in vivo antitumor activity. This chapter reviews various linkers that have been designed for ADCs and the impact of linkers on the activity and safety of ADCs as cancer therapeutics.

4.2 Sites of Attachment and Reactive Chemical Moieties on Linkers

Sites of attachment on the antibody and the nature of chemical groups on linkers to effect such attachment are important considerations in ADC design. They not only impact the conjugation efficiency and the ease of production of ADCs, but they also affect the integrity and stability of the conjugate during production/storage as well as in patients. In addition, the linker often remains as an integral part of the metabolites and plays a key role in the activity and safety of ADCs in patients.

Lysine and cysteine residues have been utilized as the sites of attachment on antibodies. Lysines are abundantly present on the antibody (80–100 lysines per antibody), and their primary amine readily reacts with *N*-hydroxysuccinimide esters, a chemical group often incorporated into linkers, to form stable amide bonds. Cysteines are present on antibodies in the form of disulfide bonds, mediating intrastrand and interstrand bridges connecting light and heavy chains. For cysteines to be used as sites for conjugation, typically the interstrand disulfide bonds need to be reduced to generate reactive thiols, which can undergo Michael reaction with a maleimide group commonly utilized in linkers to form thioether bonds. Both conjugation sites/methods allow efficient reactions and generate conjugates as a heterogeneous mixture with varying number of cytotoxic molecules per antibody. The distribution of the cytotoxic molecules per antibody can be highly reproducible from batch to batch and typically follows a binomial function (Fig. 4.2) (Singh and Erickson 2009; Wang et al. 2005).

Recent efforts to make homogeneous conjugates have explored diverse sites of attachment. An engineered cysteine on heavy and light chains of the antibody has been utilized as a site of attachment via maleimide linker chemistry (Junutula et al. 2008; Kung Sutherland et al. 2013). The reactive thiol of an engineered cysteine is

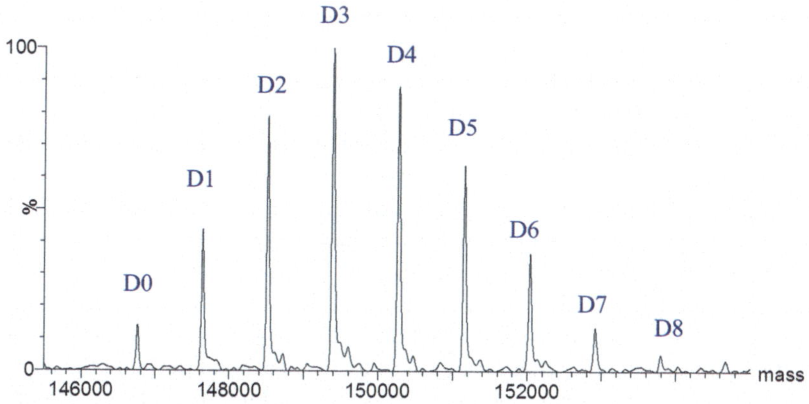

Fig. 4.2 Mass spectrometry profile of an antibody–drug conjugate. The profile depicts a maytansinoid-ADC made with an average maytansinoid-to-antibody ratio of 3.5. "n" of Dn refers to the number of cytotoxins attached to the antibody

not readily available for conjugation, because during antibody production, the sulfhydryl group of the engineered cysteine forms a mixed disulfide through exchange with cystine in the media. Before conjugation can proceed, a free thiol of the engineered cysteine must be generated by the reduction of the antibody. Reduced antibody is then subject to partial oxidation to restore the intrachain and interchain disulfides that are important for antibody integrity, while maintaining the engineered cysteine in its thiol form. The thiol can then be utilized to react with a maleimide group on linkers, a conjugation strategy used to produce ADCs from the Thiomab (Junutula et al. 2008) or a similarly engineered antibody (Kung Sutherland et al. 2013). Alternatively, the free thiol of cysteine within a specific recognition sequence can be modified by formylglycine-generating enzyme post-translationally to produce an aldehyde-bearing amino acid, which can be conjugated to a cytotoxic agent via a hydrazine linker in a process called "aldehyde tagging" (Rabuka et al. 2012).

Carbohydrate moieties that are naturally present on antibodies have been used as sites of conjugation, especially for antibodies that tend to lose binding to antigen upon conjugation via lysine residues (Walus et al. 1996). Conjugation via glycans involves oxidation of the carbohydrate using sodium periodate, followed by reaction of the resulting aldehyde with a linker bearing a hydrazide group. Although efficient, this reaction suffered from undesired oxidation of the protein. A new method for glycan modification has emerged in recent years. Mannose or galactose containing an azido substituent is introduced into the endogenous glycosylation at Asp297 of the antibody, either by expressing the antibody in the presence of azido-sugar (Ac4ManNAz or Ac4GalNAz) or by chemical remodeling using beta(1,4)-galactosyltransferase (Boeggeman et al. 2009). Use of the azido group ushered in the development of linkers with strained alkynes (e.g., cycloalkynes) that can react with the azide in copper-free Click chemistry. The use of Click chemistry gained popularity in conjugation to engineered sites, as the chemistry provides specificity without affecting the reactivity of the endogenous amino acids on antibodies.

Recently, non-natural amino acids have been introduced into specific sites in antibodies, either by the use of bacterial strains with orthogonal tRNAs that can be charged with the non-natural amino acid (Hutchins et al. 2011) or by cell-free extracts containing such tRNA and appropriate aminoacyl tRNA synthetases (Zimmerman et al. 2014). Click chemistry, along with oxime ligation, has been used for conjugation to these engineered non-natural amino acids. A non-natural amino acid containing an azido group can react with a linker containing a cycloalkyne, or a non-natural amino acid containing a hydroxylamine can react with linkers containing an aldehyde. The latter oxime ligation has been used to generate a homogeneous anti-Her2 conjugate (Axup et al. 2012).

Glutamine has been used as an acceptor for conjugation when a bacterial enzyme, transglutaminase, was employed (Dennler et al. 2014; Strop et al. 2013). Transglutaminase catalyzes the formation of a covalent bond between the acyl side chain of glutamine (e.g., endogenous Q295 of human IgG) and a free amino group (e.g., of lysine) on a linker. An antibody can be engineered by introducing a glutamine tag to scan for sites that favor transglutaminase reaction anywhere on antibodies (Strop et al. 2013), or the native glutamine residue (Q295) can be exposed for conjugation by deglycosylation at N297 that typically hinders the accessibility of glutamine (Dennler et al. 2014).

Conjugation chemistry has become sophisticated in the recent years, and different sites on the antibody and various functional groups on linkers have been explored. As a result, many claims have been made regarding the effect of these approaches (sites of conjugation and accompanying conjugation chemistry, as well as the consequence of homogeneous vs. heterogeneous conjugates) on the biological activity and potential safety of these ADCs (Junutula et al. 2008; Zimmerman et al. 2014). Thus far, there is no clinical data to support the various claims, and hence there is no clear understanding of the effect of these strategies on the activity and safety of ADCs.

4.3 Types of Linkers

Linkers are generally categorized into noncleavable or cleavable linkers. Noncleavable linkers represent linkers which remain intact during intracellular metabolism. ADCs with such linkers require lysosomal degradation of the antibody to release the cytotoxic agent. The metabolite retains the amino acid residue that served as the site of attachment to the antibody. Cleavable linkers are linkers that get cleaved during intracellular metabolism, generating metabolites that contain the cytotoxic agent with or without a portion of the linker. Cleavable linkers may be cleaved by hydrolysis, enzymatic reaction, or reduction and include acid-labile hydrazone linkers, peptide-based linkers, and disulfide linkers, respectively.

4.3.1 Noncleavable Linkers

Thioethers are a widely used format for noncleavable linkers because of the facile reaction of maleimides with thiols under mild, neutral aqueous conditions. Thioethers can also be generated by reacting thiols with halo-acetamido groups, although this reaction needs harsher conditions (e.g., higher pH and excess halo-acetamido reagents) (Alley et al. 2008). Thioether linkage has been used for microtubule-targeting agents, e.g., thiol derivatives of maytansine (Fig. 4.3) (Singh and Erickson 2009) and auristatin (MonoMethyl Auristatin F or MMAF) (Alley et al. 2008).

Ado-trastuzumab emtansine (T-DM1 or Kadcyla®) is a conjugate approved by the FDA for the treatment of HER2-positive, metastatic breast cancer in patients previously treated with trastuzumab and taxane. T-DM1 is composed of trastuzumab and DM1 linked by a heterobifunctional linker, SMCC (*N*-succinimidyl-4-(maleimidomethyl) cyclohexane-1-carboxylate) (Fig. 4.4). The SMCC–DM1 conjugate is prepared by reacting lysine residues on the antibody with the *N*-hydroxysuccinimide ester moiety of SMCC and linking the thiol of DM1 with the

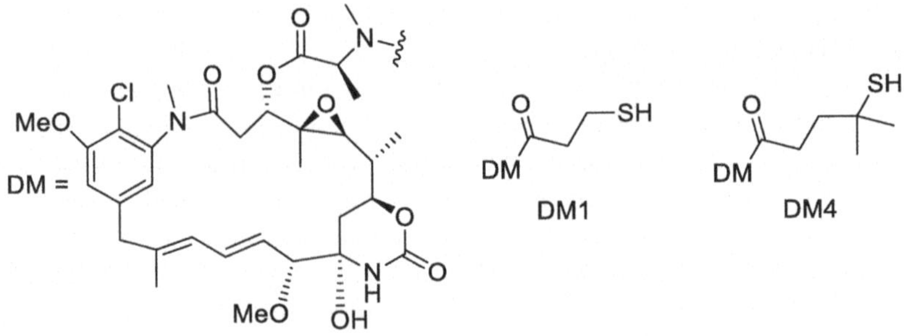

Fig. 4.3 Derivatives of maytansinoid

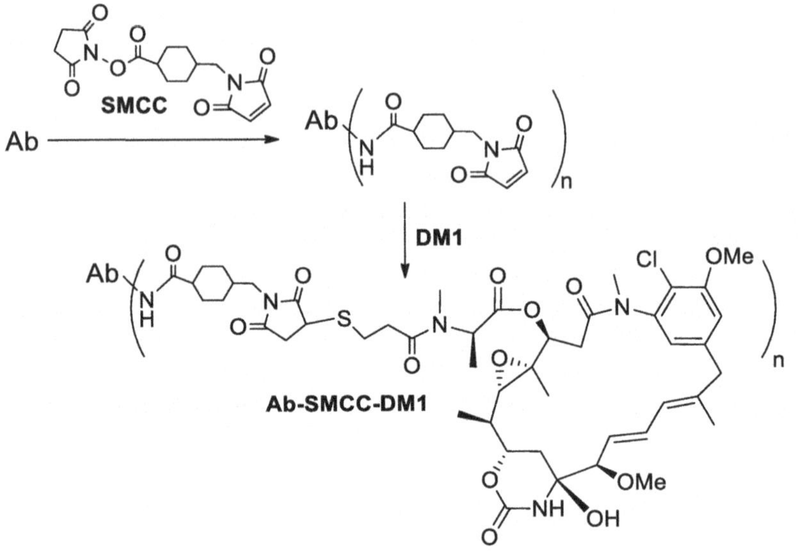

Fig. 4.4 Conjugation scheme for antibody–SMCC–DM1

maleimide group on SMCC (Fig. 4.4). The antibody–SMCC–DM1 format is currently used in other ADCs. EGFR-targeting IMGN289, CD37-targeting IMGN529, and CD70-targeting AMG172 are being evaluated in the clinic. The sole metabolite of SMCC–DM1 conjugates is Lys-SMCC–DM1 (Fig. 4.5) (Erickson et al. 2006).

An analogous thioether linker in which a hydrophilic tetraethylene glycol (PEG_4) replaced the hydrophobic cyclohexyl group of SMCC was developed (Fig. 4.6). When conjugated to an anti-EpCAM antibody, the ADC with the hydrophilic linker showed higher potency in cells expressing P-glycoprotein, which is responsible for the MDR phenotype (Kovtun et al. 2010). This PEG_4Mal–DM1 conjugate generates a single metabolite, Lys-PEG_4Mal–DM1 (Kovtun et al. 2010). A similar PEG linker was also used to conjugate DM1 to two engineered cysteines in the heavy chain of trastuzumab (Junutula et al. 2010). The homo-bifunctional BMPEO (bis-maleimido-trioxyethylene glycol) linker used contained a triethylene glycol (PEG_3) spacer with

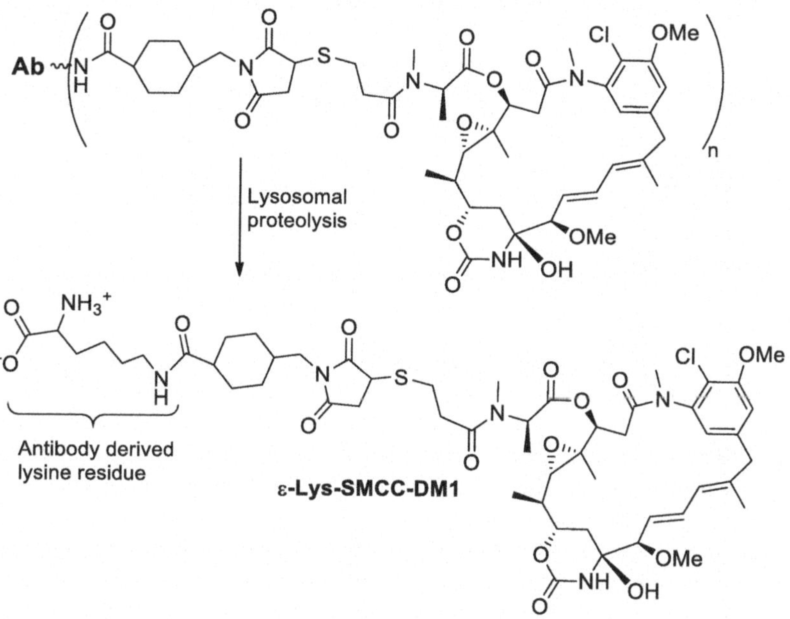

Fig. 4.5 Antibody–SMCC–DM1 generates a single metabolite Lys-SMCC–DM1

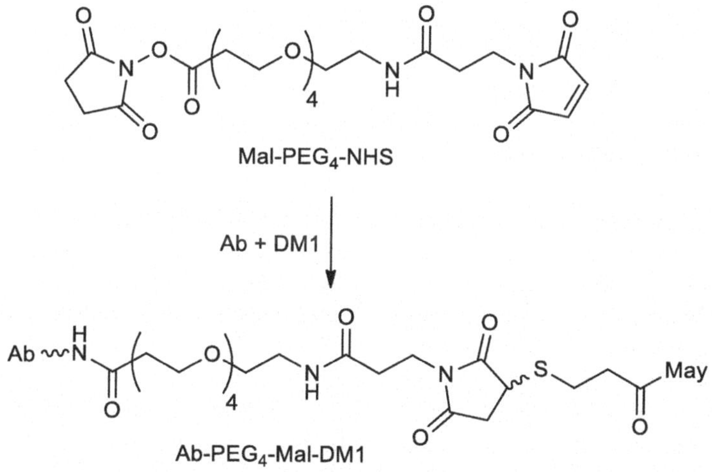

Fig. 4.6 Conjugation scheme for antibody–PEG_4–Mal–DM1

two maleimido groups, used for conjugating to the thiol of DM1 and the thiol of a reduced free cysteine, to generate a homogeneous site-specific conjugate.

A thioether linkage has also been used for the conjugation of antibodies to the tubulin inhibitor, auristatin. Anti-CD70-mc-MMAF was generated by the reaction of the thiol groups of native cysteines on the antibody with a maleimide-containing

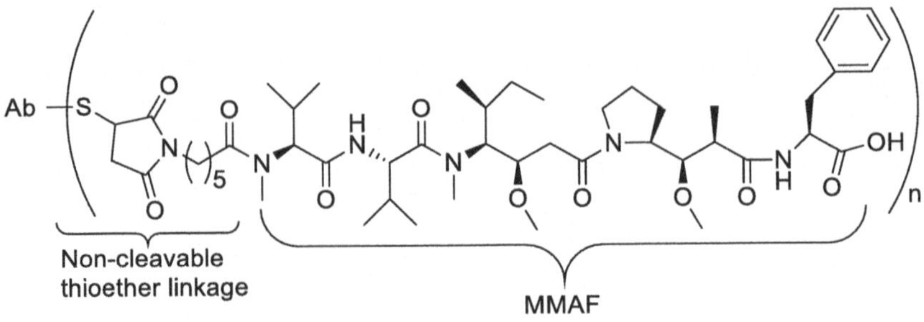

Fig. 4.7 Antibody–mc–MMAF

linker (Fig. 4.7). The anti-CD70-mc-MMAF conjugate was evaluated in the clinic for the treatment of lymphomas and renal cell carcinomas and contained an average of four MMAF molecules per antibody. The anti-CD70 antibody was evaluated preclinically with another noncleavable linker. When a bromoacetamidecaproyl (bac) group replaced the maleimidocaproyl (mc) group to generate anti-CD70-bromoacetamidecaproyl (bac)-MMAF, it showed improved plasma stability and 25% higher intratumoral drug exposure compared with a similar conjugate bearing mc-MMAF. However, despite this improvement, there was no statistically significant difference in the efficacy of these two conjugates in a subcutaneous 786-O renal cell carcinoma xenograft model (Alley et al. 2008).

4.3.2 Cleavable Linkers

4.3.2.1 Acid-Labile Linkers

Hydrazone linkers are designed to be stable in a neutral pH environment, such as in plasma, but hydrolyzed to release the cytotoxic agent in an acidic pH environment, such as in late endosomes and lysosomes. Potent DNA-targeting cytotoxic agents, e.g., calicheamicin or doxorubicin (Fig. 4.8), have been conjugated to antibodies via hydrazone linkers.

N-acetyl-γ_1^{I}-calicheamicin is a type of enediyne antibiotics that associate with the minor groove of DNA and causes double-strand breaks in a sequence-specific manner (Greenberg 2014). Calicheamicin has been evaluated in the context of antibodies to CD33, CD22, and Muc1, all of which were conjugated via hydrazone linkers (Ricart 2011; Hamann et al. 2005a).

The first FDA-approved ADC, gemtuzumab ozogamicin (Mylotarg®), was used for the treatment of patients with acute myelogenous leukemia (AML). Gemtuzumab ozogamicin is composed of anti-CD33 IgG4 conjugated to calicheamicin via a bifunctional linker 4-(4'acetylphenoxy)-butanoic acid. The *N*-hydroxysuccinimide ester of *N*-acetyl-γ-calicheamicin dimethyl hydrazide 4-(4-acetylphenoxy)-butanoic acid reacts with lysine residues on the antibody, leading to an average loading

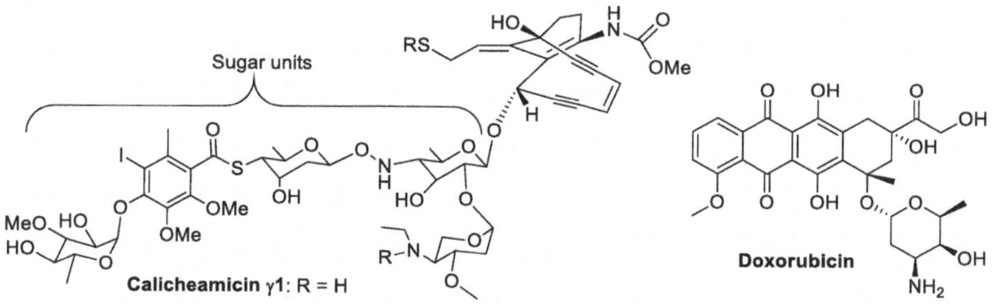

Fig. 4.8 DNA-targeting calicheamicin and doxorubicin

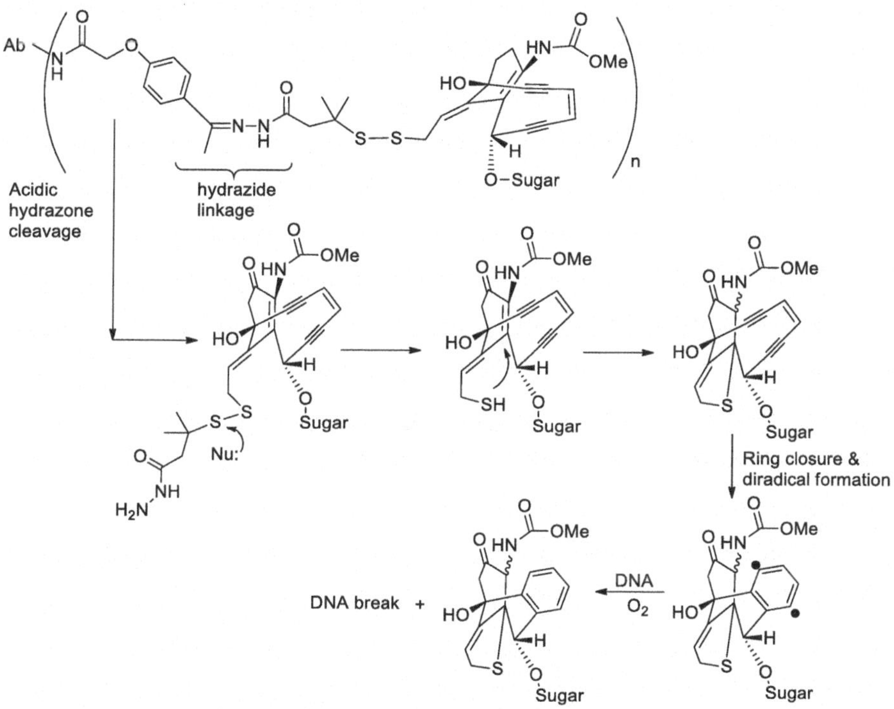

Fig. 4.9 Mechanism of release and action for antibody conjugated to calicheamicin via a hydrazone linker

of 4–6 calicheamicins per antibody, but with 50% of the antibody remaining unconjugated (Hinman et al. 1993). In an acidic environment, calicheamicin is released from the ADC upon hydrolysis of the hydrazone linker and undergoes thiol-mediated reduction, followed by Masume–Bergman cyclization. This reaction generates ρ-benzyne biradicals that abstract hydrogen atoms from the phosphodiester backbone of DNA and leads to single- and double-strand lesions (Fig. 4.9). Mylotarg was voluntarily withdrawn from the market when a confirmatory phase III clinical trial failed to demonstrate clinical benefit. A new regimen of fractionated doses for better efficacy and safety is being evaluated in the clinic (Pilorge et al. 2014). Anti-CD22 antibody was also conjugated to calicheamicin via the same acid-labile 4-(4'acetylphenoxy)-butanoic acid linker, generating inotuzumab ozogamicin (CMC-544). CMC-544 was evaluated in the clinic for the treatment of B-lymphoid malignancies (DiJoseph et al. 2004a, b; Jain et al. 2014); however, its evaluation for the treatment of CD22-positive aggressive non-Hodgkin's lymphoma was halted due to the lack of benefit in the overall survival. Another example for the hydrazone linker is highlighted in the anti-Muc1 antibody, CTM01, which was conjugated to calicheamicin via a hydrazone link to periodate-oxidized carbohydrates on the antibody (Hamann et al. 2005b).

Similarly, doxorubicin, a DNA-intercalator that elicits cytotoxicity by inhibiting topoisomerase II, has been conjugated to antibodies via a hydrazone linker. Thiols from the reduced interchain disulfides of the anti-CD74 antibody, IMMU-110, are conjugated to doxorubicin via a 4-[*N*-maleimidomethyl] cyclohexane-1 carboxylhydrazide, yielding an average of eight doxorubicin molecules per antibody. This conjugate, milatuzumab–dox, was evaluated for treatment of patients with multiple myeloma (Sapra et al. 2005) but was recently withdrawn for lack of efficacy.

4.3.2.2 Reducible Linkers

Reducible disulfide linkers take advantage of the difference in reduction potential in plasma vs. the intracellular compartment. They are designed to keep conjugates intact during systemic circulation in patients, while efficiently allowing reduction/cleavage of the disulfide bond inside cells. Reducible disulfide linkers generate neutral metabolites that can freely diffuse into neighboring cells and elicit bystander killing, aiding the tumor penetration of the cytotoxic anticancer agent (see Sect. 4.4.3).

The most abundant low-molecular thiol in blood is cysteine, the concentration of which is low at ~5 μM (Mills and Lang 1996). Serum albumin is another source of reducing potential in the blood, and its concentration is higher at ~0.6 mM (Turell et al. 2009). However, despite its relative abundance, the thiols in albumin are buried and inaccessible to thiol–disulfide exchange; thus, it is not a major source of disulfide cleavage (for more discussion, see Sect. 4.4.4). In contrast to blood, the reduction potential inside cancer cells is much higher. Reduced glutathione is present at 1–10 mM (Wu et al. 2004), and cells also contain enzymes of the protein disulfide isomerase family, which may contribute to reduction of the disulfide bond in cellular compartments. This differential allows stable circulation in plasma but efficient release inside the cells upon internalization of the ADC.

An important consideration for designing disulfide linkers is the steric hindrance of the disulfide bond and its impact on the stability of the linker. HuC242 antibody has been used as a model system to evaluate the effect of hindrance on the reactivity of disulfide linkers to thiol–disulfide exchange in vitro and in plasma (Kellogg et al. 2011). The humanized C242 antibody recognizes CanAg, which is abundantly expressed in colorectal carcinoma. Various numbers of methyl groups were

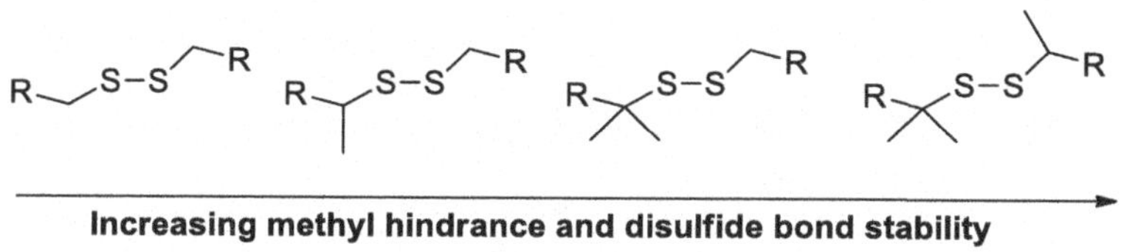

Fig. 4.10 Increasing methyl hindrance and disulfide bond stability

introduced on the carbon atoms bearing the disulfide bond (Fig. 4.10), and the rate of reduction of these disulfide bonds in ADCs by dithiothreitol (DTT) was measured. The study led to the conclusion that the stability of the disulfide linker, in the presence of a reducing agent, increased with the level of steric hindrance. Introduction of one methyl group on each side (total two methyl groups) of the disulfide bond was shown to confer greater stability than two methyl groups on one side of the disulfide. The difference in stability despite the same total number of methyl groups / hindrance highlights the importance of the position of hindrance (Kellogg et al. 2011).

Where are disulfide linkers reduced? The pKa value of the thiol of reduced glutathione is 9.65. Hence, it is expected that the reduction will be inefficient in a low pH environment such as late endosomes and lysosomes (pH ~5). By contrast, the reduction is expected to be more efficient in a higher pH environment, as in cytoplasm (pH 7.4) (Fig. 4.11). Consistent with this, no significant amount of reduction was reported to occur in lysosomes when probed with a conjugate linked to a pH-sensitive fluorophore via a disulfide linker (Austin et al. 2005; Yang et al. 2006). However, it has been demonstrated that in certain cell lines, there is some degree of reduction of the disulfide bond in endosomes, allowing an alternative pathway of metabolism before lysosomes and the consequential advantage in the cytotoxic effect (Maloney et al. 2009). Accordingly, significant reduction takes place in the endosomes for folate conjugated to FRET fluorophores via a linker with a disulfide bond (Yang et al. 2006).

4.3.2.3 Peptide Linkers

Peptide linkers are attractive for several reasons. First, peptide linkers may allow easier release of the cytotoxic molecule, as cleavage of one bond within the peptide linker is sufficient to free the metabolite within cells, as opposed to noncleavable linkers that require a cleavage of two bonds, both the N- and C-termini of the conjugated amino acid residue. Second, unlike hydrophobic synthetic linkers, peptide linkers, with the appropriate choice of amino acids, can offer hydrophilicity, allowing higher cytotoxin loading per antibody and/or better solubility in combination with unusually hydrophobic cytotoxins.

Microtubule-targeting MMAE is often conjugated to an antibody via a linker containing a valine–citrulline (Val-Cit) dipeptide and a self-immolative

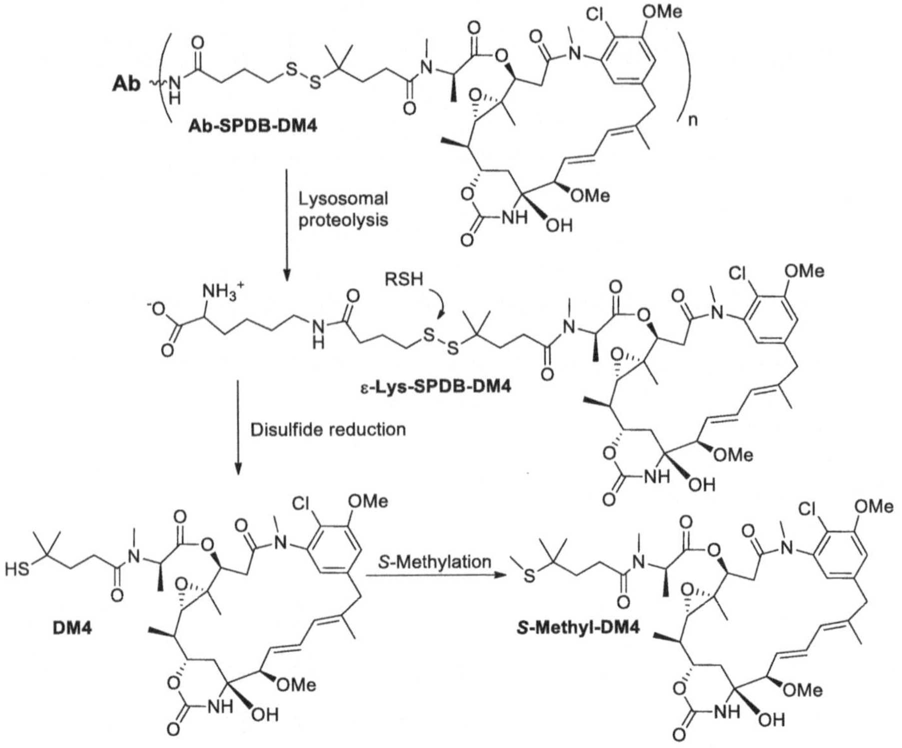

Fig. 4.11 Metabolism of antibody–SPDB–DM4

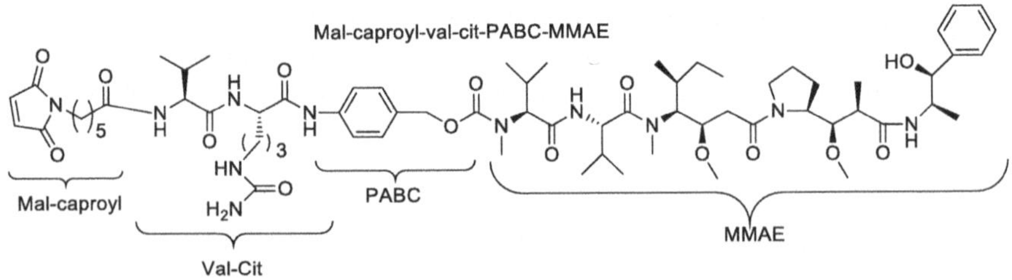

Fig. 4.12 mc-Val-Cit-PABC-MMAE

ρ-aminobenzylcarbamate spacer (PABC) (Fig. 4.12). Upon hydrolysis of the Val-Cit peptide by lysosomal proteases such as cathepsin B, PABC-MMAE is released which subsequently undergoes self-immolation at the PABC site to generate MMAE, which can diffuse into neighboring cells and elicit bystander cytotoxicity (Fig. 4.13). When compared with antibodies conjugated to MMAE via the hydrazone of 5-benzoylvaleric acid-AE ester (AEVB), the ADC with Val-Cit dipeptide linker showed better stability and greater specificity of ADC activity in vitro and in vivo (Doronina et al. 2003). Val-Cit-PABC is the linker used in brentuximab

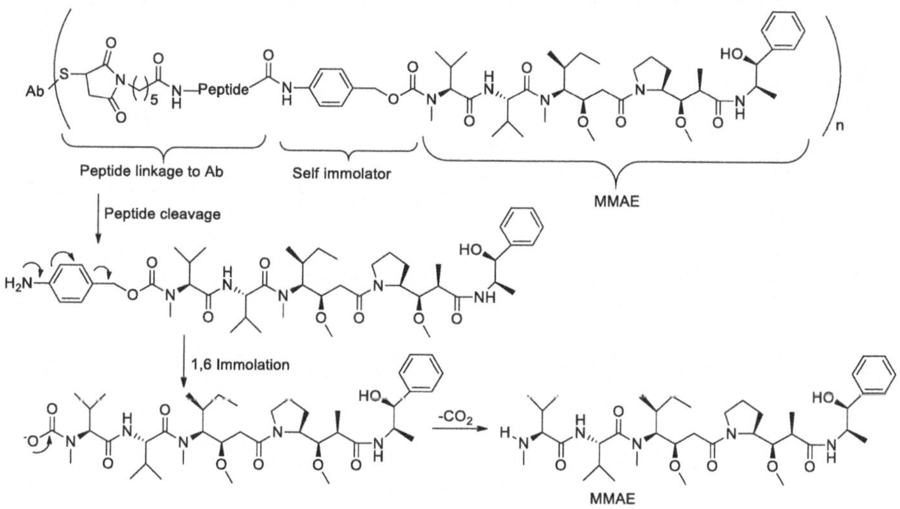

Fig. 4.13 Metabolism and self-immolation of mc-Val-Cit-PABC-MMAE

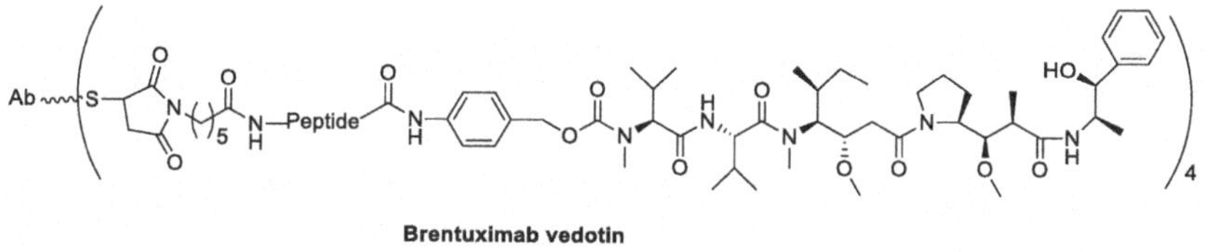

Fig. 4.14 Brentuximab vedotin

vedotin (SGN-35, Adcetris™), a CD30-targeting ADC (Fig. 4.14) (Senter and Sievers 2012). Val-Cit with PABC spacer has also been conjugated to the DNA-interacting cytotoxin, doxorubicin, in the context of mAbs c1F6 (anti-CD70) and cAC10 (anti-CD30) for targeting renal cell carcinoma and anaplastic large-cell lymphoma, respectively (Jeffrey et al. 2006).

In addition to Val-Cit, variations of the dipeptide have been evaluated. Antibodies recognizing Lewis Y and CD30 on carcinomas were conjugated to MMAE via a linker containing phenylalanine–lysine with a PABC spacer. In a head-to-head comparison using the same antibodies, the ADC with the Phe-Lys linker conferred lower stability in mouse and human plasma compared to that with a Val-Cit linker as measured ex vivo, with projected half-lives of 12.5 days vs. 30 days in mouse plasma and 80 days vs. 230 days in human plasma, respectively (Doronina et al. 2003).

Peptide linkers can be modified to confer hydrophilicity for the conjugation of hydrophobic drugs such as DNA minor-groove-binding drugs (MGBs). The valine–lysine–tetraethyleneglycol and valine–lysine–para-aminobenzyl ether self-immolative spacer allowed the conjugation of MGBs to antibodies without forming aggregates (Jeffrey et al. 2005).

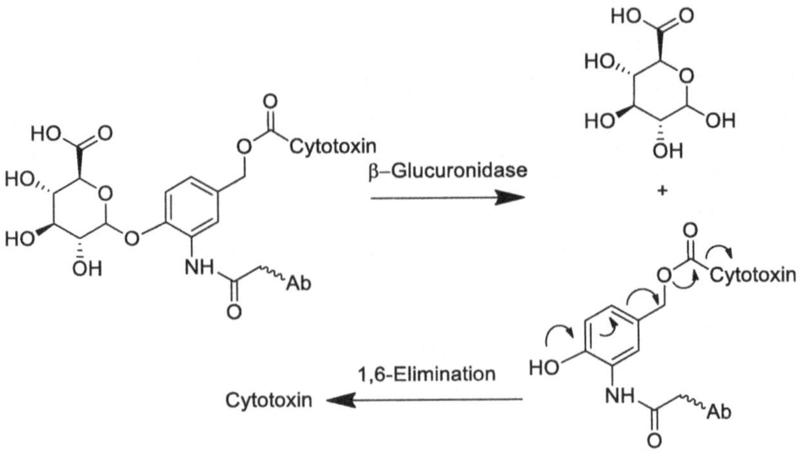

Fig. 4.15 Metabolism of antibody–β-glucuronide–cytotoxin

4.3.2.4 β-Glucuronide Linker

A β-glucuronic acid-based linker was used to conjugate MMAE, MMAF, and doxorubicin propyloxazoline individually to antibodies c1F6 (anti-CD70) and cAC10 (anti-CD30) (Jeffrey et al. 2007). β-Glucuronides are hydrolyzed by β-glucuronidase, an enzyme that is abundantly present in lysosomes and is overexpressed in some tumors (Fig. 4.15) (Albin et al. 1993) and is reported to be the main drug-metabolizing enzyme system in human breast tumors and peritumoral tissues. The hydrophilicity of this linker allows generation of monomeric ADCs with as many as eight cytotoxins per antibody. In addition, the hydrophilicity of the glucuronide linker afforded conjugation of hydrophobic drugs such as DNA minor-groove-binders (Jeffrey et al. 2005; Jeffrey et al. 2007).

4.4 Effect of Linkers on the Activity and Safety of ADC

4.4.1 Linker and Intracellular Processing/Activation

Ideally, ADCs behave as pro-drugs: they are designed to be inactive during the systemic circulation in plasma but become activated upon internalization inside cells (Fig. 4.16). In the first step, ADCs bind to antigens on the cell surface and undergo endocytosis. Upon clathrin-mediated vesicle formation, ADCs are transported to endosomes (pH ~5–6). Subsequently, endosomes fuse with lysosomes (pH ~4), the compartments that are rich in enzymes responsible for degradation of proteins, e.g., proteases and esterases. Inside lysosomes, ADCs are degraded to generate metabolites that consist of the cytotoxic agent covalently linked to the amino acid site of conjugation (Erickson et al. 2006). For example, T-DM1 generates

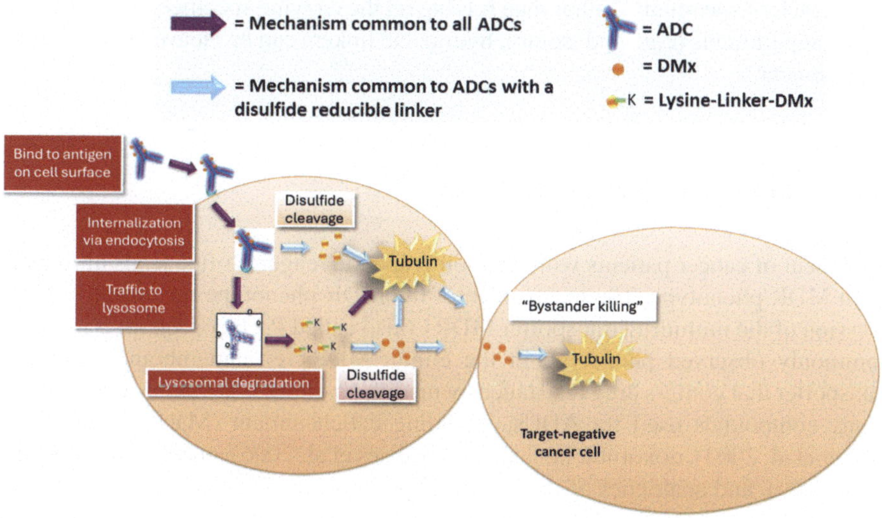

Fig. 4.16 Intracellular processing of antibody–drug conjugates

Lys-SMCC–DM1, as the conjugation occurred between the primary amine of lysine and *N*-hydroxysuccinimide ester of SMCC (Fig. 4.5) (Erickson et al. 2012).

Linkers can be modified to take advantage of different modes of activation/metabolism. For example, conjugates containing peptide linkers may allow faster rate of activation, as cleavage of one bond in the peptide linker is sufficient to release the cytotoxic agent, as opposed to ADCs with a noncleavable linker, which necessitates cleavage of two bonds at both N- and C-termini of the amino acid of attachment. Peptide linkers may also be designed so that the conjugates are metabolized in both endosomes and lysosomes, by altering the peptide sequence to match the substrate specificity of enzymes that are present in both compartments. For example, conjugates containing a peptide linker that can be cleaved by cathepsin B may be metabolized in both endosomes and lysosomes since cathepsin B is present in both compartments (Diederich et al. 2012). This may or may not be advantageous for the activity of ADC for the following reasons. Metabolism in both endosomes and lysosomes may allow faster activation that may afford an advantage in activity. However, if the linker is cleaved in the endosomes, it may also be susceptible to cleavage in the endosomal compartment during Fc recycling. Antibodies are recycled via FcRN binding in endosomes and are transported back to the cell surface, thereby avoiding degradation in lysosomes. Such recycling of antibodies via FcRN is responsible for their long half-life in plasma (Lencer and Blumberg 2005). If the linker is cleaved in endosomes during recycling, the number of molecules of cytotoxic agent linked per antibody will decrease, which leads to (i) delivery of a lower amount of cytotoxic agent per antibody to cancer cells and (ii) toxicity caused by the early release of free cytotoxin in normal tissues, with the degree of toxicity correlating to membrane permeability of the free cytotoxic agent. A similar activity and toxicity concern may apply to acid-labile hydrazone linker, with an added concern for the lack of specificity; rather than relying on the enzyme specificity and its localized compartments (e.g., endosome), hydrazone linkers can be cleaved in any acidic environment.

4.4.2 Linkers to Overcome MDR

Treatment of cancer patients with chemotherapeutic reagents often leads ultimately to an MDR phenotype. The mechanism of the MDR phenotype varies, but overexpression of the multidrug transporter MDR1 (also called P-glycoprotein) is the most commonly observed phenotype in the clinic. MDR1 is a membrane-associated transporter that confers drug resistance by mediating efflux of cytotoxic compounds. Many compounds used for ADCs, including calicheamicin (Matsui et al. 2002; Walter et al. 2003), doxorubicin, taxanes (Szakacs et al. 2006), maytansinoids (Tang et al. 2009), and analogues of dolastatin (Toppmeyer et al. 1994), are substrates of the multidrug transporter MDR1, which poses a barrier to the effective treatment of cancer patients with an ADC.

Linkers can be designed to evade the MDR1-mediated drug resistance. In a study to understand the effect of linkers on MDR1-dependent drug resistance, cell lines ranging from those naturally expressing a high level of MDR1, e.g., the colon adenocarcinoma HCT-15 and the renal adenocarcinoma UO-31, to an engineered cell line that mimics the high expression of MDR1, COLO 205^{MDR} (MDR positive; parental COLO 205 is MDR negative), were used (Kovtun et al. 2010). The presence of MDR1 led to 6–18-fold reduction in sensitivity to tubulin inhibitors including maytansine, paclitaxel, and vinblastine. Similarly, cells became resistant to anti-EpCAM–SMCC–DM1 compared to those co-treated with cyclosporin A that inhibits MDR1, indicating that MDR1 is also effective against the metabolite of the ADC, Lys-SMCC–DM1. Replacement of SMCC with PEG_4Mal (Fig. 4.6) in an anti-EpCAM ADC led to enhanced cytotoxic activity in vitro and in HCT-15 and COLO 205^{MDR} xenograft models in vivo. The sole metabolite generated by anti-EpCAM-PEG_4Mal–DM1 was Lys-PEG_4Mal-DM1, suggesting that the evasion of MDR phenotype is due to the hydrophilicity of the linker. Potency of the metabolites is expected to be the same for Lys-SMCC–DM1 and Lys-PEG_4Mal–DM1, as anti-EpCAM-SMCC–DM1 and anti-EpCAM-PEG4Mal–DM1 display similar cytotoxic potency in non-MDR cell lines (Kovtun et al. 2010).

Evasion of MDR1-mediated drug resistance is not limited to noncleavable linkers. Recently, a hydrophilic disulfide linker was generated by modifying *N*-succinimidyl-4-(2-pyridyldithio) butanoate, SPDB, with a sulfonate group positioned distal to the disulfide bond (Fig. 4.17). When evaluated as an anti-EpCAM-sulfo-SPDB-DM4 conjugate, it showed better activity in the COLO 205^{MDR} cell line and xenograft model than anti-EpCAM-SPDB-DM4, suggesting that the sulfo-SPDB linker is effective in overcoming MDR1-mediated drug resistance. Similar enhanced activity in MDR1-positive cell lines was also observed for huC242-sulfo-SPDB-DM4-targeting CanAg, a novel glycoform of Muc1 (Zhao et al. 2011).

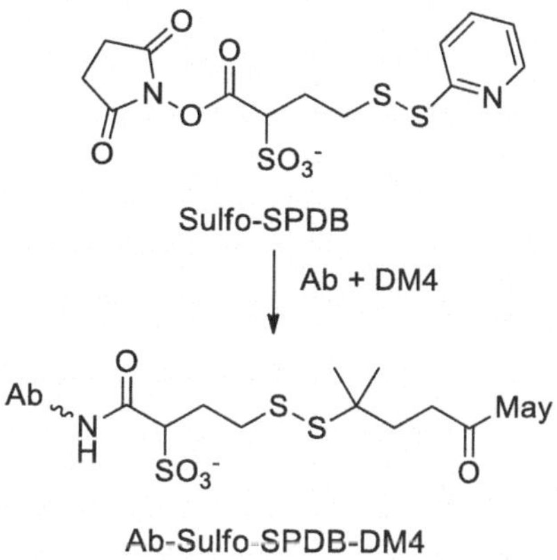

Fig. 4.17 Sulfo-SPDB linker that evades MDR resistance

In addition to MDR1-resistance, hydrophilicity of PEG_4Mal and sulfo-SPDB affords conjugation of a higher number of maytansinoid molecules per antibody (8–9 drugs per antibody (Zhao et al. 2011)). Thus, these linkers may be particularly beneficial in the conjugation of hydrophobic cytotoxins that may not have been previously feasible using other linkers. The ADC targeting the folate receptor (IMGN853) using DM4 linked to the antibody via a sulfo-SPDB linker is currently undergoing clinical evaluation in patients with ovarian carcinoma and non-small-cell lung cancer.

4.4.3 Linkers to Improve Activity in Solid Tumors

Solid tumors are architecturally complex, often heterogeneous, and composed of different tissue types. These inherent properties of solid tumors can lead to heterogeneous expression of the target antigen, which limits the population of cancer cells that can be targeted by ADC. Even for tumors expressing the target antigen homogeneously, it has been documented that tumor penetration is not efficient due in part to the large molecular size of antibody-based therapies and uneven vasculature within tumors. When anti-Her2 antibody penetration was monitored in an MDA-435/LCC6^{HER2} xenograft model, despite homogeneous expression and IHC staining of Her2, the fraction of Her2 bound by anti-Her2 antibody was patchy and localized (Baker et al. 2008). Thus, even solid tumors with a homogeneous expression of the target antigen suffer from incomplete tumor penetration, which could limit the effectiveness of ADCs.

Linkers can be designed to help compensate for this. Thus far, two types of linkers have been designed to release cell-permeable free cytotoxic agents that can diffuse into neighboring cells and cause “bystander killing”, irrespective of whether these neighboring cells express target antigen. ADCs with bystander killing have often shown better activity in vivo compared with those without the bystander effect. This advantage in vivo has not always been apparent in vitro, where the bystander killing cannot be sufficiently recapitulated due to a difference in the architecture of two-dimensional tissue culture versus three-dimensional tumors.

One type of linker that can elicit bystander killing is disulfide linkers (Kellogg et al. 2011). The disulfide bonds in the ADCs of the maytansinoid DM4 can be reduced by intracellular thiols to generate DM4 that can freely diffuse into neighboring cells, and if dividing, kill them. Furthermore, it has been demonstrated that DM4 undergoes methylation to form *S*-methyl DM4 in vitro and in vivo (Fig. 4.11) (Erickson et al. 2006; Erickson et al. 2010). Capping of DM4 through methylation may lead to an improved bystander effect: (i) methylation leads to the formation of a noncharged compound that is readily membrane-permeable and (ii) capping the free thiol of DM4 by methylation prevents possible disulfide exchange with endogenous disulfides such as cystine, which creates a hydrophilic charged compound, cysteinyl-DM4, that is not as membrane-permeable. The ADC, huC242-SPP-DM1, wherein DM1 is linked via a disulfide bond using the SPP linker, displayed bystander killing effect as a result of the reduction of the disulfide to release DM1. This ADC showed better efficacy than the conjugate with a noncleavable linker, huC242–SMCC–DM1 in COLO 205 and HT-29 xenograft models expressing the CanAg antigen, suggesting that the bystander effect plays an important role in antitumor activity in vivo (Kellogg et al. 2011).

Peptide linkers have been utilized to generate free cytotoxins that can elicit bystander killing. Conjugates containing Val-Cit-PABC-MMAE are cleaved at Val-Cit dipeptide and release *p*-aminobenzyloxycarbonyl (PABC)-MMAE, which undergoes self-immolation to generate a noncharged MMAE molecule that can penetrate neighboring cells (Fig. 4.13) (Sievers and Senter 2013).

4.4.4 Linker Stability in Plasma

Noncleavable linkers often employ a thioether bond formed by a Michael reaction between free sulfhydryl and maleimide groups, the examples of which are evident in maytansinoid and auristatin conjugates. Noncleavable linkers are considered relatively stable during circulation in plasma. However, recent data suggest that the stability of the thioether bond may vary for different linkages.

1F6-C4v2-maleimidocapropyl (mc)-MMAF, an ADC targeting the CD70 antigen on lymphomas and renal cell carcinoma, is generated by reacting a maleimido group of mc-MMAF with the reduced thiol of cysteines that normally form interchain disulfides in IgG. When compared with IF6-C4v2-bromoacetamidecaproyl (bac)-MMAF, an ADC that uses a halo-acetamido group instead of a maleimide, the IF6-C4v2-mc-MMAF conjugate showed reduced serum concentration and drug exposure, and it was found that a portion of mc-MMAF becomes conjugated to cysteine 34 of serum albumin during incubation in plasma. It is speculated that IF6-C4v2-mc-MMAF undergoes retro-Michael reaction, which releases the maleimide drug that subsequently reacts with the cysteine of serum albumin (Alley et al. 2008).

A similar retro-Michael reaction has been implicated for the instability of site-specific conjugate with a Thiomab that utilizes engineered cysteine as a reaction site for maleimide. Interestingly, a conjugation site-dependent instability of the thiol–maleimide bond was observed. When conjugated to a solvent-exposed cysteine residue S396C of Fc, mc-vc-MMAE or mc-Alexa488 was released from the antibody at a higher rate than the same chemical moiety conjugated to V205C on a light chain that is located in a positively charged environment. This site-dependent loss of conjugated moiety from the antibody was accompanied by the conjugation of the released "payload" to serum albumin, as modeled utilizing mc-Alexa488 conjugates, suggesting that a maleimide exchange has occurred between the antibody and albumin. It was speculated that the thiol–maleimide bond in a solvent-exposed environment readily undergoes maleimide exchange in plasma, whereas maleimide in a positively charged environment undergoes succinimidyl ring opening, which prevents maleimide exchange (Shen et al. 2012b).

Maytansinoid conjugates have proven to be an exception. Although the thiol-containing maytansinoid, DM1, utilizes the same thiol–maleimide chemistry, it is much more resistant to retro-Michael reaction, likely owing to the higher pKa of the thiol donor, compared to cysteine residues on the antibody. A study to further understand the stability of the SMCC–DM1 thioether linkage led to the observation that free DM1 is released only by β-elimination, followed by oxidation of the thioether bond to sulfoxide, which likely occurs in ex vivo conditions (Fishkin et al. 2011). It is suggested that thioether oxidation is a potential ex vivo artifact that is less likely to occur in vivo where the redox potential of plasma is more tightly regulated.

Stability of the disulfide linkers depends on the hindrance of the disulfide. As discussed previously (see Sect. 4.3.2.2), the greater the degree of hindrance around the disulfide bond, the lower is the propensity to reduction in vitro by dithiotreitol (DTT), and this stability correlates well with the pharmacokinetics (PK) of the ADCs in vivo. HuC242-SPDB-DM4 (two methyl groups at the carbon next to the disulfide bond) shows a longer half-life in mouse plasma than huC242-SPDB-DM3 or huC242-SPP-DM1 (both contain one methyl group at the carbon adjoining the disulfide bond) (Kellogg et al. 2011).

4.4.5 Linker Stability and Activity of ADCs

Do stable linkers provide better activity for ADCs? Preclinical findings suggest that although stable linkers may increase the exposure of tumors to ADC, it is the careful balance between the resistance to extracellular cleavage (e.g., in plasma) and the facility of intracellular cleavage (upon cellular internalization) of linkers that provides the maximal activity.

The study of reducible linkers demonstrates this point elegantly. First, an ADC with SPP-DM4 with three methyl groups around the disulfide bond shows better stability against thiol–disulfide exchange in vitro and longer half-life in plasma of 218 h compared with a SPP–DM1 conjugate, which has only one methyl group on the carbon atom adjacent to the disulfide bond (half-life at 47 h). Accordingly, the exposure for the SPP–DM4 conjugate is greater than that of SPP-DM1, with AUC being 22,712 and 5186 h μg/mL, respectively. Yet, huC242-SPP-DM1 shows better efficacy compared with the huC242 antibody conjugated to SPP-DM4 in subcutaneous COLO 205 and HT-29 xenograft models. It is hypothesized that SPP-DM1 releases catabolites more readily than SPP-DM4 inside cells, such that the higher exposure of SPP-DM4 cannot compensate for the faster intracellular activation of SPP-DM1 (Kellogg et al. 2011).

Similar results were observed for α_v integrin-targeting conjugates. CNTO365, using SPDB-DM4 with two methyl groups hindering thiol–disulfide exchange, showed better efficacy than CNTO366, using SPP-DM4 with three methyl groups hindering thiol–disulfide exchange, in HT-29 colon cancer and A-549 human lung cancer xenograft models (Chen et al. 2007).

Recent studies with T-DM1 illustrates that the linker stability alone does not predict efficacy. When compared against trastuzumab-SPP–DM1 (or T-SPP–DM1), T-DM1 (T-SMCC–DM1) showed better stability in plasma and longer half-life (Fig. 4.18). However, despite faster clearance and less total conjugate localization to tumors, T-SPP-DM1 showed similar amount of metabolites generated at tumors (Fig. 4.18). As such, T-SPP–DM1 demonstrated a similar efficacy as T-SMCC–DM1 in the BT474-EEI xenograft model (Erickson et al. 2012). It is clear that the rate of activation inside the tumors and the total amount of metabolites are important predictive factors in antitumor activity. Moreover, different mechanisms of action for cell killing, i.e., bystander activity for T-SPP-DM1, but not for T-DM1, should also be considered in interpreting the efficacy data.

In conclusion, efficacy conferred by the linker cannot be predicted based on one aspect of ADC behavior, such as pharmacokinetics, as a number of different factors can contribute to the activity of ADC. Thus, each ADC with different linkers must be tested empirically to determine the combined effect of PK, exposure, rate of intracellular activation, and mode of killing (e.g., bystander, etc.) in the context of each target.

4.4.6 Linker Stability and Safety

4.4.6.1 Effect of Linker on Liver Detoxification of ADC and Biodistribution

Liver is the primary site of antibody metabolism, and indeed, significant amounts of metabolites from ADCs are found in liver. The anti-CD56 antibody, huN901, conjugated to maytansinoid via various linkers was used to study the effect of linker on the detoxification of ADC (Sun et al. 2011). A radioactive tracer, tritium, was incorporated to the C-20 methoxy group of maytansinoid to allow for the detection of metabolites. HuN901-SMCC–[^{3}H]DM1 with a noncleavable linker was

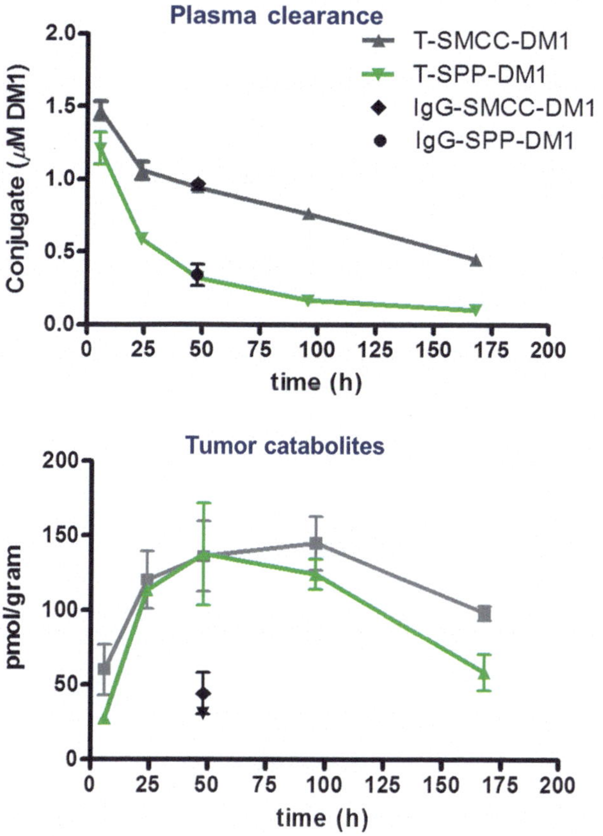

Fig. 4.18 Pharmacokinetics of trastuzumab–SPP–DM1 (T-SPP–DM1) and T-DM1 (T-SMCC–DM1) and accumulation of the metabolites in the BT474 EEI-derived tumor xenograft model

metabolized in liver to Lys-SMCC–[^{3}H]DM1, which is more than 50-fold less cytotoxic than the parental compound due to poor cell penetration. Both huN901-SPP-[^{3}H]DM1 and huN901-SPDB-[^{3}H]DM4 containing disulfide linkers also generate initially the analogous lysine-linked maytansinoid. However, subsequent reduction, *S*-methylation, and NADPH-dependent oxidation in the liver lead to the formation of *S*-methyl sulfoxide and *S*-methyl sulfone derivatives of maytansinoid (Fig. 4.19). When tested in vitro, these oxidized maytansinoids were found to be 5–50-fold less cytotoxic than parental maytansine in many human cancer cell lines, illustrating efficient detoxification of ADCs in liver (Sun et al. 2011).

Biodistribution of ADC and the effect of linkers on tissue distribution were assessed using huC242 antibody targeting CanAg antigen. Unconjugated antibody, antibody–SPP–DM1, and antibody–SPDB–DM4 were labeled with ^{125}I on the antibody backbone to track the localization of ADC to various tissues. Following a

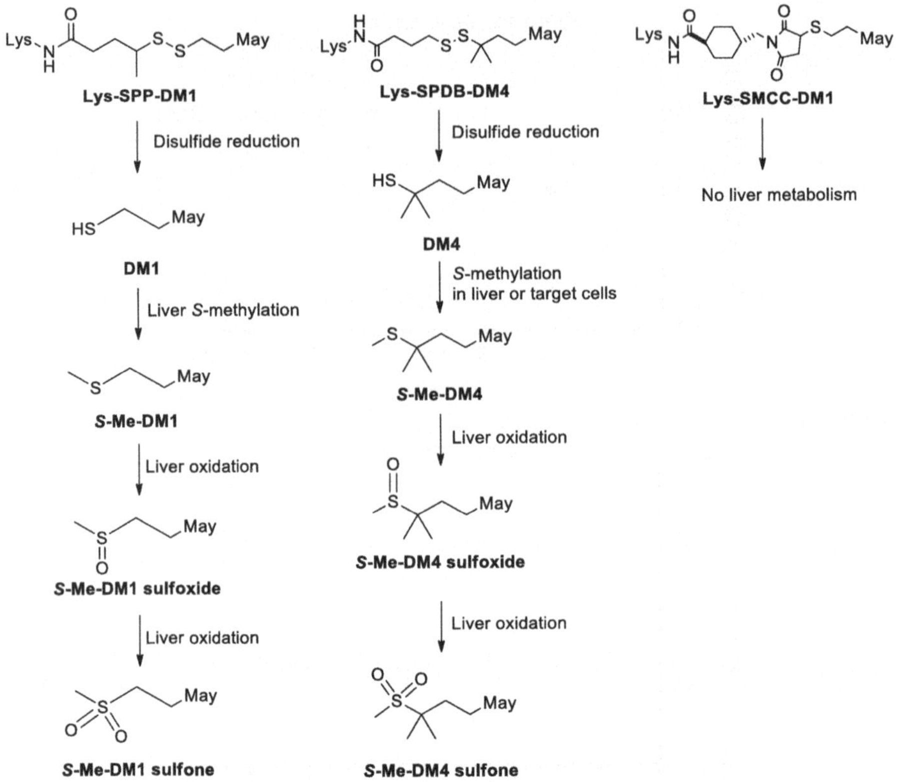

Fig. 4.19 Liver detoxification of antibody–maytansinoid conjugates

single bolus injection of 4.16 mg/kg, it was found that the biodistribution profile is similar among all conjugates and the unconjugated antibody, demonstrating that ADC distribution is dictated by the antibody component (Xie et al. 2004; Xie and Blattler 2006). A closer look at huC242-SPDB-DM4 with tritium label at the C-20 methoxy group of maytansinoid to follow the drug portion of ADC led to the finding that 30–50% of injected dose/g was recovered in gall bladder from 2 h to 2 days post-dosing, which is consistent with the hepatobiliary elimination of maytansinoid (Erickson and Lambert 2012). Similar observations were made when rats were administered with a single bolus injection of T-[^{3}H]DM1; up to 80% of radioactivity was recovered in the feces over 7 days, consistent with the hepatobiliary elimination of maytansinoid (Shen et al. 2012a). Biodistribution profile of huC242-SMCC–[^{3}H]DM1 resembled the profile of huC242-SPDB-[^{3}H]DM4 (Erickson and Lambert 2012), suggesting a lack of significant contribution of linkers on biodistribution for the linkers tested. Thus, the different degree of toxicity conferred by SPP, SPDB, and SMCC linkers in mice may be due to the cell permeability and subsequent potency of the metabolites, rather than the linker-dependent distribution of ADCs.

4.4.6.2 Effect of Linkers on Safety in Clinic: Inverse Correlation of Stability and Safety

There are many ADCs with different linkers in the clinic today. The pharmacokinetics of the ADC in humans has been a reflection of both the stability of the linker and antigen-mediated clearance. For maytansinoid conjugates containing disulfide linkers, the pharmacokinetics in humans has been consistent with the susceptibility of the linker to reduction by DTT and the pharmacokinetics observed in preclinical animals. Cantuzumab or anti-huC242 targeting CanAg antigen provides an ideal example, in which the same antibody has been conjugated with two linkers, and the resulting conjugates have been evaluated in phase I clinical trials. Cantuzumab mertansine (huC242-SPP-DM1), which contains a mildly hindered disulfide, has a half-life of 2 days in human plasma (Rodon et al. 2008). Cantuzumab ravtansine (huC242-SPDB-DM4), which contains a highly hindered disulfide, shows a half-life of 4.6 days (Qin et al. 2008). The pharmacokinetics of these conjugates in humans reflects the difference in the linker stability of SPP-DM1 and SPDB-DM4 in the context of CanAg. SAR3419 (huB4-SPDB-DM4) targeting CD19 shows 7.9 days of half-life in human plasma (Younes et al. 2009; Ribrag et al. 2014), indicating that the faster clearance of huC242-SPDB-DM4 (half-life of 4.6 days) is likely due to some contribution of antigen-mediated clearance rather than the inherent instability of the SPDB-DM4 linker–drug combination. Ado-trastuzumab emtansine has a half-life of 4.4 days in human plasma (Krop et al. 2010), reflecting largely the antigen-mediated clearance.

The maximum tolerated dose (MTD) can be affected by target-dependent toxicity, i.e., target expression on normal tissues could contribute to the final tolerable level of dose. However, different ADCs with the same linker–cytotoxic agent pairing directed against unrelated targets with diverse expression in tissues demonstrate that pairings can create an upper limit for the highest administered dose. For example, the MTD for auristatin conjugates containing the dipeptide Val-Cit linker is typically close to 2 mg/kg (Younes et al. 2010) due to neutropenia and/or peripheral neuropathy. Maytansinoid conjugates also show a strong linker impact on tolerability. Both T-DM1 and AMG595, which use the SMCC–DM1 pairing, have dose-limiting toxicity (DLT) of reversible thrombocytopenia; MTD for T-DM1 is 3.6 mg/kg (Q3W), and AMG595 has been dosed to 3.0 mg/kg (Q3W) (Krop et al. 2010). SAR3419 or huB4-SPDB-DM4 with a cleavable disulfide linker shows an MTD of 4.3 mg/kg (Q3W) with the DLT of reversible ocular toxicity (Younes et al. 2009). Cantuzumab mertansine or anti-huC242-SPP-DM1 with the most readily cleavable disulfide linker had an MTD of 6.4 mg/kg (Q3W) with the DLT of reversible elevation of liver transaminases (Rodon et al. 2008). Interestingly, there is an inverse correlation of tolerability and stability of linkers (Fig. 4.20). The chemical stability of the linkers in plasma (in vivo) can be ranked as SMCC>SPDB>SPP, with SMCC being the most stable linker and SPP being the least stable linker. In contrast to the chemical stability in plasma, the tolerability as demonstrated by the MTDs in the clinic for the maytansinoid conjugates listed above can be ranked as SPP>SPDB>SMCC. More clinical data are needed to determine whether (i) this

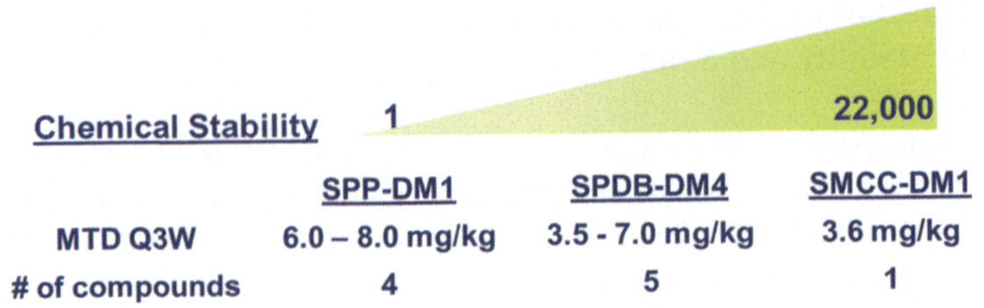

Fig. 4.20 Inverse correlation of the linker stability and MTD in human. # of compounds refers to those that have been evaluated in the clinic

stands true for all maytansinoid conjugates that may yield different small-molecular-weight metabolites during their eventual elimination and (ii) whether a similar trend can be found for other cytotoxin payloads. These findings suggest that it should not be hastily concluded that the most stable linker is the best linker for clinical development, and as has been discussed above, stability, efficacy, and safety must be all factored in for consideration of the optimal linker for ADCs.

References

Albin N, Massaad L, Toussaint C, Mathieu MC, Morizet J, Parise O, Gouyette A, Chabot GG (1993) Main drug-metabolizing enzyme systems in human breast tumors and peritumoral tissues. Cancer Res 53(15):3541–3546

Alley SC, Benjamin DR, Jeffrey SC, Okeley NM, Meyer DL, Sanderson RJ, Senter PD (2008) Contribution of linker stability to the activities of anticancer immunoconjugates. Bioconjug Chem 19(3):759–765. https://doi.org/10.1021/bc7004329

Austin CD, Wen X, Gazzard L, Nelson C, Scheller RH, Scales SJ (2005) Oxidizing potential of endosomes and lysosomes limits intracellular cleavage of disulfide-based antibody-drug conjugates. Proc Natl Acad Sci USA 102(50):17987–17992. https://doi.org/10.1073/pnas.0509035102

Axup JY, Bajjuri KM, Ritland M, Hutchins BM, Kim CH, Kazane SA, Halder R, Forsyth JS, Santidrian AF, Stafin K, Lu Y, Tran H, Seller AJ, Biroc SL, Szydlik A, Pinkstaff JK, Tian F, Sinha SC, Felding-Habermann B, Smider VV, Schultz PG (2012) Synthesis of site-specific antibody-drug conjugates using unnatural amino acids. Proc Natl Acad Sci USA 109(40):16101–16106. https://doi.org/10.1073/pnas.1211023109

Baker JH, Lindquist KE, Huxham LA, Kyle AH, Sy JT, Minchinton AI (2008) Direct visualization of heterogeneous extravascular distribution of trastuzumab in human epidermal growth factor receptor type 2 overexpressing xenografts. Clin Cancer Res Off J Am Assoc Cancer Res 14(7):2171–2179. https://doi.org/10.1158/1078-0432.CCR-07-4465

Boeggeman E, Ramakrishnan B, Pasek M, Manzoni M, Puri A, Loomis KH, Waybright TJ, Qasba PK (2009) Site specific conjugation of fluoroprobes to the remodeled Fc N-glycans of monoclonal antibodies using mutant glycosyltransferases: application for cell surface antigen detection. Bioconjug Chem 20(6):1228–1236. https://doi.org/10.1021/bc900103p

Chen Q, Millar HJ, McCabe FL, Manning CD, Steeves R, Lai K, Kellogg B, Lutz RJ, Trikha M, Nakada MT, Anderson GM (2007) Alphav integrin-targeted immunoconjugates regress established human tumors in xenograft models. Clin Cancer Res Off J Am Assoc Cancer Res 13(12):3689–3695. https://doi.org/10.1158/1078-0432.CCR-07-0026

Dennler P, Chiotellis A, Fischer E, Bregeon D, Belmant C, Gauthier L, Lhospice F, Romagne F, Schibli R (2014) Transglutaminase-based chemo-enzymatic conjugation approach yields homogeneous antibody-drug conjugates. Bioconjug Chem 25(3):569–578. https://doi.org/10.1021/bc400574z

Diederich S, Sauerhering L, Weis M, Altmeppen H, Schaschke N, Reinheckel T, Erbar S, Maisner A (2012) Activation of the Nipah virus fusion protein in MDCK cells is mediated by cathepsin B within the endosome-recycling compartment. J Virol 86(7):3736–3745. https://doi.org/10.1128/JVI.06628-11

DiJoseph JF, Armellino DC, Boghaert ER, Khandke K, Dougher MM, Sridharan L, Kunz A, Hamann PR, Gorovits B, Udata C, Moran JK, Popplewell AG, Stephens S, Frost P, Damle NK (2004a) Antibody-targeted chemotherapy with CMC-544: a CD22-targeted immunoconjugate of calicheamicin for the treatment of B-lymphoid malignancies. Blood 103(5):1807–1814. https://doi.org/10.1182/blood-2003-07-2466

DiJoseph JF, Goad ME, Dougher MM, Boghaert ER, Kunz A, Hamann PR, Damle NK (2004b) Potent and specific antitumor efficacy of CMC-544, a CD22-targeted immunoconjugate of calicheamicin, against systemically disseminated B-cell lymphoma. Clin Cancer Res Off J Am Assoc Cancer Res 10(24):8620–8629. https://doi.org/10.1158/1078-0432.CCR-04-1134

Doronina SO, Toki BE, Torgov MY, Mendelsohn BA, Cerveny CG, Chace DF, DeBlanc RL, Gearing RP, Bovee TD, Siegall CB, Francisco JA, Wahl AF, Meyer DL, Senter PD (2003) Development of potent monoclonal antibody auristatin conjugates for cancer therapy. Nat Biotechnol 21(7):778–784. https://doi.org/10.1038/nbt832. nbt832 [pii]

Erickson HK, Lambert JM (2012) ADME of antibody-maytansinoid conjugates. Aaps J 14(4):799–805. https://doi.org/10.1208/s12248-012-9386-x

Erickson HK, Park PU, Widdison WC, Kovtun YV, Garrett LM, Hoffman K, Lutz RJ, Goldmacher VS, Blattler WA (2006) Antibody-maytansinoid conjugates are activated in targeted cancer cells by lysosomal degradation and linker-dependent intracellular processing. Cancer Res 66(8):4426–4433. https://doi.org/10.1158/0008-5472.CAN-05-4489

Erickson HK, Widdison WC, Mayo MF, Whiteman K, Audette C, Wilhelm SD, Singh R (2010) Tumor delivery and in vivo processing of disulfide-linked and thioether-linked antibody-maytansinoid conjugates. Bioconjug Chem 21(1):84–92. https://doi.org/10.1021/bc900315y

Erickson HK, Lewis Phillips GD, Leipold DD, Provenzano CA, Mai E, Johnson HA, Gunter B, Audette CA, Gupta M, Pinkas J, Tibbitts J (2012) The effect of different linkers on target cell catabolism and pharmacokinetics/pharmacodynamics of trastuzumab maytansinoid conjugates. Mol Cancer Ther 11(5):1133–1142. https://doi.org/10.1158/1535-7163.MCT-11-0727

Fishkin N, Maloney EK, Chari RV, Singh R (2011) A novel pathway for maytansinoid release from thioether linked antibody-drug conjugates (ADCs) under oxidative conditions. Chem Commun (Camb) 47(38):10752–10754. https://doi.org/10.1039/c1cc14164c

Greenberg MM (2014) Abasic and oxidized abasic site reactivity in DNA: enzyme inhibition, cross-linking, and nucleosome catalyzed reactions. Acc Chem Res 47(2):646–655. https://doi.org/10.1021/ar400229d

Hamann PR, Hinman LM, Beyer CF, Greenberger LM, Lin C, Lindh D, Menendez AT, Wallace R, Durr FE, Upeslacis J (2005a) An anti-MUC1 antibody-calicheamicin conjugate for treatment of solid tumors. Choice of linker and overcoming drug resistance. Bioconjug Chem 16(2):346–353. https://doi.org/10.1021/bc049795f

Hamann PR, Hinman LM, Beyer CF, Lindh D, Upeslacis J, Shochat D, Mountain A (2005b) A calicheamicin conjugate with a fully humanized anti-MUC1 antibody shows potent antitumor effects in breast and ovarian tumor xenografts. Bioconjug Chem 16(2):354–360. https://doi.org/10.1021/bc049794n

Hinman LM, Hamann PR, Wallace R, Menendez AT, Durr FE, Upeslacis J (1993) Preparation and characterization of monoclonal antibody conjugates of the calicheamicins: a novel and potent family of antitumor antibiotics. Cancer Res 53(14):3336–3342

Hutchins BM, Kazane SA, Staflin K, Forsyth JS, Felding-Habermann B, Schultz PG, Smider VV (2011) Site-specific coupling and sterically controlled formation of multimeric antibody fab

fragments with unnatural amino acids. J Mol Biol 406(4):595–603. https://doi.org/10.1016/j.jmb.2011.01.011
Jain N, O'Brien S, Thomas D, Kantarjian H (2014) Inotuzumab ozogamicin in the treatment of acute lymphoblastic leukemia. Front Biosci (Elite Ed) 6:40–45
Jeffrey SC, Torgov MY, Andreyka JB, Boddington L, Cerveny CG, Denny WA, Gordon KA, Gustin D, Haugen J, Kline T, Nguyen MT, Senter PD (2005) Design, synthesis, and in vitro evaluation of dipeptide-based antibody minor groove binder conjugates. J Med Chem 48(5):1344–1358. https://doi.org/10.1021/jm040137q
Jeffrey SC, Nguyen MT, Andreyka JB, Meyer DL, Doronina SO, Senter PD (2006) Dipeptide-based highly potent doxorubicin antibody conjugates. Bioorg Med Chem Lett 16(2):358–362. https://doi.org/10.1016/j.bmcl.2005.09.081
Jeffrey SC, Nguyen MT, Moser RF, Meyer DL, Miyamoto JB, Senter PD (2007) Minor groove binder antibody conjugates employing a water soluble beta-glucuronide linker. Bioorg Med Chem Lett 17(8):2278–2280. https://doi.org/10.1016/j.bmcl.2007.01.071
Junutula JR, Raab H, Clark S, Bhakta S, Leipold DD, Weir S, Chen Y, Simpson M, Tsai SP, Dennis MS, Lu Y, Meng YG, Ng C, Yang J, Lee CC, Duenas E, Gorrell J, Katta V, Kim A, McDorman K, Flagella K, Venook R, Ross S, Spencer SD, Lee Wong W, Lowman HB, Vandlen R, Sliwkowski MX, Scheller RH, Polakis P, Mallet W (2008) Site-specific conjugation of a cytotoxic drug to an antibody improves the therapeutic index. Nat Biotechnol 26(8):925–932. https://doi.org/10.1038/nbt.1480
Junutula JR, Flagella KM, Graham RA, Parsons KL, Ha E, Raab H, Bhakta S, Nguyen T, Dugger DL, Li G, Mai E, Lewis Phillips GD, Hiraragi H, Fuji RN, Tibbitts J, Vandlen R, Spencer SD, Scheller RH, Polakis P, Sliwkowski MX (2010) Engineered thio-trastuzumab-DM1 conjugate with an improved therapeutic index to target human epidermal growth factor receptor 2-positive breast cancer. Clin Cancer Res Off J Am Assoc Cancer Res 16(19):4769–4778. https://doi.org/10.1158/1078-0432.CCR-10-0987
Kellogg BA, Garrett L, Kovtun Y, Lai KC, Leece B, Miller M, Payne G, Steeves R, Whiteman KR, Widdison W, Xie H, Singh R, Chari RV, Lambert JM, Lutz RJ (2011) Disulfide-linked antibody-maytansinoid conjugates: optimization of in vivo activity by varying the steric hindrance at carbon atoms adjacent to the disulfide linkage. Bioconjug Chem 22(4):717–727. https://doi.org/10.1021/bc100480a
Kovtun YV, Audette CA, Mayo MF, Jones GE, Doherty H, Maloney EK, Erickson HK, Sun X, Wilhelm S, Ab O, Lai KC, Widdison WC, Kellogg B, Johnson H, Pinkas J, Lutz RJ, Singh R, Goldmacher VS, Chari RV (2010) Antibody-maytansinoid conjugates designed to bypass multidrug resistance. Cancer Res 70(6):2528–2537. https://doi.org/10.1158/0008-5472.CAN-09-3546
Krop IE, Beeram M, Modi S, Jones SF, Holden SN, Yu W, Girish S, Tibbitts J, Yi JH, Sliwkowski MX, Jacobson F, Lutzker SG, Burris HA (2010) Phase I study of trastuzumab-DM1, an HER2 antibody-drug conjugate, given every 3 weeks to patients with HER2-positive metastatic breast cancer. J Clin Oncol 28(16):2698–2704. https://doi.org/10.1200/JCO.2009.26.2071
Kung Sutherland MS, Walter RB, Jeffrey SC, Burke PJ, Yu C, Kostner H, Stone I, Ryan MC, Sussman D, Lyon RP, Zeng W, Harrington KH, Klussman K, Westendorf L, Meyer D, Bernstein ID, Senter PD, Benjamin DR, Drachman JG, McEarchern JA (2013) SGN-CD33A: a novel CD33-targeting antibody-drug conjugate utilizing a pyrrolobenzodiazepine dimer is active in models of drug-resistant AML. Blood 122:1455. https://doi.org/10.1182/blood-2013-03-491506
Lambert JM (2013) Drug-conjugated antibodies for the treatment of cancer. Br J Clin Pharmacol 76(2):248–262. https://doi.org/10.1111/bcp.12044
Lencer WI, Blumberg RS (2005) A passionate kiss, then run: exocytosis and recycling of IgG by FcRn. Trends Cell Biol 15(1):5–9. https://doi.org/10.1016/j.tcb.2004.11.004
Maloney E, Fishkin N, Chari R, Singh R (2009) Abstract B120: designing potent antibody-drug conjugates: the impact of lysosomal processing efficiency and conjugate linker selection on anticancer activity. Mol Cancer Ther 8(Supplement 1):B120. https://doi.org/10.1158/1535-7163.targ-09-b120

Matsui H, Takeshita A, Naito K, Shinjo K, Shigeno K, Maekawa M, Yamakawa Y, Tanimoto M, Kobayashi M, Ohnishi K, Ohno R (2002) Reduced effect of gemtuzumab ozogamicin (CMA-676) on P-glycoprotein and/or CD34-positive leukemia cells and its restoration by multidrug resistance modifiers. Leukemia 16(5):813–819. https://doi.org/10.1038/sj.leu.2402459

Mills BJ, Lang CA (1996) Differential distribution of free and bound glutathione and cyst(e)ine in human blood. Biochem Pharmacol 52(3):401–406. https://doi.org/10.1016/0006-2952(96)00241-9. [pii]

Pilorge S, Rigaudeau S, Rabian F, Sarkozy C, Taksin AL, Farhat H, Merabet F, Ghez S, Raggueneau V, Terre C, Garcia I, Renneville A, Preudhomme C, Castaigne S, Rousselot P (2014) Fractionated gemtuzumab ozogamicin and standard dose cytarabine produced prolonged second remissions in patients over the age of 55 years with acute myeloid leukemia in late first relapse. Am J Hematol 89(4):399–403. https://doi.org/10.1002/ajh.23653

Qin A, Watermill J, Mastico RA, Lutz RJ, O'Keeffe J, Zildjian S, Mita AC, Phan AT, Tolcher AW (2008) The pharmacokinetics and pharmacodynamics of IMGN242 (huC242-DM4) in patients with CanAg-expressing solid tumors. Paper presented at the ASCO Meet Abstr

Rabuka D, Rush JS, deHart GW, Wu P, Bertozzi CR (2012) Site-specific chemical protein conjugation using genetically encoded aldehyde tags. Nat Protoc 7(6):1052–1067. https://doi.org/10.1038/nprot.2012.045

Ribrag V, Dupuis J, Tilly H, Morschhauser F, Laine F, Houot R, Haioun C, Copie C, Varga A, Lambert J, Hatteville L, Ziti-Ljajic S, Caron A, Payrard S, Coiffier B (2014) A dose-escalation study of SAR3419, an anti-CD19 antibody maytansinoid conjugate, administered by intravenous infusion once weekly in patients with relapsed/refractory B-cell non-Hodgkin lymphoma. Clin Cancer Res Off J Am Assoc Cancer Res 20(1):213–220. https://doi.org/10.1158/1078-0432.CCR-13-0580

Ricart AD (2011) Antibody-drug conjugates of calicheamicin derivative: gemtuzumab ozogamicin and inotuzumab ozogamicin. Clin Cancer Res Off J Am Assoc Cancer Res 17(20):6417–6427. https://doi.org/10.1158/1078-0432.CCR-11-0486

Rodon J, Garrison M, Hammond LA, de Bono J, Smith L, Forero L, Hao D, Takimoto C, Lambert JM, Pandite L, Howard M, Xie H, Tolcher AW (2008) Cantuzumab mertansine in a three-times a week schedule: a phase I and pharmacokinetic study. Cancer Chemother Pharmacol 62(5):911–919. https://doi.org/10.1007/s00280-007-0672-8

Sapra P, Stein R, Pickett J, Qu Z, Govindan SV, Cardillo TM, Hansen HJ, Horak ID, Griffiths GL, Goldenberg DM (2005) Anti-CD74 antibody-doxorubicin conjugate, IMMU-110, in a human multiple myeloma xenograft and in monkeys. Clin Cancer Res Off J Am Assoc Cancer Res 11(14):5257–5264. https://doi.org/10.1158/1078-0432.CCR-05-0204

Senter PD, Sievers EL (2012) The discovery and development of brentuximab vedotin for use in relapsed Hodgkin lymphoma and systemic anaplastic large cell lymphoma. Nat Biotechnol 30(7):631–637. https://doi.org/10.1038/nbt.2289

Shen BQ, Bumbaca D, Saad O, Yue Q, Pastuskovas CV, Khojasteh SC, Tibbitts J, Kaur S, Wang B, Chu YW, LoRusso PM, Girish S (2012a) Catabolic fate and pharmacokinetic characterization of trastuzumab emtansine (T-DM1): an emphasis on preclinical and clinical catabolism. Curr Drug Metab 13(7):901–910

Shen BQ, Xu K, Liu L, Raab H, Bhakta S, Kenrick M, Parsons-Reponte KL, Tien J, Yu SF, Mai E, Li D, Tibbitts J, Baudys J, Saad OM, Scales SJ, McDonald PJ, Hass PE, Eigenbrot C, Nguyen T, Solis WA, Fuji RN, Flagella KM, Patel D, Spencer SD, Khawli LA, Ebens A, Wong WL, Vandlen R, Kaur S, Sliwkowski MX, Scheller RH, Polakis P, Junutula JR (2012b) Conjugation site modulates the in vivo stability and therapeutic activity of antibody-drug conjugates. Nat Biotechnol 30(2):184–189. https://doi.org/10.1038/nbt.2108

Sievers EL, Senter PD (2013) Antibody-drug conjugates in cancer therapy. Annu Rev Med 64:15–29. https://doi.org/10.1146/annurev-med-050311-201823

Singh R, Erickson HK (2009) Antibody-cytotoxic agent conjugates: preparation and characterization. Methods Mol Biol 525:445–467, xiv. https://doi.org/10.1007/978-1-59745-554-1_23

Strop P, Liu SH, Dorywalska M, Delaria K, Dushin RG, Tran TT, Ho WH, Farias S, Casas MG, Abdiche Y, Zhou D, Chandrasekaran R, Samain C, Loo C, Rossi A, Rickert M, Krimm S, Wong T, Chin SM, Yu J, Dilley J, Chaparro-Riggers J, Filzen GF, O'Donnell CJ, Wang F, Myers JS, Pons J, Shelton DL, Rajpal A (2013) Location matters: site of conjugation modulates stability and pharmacokinetics of antibody drug conjugates. Chem Biol 20(2):161–167. https://doi.org/10.1016/j.chembiol.2013.01.010

Sun X, Widdison W, Mayo M, Wilhelm S, Leece B, Chari R, Singh R, Erickson H (2011) Design of antibody-maytansinoid conjugates allows for efficient detoxification via liver metabolism. Bioconjug Chem 22(4):728–735. https://doi.org/10.1021/bc100498q

Szakacs G, Paterson JK, Ludwig JA, Booth-Genthe C, Gottesman MM (2006) Targeting multidrug resistance in cancer. Nat Rev Drug Discov 5(3):219–234. https://doi.org/10.1038/nrd1984

Tang R, Cohen S, Perrot JY, Faussat AM, Zuany-Amorim C, Marjanovic Z, Morjani H, Fava F, Corre E, Legrand O, Marie JP (2009) P-gp activity is a critical resistance factor against AVE9633 and DM4 cytotoxicity in leukaemia cell lines, but not a major mechanism of chemoresistance in cells from acute myeloid leukaemia patients. BMC Cancer 9:199. https://doi.org/10.1186/1471-2407-9-199

Toppmeyer DL, Slapak CA, Croop J, Kufe DW (1994) Role of P-glycoprotein in dolastatin 10 resistance. Biochem Pharmacol 48(3):609–612

Turell L, Carballal S, Botti H, Radi R, Alvarez B (2009) Oxidation of the albumin thiol to sulfenic acid and its implications in the intravascular compartment. Braz J Med Biol Res 42(4):305–311

Walter RB, Raden BW, Hong TC, Flowers DA, Bernstein ID, Linenberger ML (2003) Multidrug resistance protein attenuates gemtuzumab ozogamicin-induced cytotoxicity in acute myeloid leukemia cells. Blood 102(4):1466–1473. https://doi.org/10.1182/blood-2003-02-0396

Walus LR, Pardridge WM, Starzyk RM, Friden PM (1996) Enhanced uptake of rsCD4 across the rodent and primate blood-brain barrier after conjugation to anti-transferrin receptor antibodies. J Pharmacol Exp Ther 277(2):1067–1075

Wang L, Amphlett G, Blattler WA, Lambert JM, Zhang W (2005) Structural characterization of the maytansinoid-monoclonal antibody immunoconjugate, huN901-DM1, by mass spectrometry. Protein Sci 14(9):2436–2446. https://doi.org/10.1110/ps.051478705

Wu G, Fang YZ, Yang S, Lupton JR, Turner ND (2004) Glutathione metabolism and its implications for health. J Nutr 134(3):489–492

Xie H, Blattler WA (2006) In vivo behaviour of antibody-drug conjugates for the targeted treatment of cancer. Expert Opin Biol Ther 6(3):281–291. https://doi.org/10.1517/14712598.6.3.281

Xie H, Audette C, Hoffee M, Lambert JM, Blattler WA (2004) Pharmacokinetics and biodistribution of the antitumor immunoconjugate, cantuzumab mertansine (huC242-DM1), and its two components in mice. J Pharmacol Exp Ther 308(3):1073–1082. https://doi.org/10.1124/jpet.103.060533

Yang J, Chen H, Vlahov IR, Cheng JX, Low PS (2006) Evaluation of disulfide reduction during receptor-mediated endocytosis by using FRET imaging. Proc Natl Acad Sci USA 103(37):13872–13877. https://doi.org/10.1073/pnas.0601455103

Younes A, Gordon L, Kim S, Romaguera J, Copeland AR, de Castro Farial S, Kwak L, Fayad L, Hagemeister F, Fanale M, Lambert J (2009) Phase I multi-dose escalation study of the anti-CD19 maytansinoid immunoconjugate SAR3419 administered by intravenous (IV) infusion every 3 weeks to patients with relapsed/refractory B-Cell non-Hodgkin's lymphoma (NHL). Paper presented at the ASH Annu Meet Abstr

Younes A, Bartlett NL, Leonard JP, Kennedy DA, Lynch CM, Sievers EL, Forero-Torres A (2010) Brentuximab vedotin (SGN-35) for relapsed CD30-positive lymphomas. N Engl J Med 363(19):1812–1821. https://doi.org/10.1056/NEJMoa1002965

Zhao RY, Wilhelm SD, Audette C, Jones G, Leece BA, Lazar AC, Goldmacher VS, Singh R, Kovtun Y, Widdison WC, Lambert JM, Chari RV (2011) Synthesis and evaluation of hydrophilic linkers for antibody-maytansinoid conjugates. J Med Chem 54(10):3606–3623. https://doi.org/10.1021/jm2002958

Zimmerman ES, Heibeck TH, Gill A, Li X, Murray CJ, Madlansacay MR, Tran C, Uter NT, Yin G, Rivers PJ, Yam AY, Wang WD, Steiner AR, Bajad SU, Penta K, Yang W, Hallam TJ, Thanos CD, Sato AK (2014) Production of site-specific antibody-drug conjugates using optimized non-natural amino acids in a cell-free expression system. Bioconjug Chem 25(2):351–361. https://doi.org/10.1021/bc400490z

Part III
Development Issues

Chapter 5
Formulation Development for Antibody–Drug Conjugates

Junyan A. Ji, Jun Liu, and Y. John Wang

Abstract Antibody drug conjugates (ADCs) are complex molecules designed to provide targeted therapy to cancer patients. An ADC is formed commonly by conjugating an antibody with small cytotoxic drugs generally either through lysine ε-amino groups to form a lysine-linked ADC or through sulfhydryl groups of reducing interchain cystine. THIOMAB®–drug conjugates (TDCs) are a new subclass of ADCs, in which the engineered free cysteine residues at specific sites of the antibody are conjugated with cytotoxins. The majority of ADCs currently on the market and under development contains a common moiety: a thio-succinimide linker that is susceptible to the hydrolysis of succinimide and cleavage of the thio–ether bond. ADC contains a heterogeneous mixture of different drug-loaded antibody species, whereas TDC exhibits a more homogeneous distribution of linker drugs. Linker stability, drug load distribution, and drug-to-antibody ratio (DAR) are believed to play critical roles in the efficacy and safety of ADC products. This chapter describes analytical methods and formulation strategies to understand and characterize these critical quality attributes in order to develop a stable formulation for ADC products. The effect of DAR on ADC aggregation is presented. Additionally, formulation parameters such as pH and buffers, antioxidants, and surfactants are described. Finally, the advantages and drawbacks of a lyophilized formulation are discussed.

Keywords Formulations · Drug–antibody ratio · Stability · ADC aggregation · Lyophilization

J. A. Ji (✉) · J. Liu · Y. J. Wang
Genentech Late Stage Pharmaceutical Development, South San Francisco, CA, USA
e-mail: Ji.junyan@gene.com

J. L. Zaro, J. Wang (eds.), *Antibody-Drug Conjugates*, AAPS Advances in the Pharmaceutical Sciences Series 17, https://doi.org/10.1007/978-3-031-83005-1_5

5.1 Introduction

Antibody–drug conjugates (ADCs) offer the promise of delivering more powerful tumor-killing activity while resulting in diminished side effects for cancer patients. Delivery of ADCs is precise and selective by combining a targeted monoclonal antibody (or antibody fragment) linked to a potent anticancer therapeutic. A variety of cytotoxins such as auristatins, maytansinoids, calicheamicin, duocarmycin (Tse et al. 2006; Phillip et al. 2008; Naito et al. 2000; Terrett et al. 2007), and other small molecules (Kim 2009) have been conjugated to monoclonal antibodies to generate ADCs. A variety of linker chemistries have been developed in the past 20 years including acid-labile hydrozones, sterically hindered (SPDB) and unhindered (SPP) disulfides, cleavable (vc) and noncleavable peptides, noncleavable thio-ethers (SMCC), etc. (Nolting 2013). Cysteine and lysine are the two most common naturally occurring amino acids that are used to attach the drugs through the linker to the antibody. Due to its specific reactivity, fast conjugation reaction, and lack of by-products, thiol-based Michael-type addition reactions have been widely employed for covalent conjugation of proteins, peptides, and drugs by the reaction of free cysteine or thiols with maleimides (Mather et al. 2006). In the case of brentuximab vedotin (ADCETRIS®) which was approved by FDA in 2011, brentuximab is conjugated to maleimide-VC-PAB-MMAE with the sulfhydryl groups of cysteine from the reduced interchain disulfides of the antibody (for the structure, refer to Chap. 4, Fig. 4.14 and Chap. 11, Fig. 11.1) (Lyon et al. 2012). The same reaction is also used in the newly developed TDCs in which the engineered free cysteine residues at specific sites of the antibody are conjugated with cytotoxins (Junutula et al. 2008, 2010). For trastuzumab emtansine (KADCYLA®) which was approved in 2013, the lysine residues on trastuzumab (Tmab) are first modified with the linker species SMCC to form the intermediate Tmab- MCC which is then reacted with the cytotoxin DM1 to form the antibody–drug conjugate T-DM1 (for structure, refer to Chap. 4, Fig. 4.4, and Chap. 12, Fig. 12.1) (Wakankar et al. 2010). Although the synthetic routes vary, the linker of these conjugates contains the same thio-succinimide moiety, of which the stability is critical for the product quality, safety, and efficacy (Alley et al. 2008). While the TDCs exhibit a more homogeneous distribution of linker drugs, both lysine-linked and cysteine-linked ADCs often lead to a heterogeneous distribution of small-molecule drugs conjugated to antibodies.

Linker stability, drug-to-antibody ratio (DAR), and drug load distribution are critical to product quality, safety, and efficacy. This chapter focuses on the analytical methods and formulation strategies to address these critical quality attributes. . It discusses the degradation pathways of the thio-succinimide linker, the effect of DAR on ADC aggregation propensity, and formulation development considerations, including the choice of pH and buffers, antioxidants, surfactants, and issues regarding liquid and lyophilized formulations.

5.2 Stability of the Thio-Succinimide Linker

5.2.1 Mechanism of Degradation Pathways

Studying linker stability alone in the presence of a myriad of other degradation pathways in the antibody is difficult due to the complexity of the ADC structure and degradation products. The degradation of the thio-succinimide linker using a synthetic small-molecule model compound *S-(N-ethylsuccinimido)-N-acetyl-L-cysteine* (NEM-NACys) (Fig. 5.1) was studied (Baldwin and Kiick 2011). Two primary degradation pathways were identified: (i) hydrolysis of succinimide to form an open-ring structure and (ii) retro-Michael reaction in the presence of thiol compounds. Succinimide is the key intermediate for deamidation of asparagine (Asn) and isomerization of aspartic acid (Asp) residues in a protein or peptide, either in vivo or in vitro (Wakankar and Borchardt 2006). Formation of succinimide is the rate-determining step in these reactions. Deamidation of Asn is dependent on the hydroxide-ion [OH^-] concentration, even as low as pH 3–4, and it increases rapidly under neutral and basic pH as the rate of succinimide hydrolysis accelerates. Unlike succinimide hydrolysis, the thio–ether bond is relatively stable in the absence of Michael donors such as thiol-containing compounds. The kinetics of the retro-Michael reaction and the extent of thiol exchange are modulated by the reactivity of the Michael donor. It should be noted that a thio–ether bond is stabilized upon succinimide ring opening, and thus no further thiol exchange is possible even in the presence of a Michael donor (Baldwin and Kiick 2011).

Studying the degradation mechanism of the thio-succinimide moiety in the model compound facilitates an understanding of the thio-succinimide linker stability in ADCs. The free drug release from a cysteine-linked ADC involving a retro-Michael exchange reaction to form an albumin adduct was reported (Alley et al. 2008). Using radioisotopes and liquid chromatography–mass spectrometry (LCMS), it was determined that the first step involves a reversible dissociation reaction, in which the thio-succinimide-linked cytotoxic drug is cleaved from the antibody.

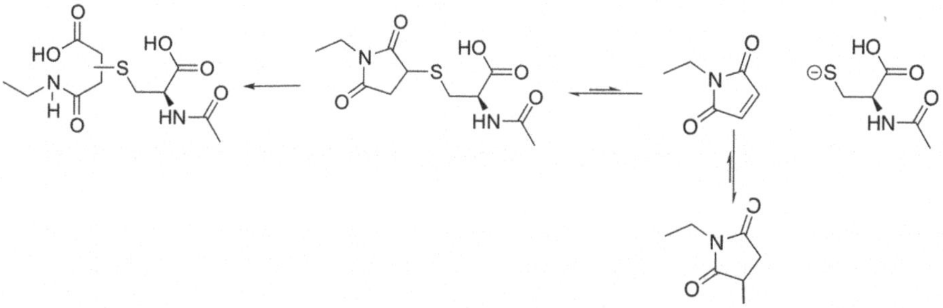

Fig. 5.1 Degradation pathways of a model compound *S-(N-ethylsuccinimido)-N-acetyl-L-cysteine* (NEM-NACys): (**i**) succinimide hydrolysis and (**ii**) retro-Michael reaction in the presence of a Michael donor (GSH glutathione)

Albumin, the predominant thiol-containing protein in plasma, has a reactive cysteine residue at position 34. Albumin reacts with the maleimide attached with the cytotoxin to form an albumin–drug conjugate. However, the retro-Michael reaction was not observed in phosphate-buffered saline because of the absence of a Michael donor.

Linker stability is critical to the safety and efficacy of ADCs. Using TDC as an example, three variants of monoclonal antibody are engineered, each having cysteine placed at a different site: light chain (Lc), heavy chain (Hc), and Fc region (Fc), differing in the local charge environment and solvent accessibility to form conjugates via the Michael addition reaction (Shen et al. 2012). The environment of the Lc conjugation site is the most positively charged, and the Fc conjugation site has the most solvent accessibility. The extent of succinimide hydrolysis and albumin adduct formation was compared among these three variants. Due to the high local concentration of hydroxide ion for the positive-charge environment in the Lc domain, succinimide hydrolysis appears to be much faster in the Lc TDC variant than in the Hc and the Fc TDC variants. Consequently, the Lc TDC variant has the least amount of albumin adduct formation because the thio–ether bond is stabilized upon the ring-opening of succinimide, thereby preventing further thio–maleimide exchange. On the other hand, the Fc TDC variant has the most solvent accessibility, and thus the most albumin adduct forms via the retro-Michael reaction. Bioactivity loss and toxicity of protein conjugate have been observed in studies using cyno monkeys (*Macaca Fascicularis*) involving the retro-Michael reaction, which is due to the solvent-accessible sites undergoing maleimide exchange in the presence of albumin or another Michael donor, such as GSH (Lin et al. 2008).

Both in vitro and in vivo stabilities, and the potentially therapeutic activity of an ADC, could be greatly improved with a stable succinimide ring-opened linker, which would not result in a retro-Michael exchange reaction. To fulfill this need, a modified thio-succinimide linker was developed by incorporating a basic amino group adjacent to the maleimide group to introduce a rapid intramolecular base catalysis of the succinimide hydrolysis at neutral pH and room temperature (Lyon et al. 2013a; Challener 2014). The hydrolysis was completed in less than 2 h, which is suitable for manufacturing processes. The antitumor activity and safety were greatly improved with this type of self-stabilized ADC, which represents a potential advancement over the current ADC technology (Lyon et al. 2013b).

5.2.2 Characterization of Thio-Succinimide Hydrolysis of ADC

Linker stability is one of the critical quality attributes for ADCs. For the thio-succinimide linker, succinimide hydrolysis creates additional charges from the ring-opened form, changing its charge state and hydrophobicity. Analytical methods that have been widely used for the identification and quantification of succinimide in proteins are also applicable to ADCs. Reverse-phase liquid chromatography with mass spectrometric detection (rpHPLC-MS) has been used for the analysis of thio-succinimide linker hydrolysis of intact ADCs in serum plasma (Xu et al. 2011). A fast FabRICATOR® (Genovis AB) rpHPLC mapping method was recently developed utilizing the unique specificity of a novel proteolytic enzyme that cleaves at the hinge region, creating pure F(ab′)2 fragments and Fc without any further degradation at mild physiological pH conditions within just a few minutes (Schneiderheinze 2011; Wagner-Rousset et al. 2014). With mass spectrometry detection, the FabRICATOR® rpHPLC method can also monitor site-specific thio-succinimide hydrolysis by separating and quantifying the ring-opened and ring-closed forms. Another alternative analytical method is imaged capillary isoelectric focusing (iCIEF), a common high-resolution separation method for charge variants of proteins. To accurately determine the level of succinimide hydrolysis of thio-succinimide linker by iCIEF, the contribution from the deamidation of the antibody moiety of an ADC must be subtracted. Figure 5.2 shows an overlay of iCIEF profiles of a cysteine-linked ADC and its unconjugated monoclonal antibody precursor. The heterogeneity of the iCIEF profile in the acidic region is due to various negative charges of the ADC gained from both deamidation and succinimide hydrolysis of the linker. Assuming that (i) charge variants are only caused by deamidation and thio-succinimide linker hydrolysis and (ii) each peak shift to the acidic region from the main peak represents a gain of one charge, an approach using weighted peak area which considers the charge state and drug load to quantify the thio-succinimide hydrolysis of ADCs was utilized (Zheng et al. 2013).

5.3 Effect of DAR on ADC Stability

5.3.1 Determination of DAR and Drug-Load Distribution

Various analytical methods have been used to determine DAR (Ducry 2013). A simple and convenient UV/vis spectrometric method can be used, provided that the spectra of the antibody and the drug have distinct maximum absorbance wavelength, so that the concentration of both can be separately determined, and thus allow the calculation of an average DAR. The method works for both cysteine-linked and lysine-linked ADCs. However, it is important to note that the absorptivity of the antibody and the drug may be affected in different buffers, ionic strengths, or pH. Hydrophobic interaction chromatography (HIC) has also been used for the determination of DAR for cysteine-linked ADCs, based on the increasing hydrophobicity with the least hydrophobic, unconjugated form eluting first and the most hydrophobic species with the highest drug load eluting last. The relatively mild and nondenaturing conditions of HIC, which do not interrupt the structure of an ADC, enable the measurement of the drug-load distribution of an intact ADC with different DARs. The average DAR can be calculated based on the percentage of each peak area and its respective drug load. Similar results can be obtained by using rpHPLC. rpHPLC is an alternative method to provide better resolution than HIC for

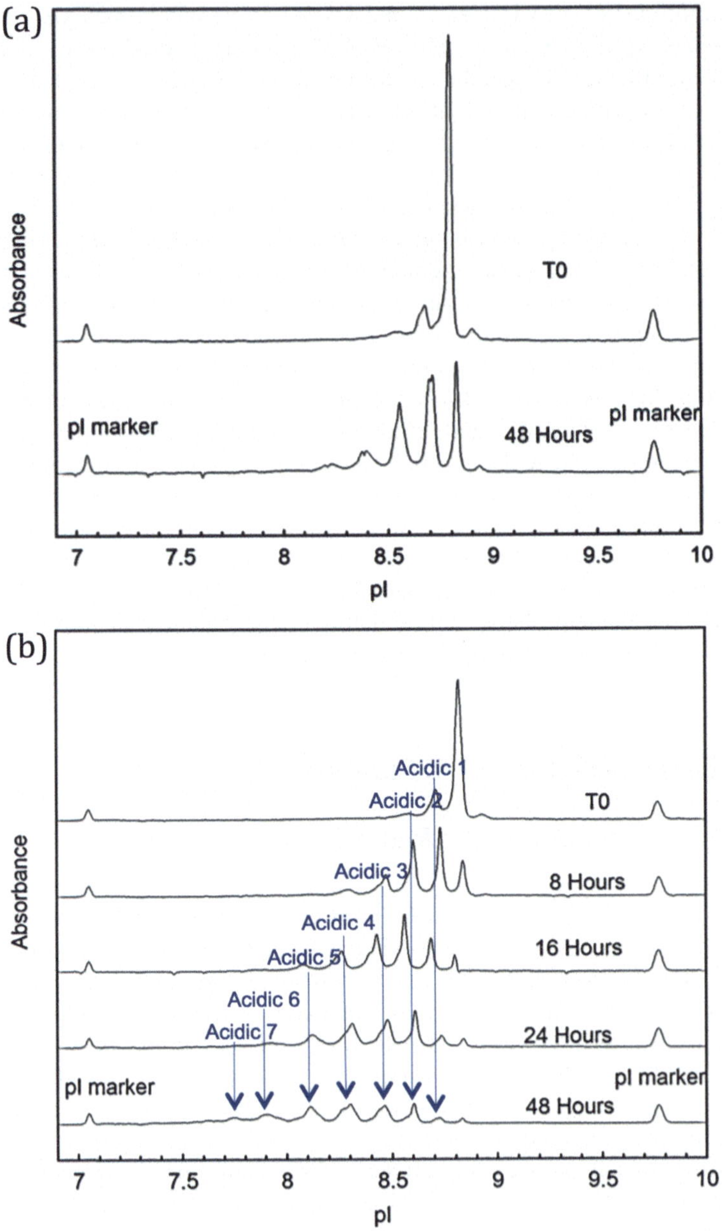

Fig. 5.2 ICIEF profiles of (**a**) an unconjugated Mab solution incubated at pH 9 for 48 h: negatively charged variants generated from deamidation and (**b**) a cysteine-linked ADC incubated at pH 9 for 8, 16, 24, and 48 h: negatively charged variants generated from both deamidation and succinimide hydrolysis (Zheng et al. 2012)

the determination of DAR. However, the presence of organic solvent and the small amount of organic acid present in the mobile phase of RP-HPLC can be disruptive to the intact cysteine-linked ADC. Treatment with a reductant, such as dithiothreitol (DTT), can result in fully reduced interchain disulfides and yields a stable light chain, a light chain loaded with one drug, a heavy chain, and a heavy chain with 1 to 3 drugs attached. The average DAR is calculated based on the percentage of each peak area and its associated drug load of each peak. The rpHPLC method has been recently improved by the FabRICATOR® rpHPLC method (Schneiderheinze 2011; Wagner-Rousset et al. 2014). Another fast and direct method for the determination of DAR is LC-ESI-MS, which identifies the drug-load distribution by intact mass measurement and determines the DAR using the peak area of each species. Deglycosylation and removal of the C-terminal lysine heterogeneity are often needed to reduce the complexity of the mass spectra (Rosati et al. 2014).

5.3.2 *Aggregation*

Both cysteine-linked and lysine-linked ADCs often lead to the heterogeneous distribution of drug molecules conjugated to antibodies. The cysteine-linked ADC with the conjugation sites at the sulfhydryl groups generated from the nonspecific reduction of interchain disulfide bonds of the antibody such as brentuximab vedotin typically generates a mixture of ADCs with 0–8 drugs per antibody (Ducry 2013; Valliere-Douglass et al. 2012). The DARs of lysine-linked ADC such as trastuzumab emtansine being conjugated through the lysine residues of the antibody range from DAR 0 to DAR 9 (Flygare et al. 2013). Studies have shown that both types of ADCs are more prone to aggregation compared with the unconjugated monoclonal antibodies because of the changes of the surface property resulting from the hydrophobic cytotoxic drug attachment or an altered higher order structure of the antibody due to the drug load inducing new modes of inter- or intramolecular interactions (Wakankar et al. 2010; Adem et al. 2014; Acchione et al. 2012). For instance, higher DAR species such as DAR 6 and DAR 8 of an ADC are the major contributors to the formation of high-molecular-weight species (HMWS) under thermal stress for a cysteine-linked ADC with an average DAR of 3.5 (Beckley et al. 2013). These high-DAR species are found to be irreversible, noncovalent, and structurally altered forms of the ADC. Based on differential scanning calorimetry (DSC), it is deduced that destabilization of the hinge region/CH2 domain of the ADC molecule occurs upon thermal stress (Fig. 5.3). The melting temperature onset of the CH2 domain decreases with the increase of the average DAR value of the antibody, resulting in multiple CH2 melting transitions. Similar observations were reported for trastuzumab emtansine in which modification and conjugation of the antibody significantly impacted the thermal stability, primarily the CH2 domain (Wakankar et al. 2010).

Ionic strength may also affect the stability of ADC, making it more prone to aggregation than unconjugated antibodies. A recent example of an ADC showed

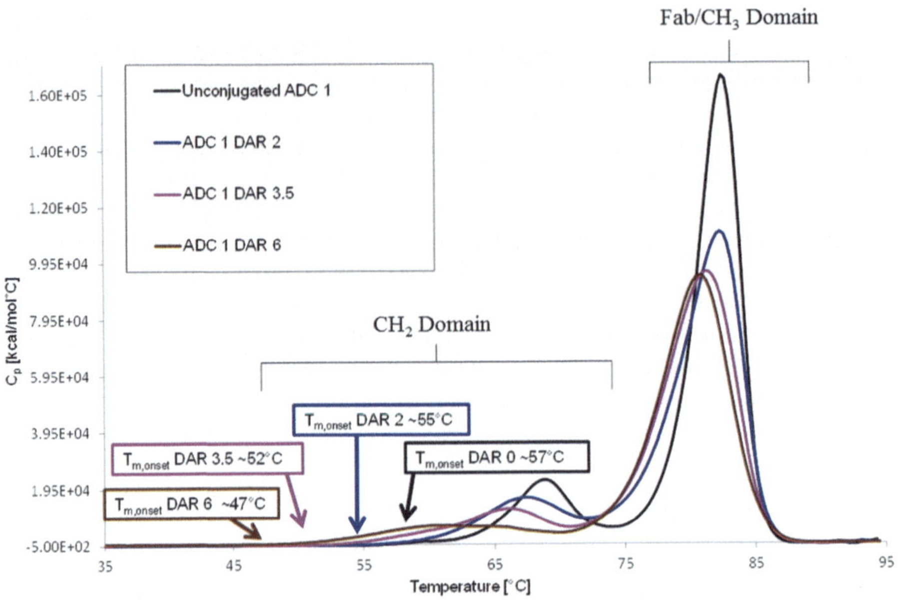

Fig. 5.3 DSC thermograms of unconjugated ADC 1 and ADC 1 batches with average DAR 2, 3.5, and 6, respectively (Beckley et al. 2013)

that a significant amount of aggregates formed with an increase of ionic strength, while the thermal stability of the monoclonal antibody was not affected in solutions up to 500 mM NaCl solution (Adem et al. 2014). The effect of DAR on the physical stability of ADCs can be exacerbated with the increase of the ionic strength of the solution. The correlations between thermal-induced unfolding and drug payload by differential scanning calorimetry showed that the melting point of the same DAR species was significantly lower in the solution with a high ionic strength (20 mM histidine acetate, pH 5.5, 100 mM NaCl) than in that with a low ionic strength (20 mM histidine acetate, pH 5.5) (Fig. 5.4). Other forms of stress can also induce aggregation such as agitation, light exposure, oxidative stress, or dilution into saline during intravenous administration, which is similar to how most of the monoclonal antibodies respond to these stresses (Kumru et al. 2012; Nakajima and Suzuki 2013; Zheng et al. 2012).

Aggregation could potentially affect the safety and efficacy profile of the ADC. Controlling DAR and drug load distribution may facilitate the analytical characterization and formulation development. Antibodies with engineered free cysteine residues at specific sites, such as the TDCs, yielding predominantly homogeneous 2-DAR species present a potential approach to the development of novel ADCs with improved quality and stability (Panowksi et al. 2014).

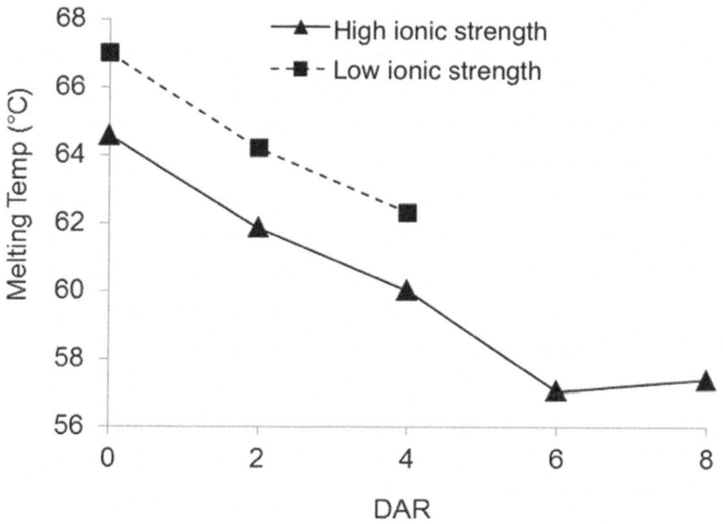

Fig. 5.4 Effect of ionic strength on the thermal stability of a cysteine-linked ADC in a low ionic strength solution (20 mM histidine acetate pH 5.5) and a high ionic strength solution (20 mM histidine acetate pH 5.5, 100 mM NaCl) (Adem et al. 2014)

5.4 Formulation Considerations

The formulation development of ADCs is similar to that of antibodies. Additional critical quality attributes of an ADC product such as DAR and the release of free drug need to be controlled to provide a stable, safe, and efficacious product to patients. A balanced consideration among the stability of antibody, cytotoxic drug, linker, and the specific attributes of a conjugate is required for ADC formulation development.

5.4.1 pH and Buffer

Similar to the controlling degradation pathways of deamidation, fragmentation, isomerization, etc. of the monoclonal antibodies, the choice of formulation pH and buffer plays a critical role in controlling the rate of succinimide hydrolysis of the thio-succinimide ether linker of ADCs. The pH–rate profile of deamidation, isomerization, and succinimide hydrolysis was established using a model tetra-peptide, as shown in Fig. 5.5 (Gieger and Clarke 1987; Capasso et al. 1993, 1995), which illustrates the dependency of the reactions on hydroxide-ion concentration. Similar observation can be found in the thio-succinimide reaction. Examples of deamidation occurring at a lower pH have been reported, however primarily through a mechanism independent of succinimide formation (Goswami et al. 2013). While the rate

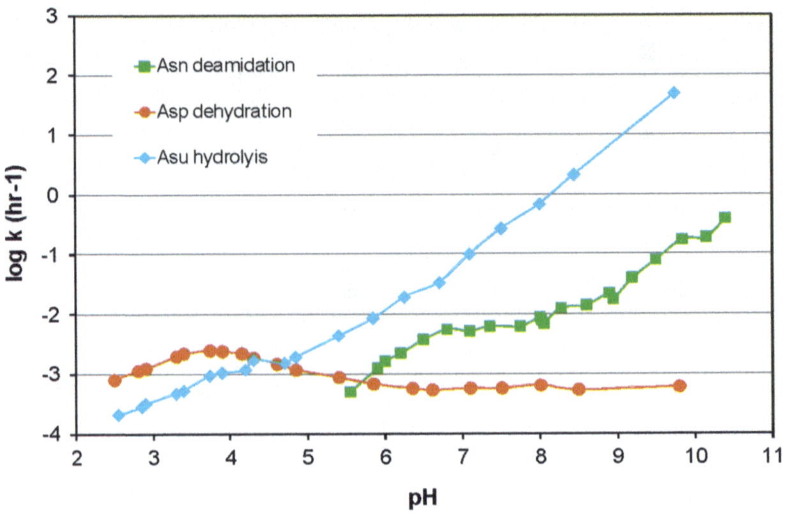

Fig. 5.5 pH–rate profiles of a tetra-peptide (Ac-Gly-x-Gly-NHMe): x = asparagine (Asn) or aspartic acid (Asp) or succinimide (Asu). (Data reconstructed from refs. (Gieger and Clarke 1987; Capasso et al. 1993, 1995)

of deamidation increases rapidly with the increase of pH, the pH–rate profiles of Asp show its pH dependency in the range of pH 3–6 and then no effect above pH 7 with a sharp maximum at a pH close to the apparent pKa of the corresponding carboxyl group. Although the increased alkalinity increases the intramolecular nucleophilic attack by the deprotonated peptide nitrogen to facilitate succinimide formation in the Asp residue, the OH or O moiety needs to leave in order to complete the ring-closure reaction. A faster rate of attack by nitrogen at a higher pH is canceled out because O-, formed readily at a high pH, is a poor leaving group. As a result, the ring-closure reaction of aspartic acid is pH-independent from pH 5 or above (Capasso et al. 1995). Factors other than pH such as sequence and steric effect of the side chain of neighboring amino acids may also influence the rate of deamidation and isomerization in proteins (Tyler-Cross and Schirch 1991; Son and Kwon 1995). As the rate-determining step is the formation of succinimide intermediate for deamidation of Asn and isomerization of Asp, succinimide hydrolysis of the thiosuccinimide linker is expected to be much faster. As illustrated in Fig. 5.5, the rate of succinimide hydrolysis of the tetra-peptide can be up to 10 to 100 times faster than the deamidation of Asn or isomerization of Asp rate. Therefore, the choice of pH range for ADC formulation is limited due to its rapid rate of succinimide hydrolysis even in the range of pH 5–7 which is typical for most of the monoclonal antibody formulations. For this reason, it is recommended that a slightly acidic pH be used to minimize deamidation and succinimide hydrolysis; however, the pH should not be so low that the Asp isomerization, if present, becomes noticeable. Alternatively, to avoid these hydrolytic degradations altogether, the formulation of a lyophilized product could be considered.

Besides pH, buffer components and ionic strength also play an important role in ADC formulation as they do in monoclonal antibody formulations. Due to the dependency upon hydroxide-ion concentration as in the deamidation of Asn, isomerization of Asp, or hydrolysis of succinimide, one needs to be cognizant of the temperature effect on the hydroxide-ion concentration, $[OH^-]$, in different buffer solutions. At different temperatures, the buffering species will affect degradation rate trends because of the changing $[OH^-]$ in the solution. At an elevated temperature, base-type buffers such as histidine, Tris, etc. will generate less hydroxide ions than the acid-type buffer such as acetate. A buffer species such as acetate that may not seem advantageous in reducing degradation under accelerated temperature could be a better choice at the long-term storage temperature (Al et al. 2013). Histidine buffer, one of the most commonly used buffer components in commercial formulations of monoclonal antibodies, may also be perceived to provide better protection against the aggregation of ADCs. However, caution needs to be practiced when considering histidine buffer because of potential oxidative degradation (Stroop et al. 2011; Qi et al. 2009). Proper choice of a buffering agent can potentially achieve both pH control and stabilization of antibodies.

5.4.2 *Antioxidants*

Oxidation is a major concern in the development of stable formulations for the biopharmaceuticals (Nguyen 1994; Li et al. 1995; Hovorka and Schoneich 2001; Teh et al. 1987; Pearlman and Bewley 1993; Zhao et al. 1997; Sasaoki et al. 1989). The oxidizable amino acid residues in proteins include Met, Tyr, Trp, His, and Cys. More recently, Leu was also found to be subject to oxidation vial free-radical reactions (Steimann et al. 2012). H_2O_2, t-butyl hydroperoxide (tBHP), ozone, or UV radiation have been used to stress protein to predict the oxidative potential of Met or Trp residues (Wei et al. 2007; Liu et al. 2008a, b; Harmon et al. 2006). A specific and reproducible stress model using a free-radical generator 2, 2′- azobis- 2- methylpropanimidamide, dihydrochloride (AAPH) was developed to assess the propensity of both Met and Trp oxidation (Ji 2009). The same approach can be applied to investigate the oxidative potential of ADCs since AAPH stress can be applied to both the protein and cytotoxic drug (Ji 2009). To protect the protein against oxidation, methionine is commonly used as a stabilizer in protein formulations (Lam et al. 1997; Oeswein et al. 1998). Several other antioxidants have been reviewed (Nema et al. 1997). However, some are unsuitable for protein formulation because of the following reasons: BHA and BHT are limited to lipid- or surfactant-containing formulations; thio compounds (thioglycerol, Cys, N-acetyl cysteine, and glutathione) may cause disulfide exchange; and reducing agents (sulfites) and ascorbic acid have been shown to cause undesirable reactions. Chelators such as EDTA or DPTA have been used to prevent metal-ion-induced oxidation, although EDTA has been found to be ineffective at preventing metal-induced oxidation in a couple of incidents (Ji et al. 2009a). Other pharmaceutical excipients such as mannitol or sucrose also show relative antioxidant effects (Ji et al. 2009b). Polyols are considered a universal stabilizer against both physical and chemical degradation. Mannitol is a well-known hydroxyl free-radical scavenger but not for alkyl peroxides or H_2O_2. Sucrose has generated similar results as mannitol, except that it is less effective. The utility of antioxidants as stabilizers should be further explored in various dosage forms with drug candidates including ADC molecules.

5.4.3 Surfactants

Nonionic surfactants such as polysorbate (PS) 20 and PS 80 have been widely used in the biopharmaceutical formulations as a stabilizer to prevent surface adsorption and protein aggregation during manufacturing processes, shipping, storage, and even during administration, due to their effectiveness at low concentrations, inert nature, and long history of safety record when mixed with proteins (Kumru et al. 2012; Carpenter et al. 1999; Wang 1999; Kreilgaard et al. 1998; Chang et al. 1996). The higher propensity of ADC toward aggregation as compared with the antibody, as a result of heterogeneous DAR distribution and hydrophobicity of the cytotoxic drug in nature may require more PS in the ADC formulation. It is also important to evaluate the in-use stability for administration. For IV delivery, the volume, material, headspace of the IV bag, diluent and dilution factor, dosing strength, dosing duration, and simulation of shipping post-dilution all need to be assessed to determine the appropriate PS content in the drug product (Alavattam et al. 2012).

Although highly effective in stabilizing proteins from aggregation and adsorption, PS is not innocuous. One needs to be cognizant of the oxidation potential when PS is used. Significant amounts of peroxides, which cause PS and protein oxidation, can form under the stress of light, elevation of temperature, or catalyst upon storage (Ha et al. 2002). There have been concerns about PS degradation recently (Kerwin 2008). The detailed degradation mechanisms of PS were studied by Mahler et al. (Kishore et al. 2011; Borisov et al. 2011). PS cleaves at both the polyethylene oxide ether bond and fatty acid ester bond. The degradants from PS can cause aggregation, visible, and subvisible particle formation (Maggio 2012). A new mechanism of hydrolytic degradation of PS by enzymatic catalysis was reported by LaBrenz in 2014 (Labrenz 2014). A lipase-like enzyme present in an antibody bulk solution was alleged to cause significant PS 80 degradation, with a concurrent increase in particle counts. When PS cannot be used as an effective stabilizer, alternative surfactants can be explored (Patel et al. 2008; Ji et al. 2013).

5.4.4 Liquid Versus Lyophilized Formulation

A successful formulation ensures the delivery of a manufacturable, stable, safe, efficacious, and elegant as well as marketable product to the patients. The selection of a formulation for clinical and commercial applications is dictated by many factors. Product stability is one of the key drivers. In addition, it is also important to consider the manufacturing capability, cost, convenience of dosing, and competitive landscape. For both liquid and lyophilized products, it is essential to understand the chemistry and level of major degradation products during the early and late stages of product development. To this end, one needs to design appropriate preclinical and clinical studies to address the potential impact of product-related degradants and impurity.

Due to the complexity of ADC, formulation development requires an understanding of not only the physicochemical stability of the monoclonal antibody but also that of small-molecule cytotoxic drug, linker moiety, as well as the intrinsic instability of the ADC itself. In addition to the high aggregation propensity of an ADC, one of the major concerns for efficacy, safety and stability, is the presence of excessive levels of free drug in an ADC product that could potentially produce significant adverse effects (Chih et al. 2011). Although a liquid formulation is desirable for the manufacturing and convenience of drug delivery, the lyophilized formulation is often chosen due to the concerns of product stability in aqueous solution, despite the complexity of the manufacturing process, as evidenced by the many clinical and commercial products including both brentuximab vedotin and trastuzumab emtansine (Goswami et al. 2013). However, lyophilization is not a panacea for all stability problems (Roy and Gupta 2004). The rational design of stable lyophilization formulations has been extensively reviewed (Carpenter and Manning 2002; Rey and May 2010; Nireesha et al. 2013; Patel et al. 2013; Johnson and Lewis 2011). While it is typically assumed that enhanced stability can be achieved at a low moisture level, work has also been conducted on the effect of surface area on the physical and chemical stabilities of the proteins (Johnson and Lewis 2011; Abdul-Fattah et al. 2007a, b, c, 2008; Luthra et al. 2008; Chang et al. 2005). Surface-exposed proteins are more susceptible to damage, and improved stability appears to correlate with lower amounts of surface exposure, particularly in lyophilized products with low levels of proteins. In some cases, products having equivalent residual moisture contents could display markedly different stability behaviors during storage due to the different surface areas (Yu and Anchordoquy 2009). In addition to the critical quality attributes such as soluble aggregates, visible and subvisible particles, moisture, cake appearance, reconstitution time, and free drug which are commonly evaluated during stability testing, formulation and lyophilization cycle development, specific surface area measurement of a lyo-product using Brunauer–Emmett–Teller (BET) analysis, based on nitrogen multilayer adsorption as a function of relative pressure, could be recommended for characterization purpose. These studies are essential for the development of a robust lyophilization cycle to achieve a successful lyophilized formulation.

Acknowledgments We thank Drs. Jamie Moore, Dana Anderson, Sreedhara Alavattam, Kai Zheng, and Lynn Gennaro for their comments and help in preparing this chapter.

References

Abdul-Fattah AM, Truong-Le V, Yee L, Nguyen L, Kalonia DS, Cicerone MT, Pikal MJ (2007a) Drying-induced variations in physico-chemical properties of amorphous pharmaceuticals and their impact on stability (I): stability of a monoclonal antibody. J Pharm Sci 96:1983–2008

Abdul-Fattah AM, Truong-Le V, Yee L, Pan E, Ao Y, Kalonia DS, Pikal MJ (2007b) Drying-induced variations in physico-chemical properties of amorphous pharmaceuticals and their impact on stability II: stability of a vaccine. Pharm Res 24:715–727

Abdul-Fattah AM, Dellerman KM, Bogner RH, Pikal MJ (2007c) The effect of annealing on the stability of amorphous solids: chemical stability of freeze-dried moxalactam. J Pharm Sci 96:1237–1250

Abdul-Fattah AM, Lechuga-Ballesteros D, Kalonia DS, Pikal MJ (2008) The impact of drying method and formulation on the physical properties and stability of methionyl human growth hormone in the amorphous solid state. J Pharm Sci 97:163–184

Acchione M, Kwon H, Jochheim CM, Atkins WM (2012) Impact of linker and conjugation chemistry on antigen binding, Fc receptor binding and thermal stability of model antibody-drug conjugates. mAbs 4:362–372

Adem YT, Schwarz KA, Duenas E, Patapoff TW, Galush WJ, Esue O (2014) Auristatin antibody drug conjugate physical instability and the role of drug payload. Bioconjug Chem 25(4):656–664

Al P, Wong RL, Zhang TY, Kao Y, Wang YJ (2013) Asparagine deamidation dependence on buffer type, pH, and temperature. J Pharm Sci 102(6):1712–1723

Alavattam S, Glover ZK, Piros LN, Xiao N, Patel A, Kabakoff B (2012) Stability of IgG1 monoclonal antibodies in intravenous infusion bags under clinical in-use conditions. J Pharm Sci 101(1):21–30

Alley SC, Benjamin DR, Jeffrey SC, Okeley NM, Meyer DL, Sanderson RJ, Senter PD (2008) Contribution of linker stability to the activities of anticancer immunoconjugates. Bioconjug Chem 19(3):759–765

Baldwin AD, Kiick KL (2011) Tunable degradation of maleimide-thiol adducts in reducing environments. Bioconjug Chem 22:1946–1953

Beckley NS, Lazzareschi KP, Chih H, Sharma VK, Flores HL (2013) Investigation into temperature-induced aggregation of an antibody rrug conjugate. Bioconjug Chem 24(10):1674–1683

Borisov OV, Ji JA, Wang YJ, Vega F, Ling VT (2011) Toward understanding molecular heterogeneity of polysorbates by application of liquid chromatography-mass spectrometry with computer-aided data analysis. Anal Chem 83:3934–3942

Capasso S, Mazzarella L, Sica F, Zagari A, Salcadori S (1993) Kinetics and mechanism of succinimide ring formation in the deamidation process of asparagine residues. J Chem Soc Perkin Trans 2:679–682

Capasso S, Kirby AJ, Salvadori S, Zagari A (1995) Kinetics and mechanism of the reversible isomerization of aspartic acid residues in tetrapeptides. J Chem Soc Perkin Trans 2:437–442

Carpenter JF, Manning MC (eds) (2002) Rational design of stable proyein formulations: Theory and practice. Springer Science & Business Media

Carpenter JF, Kendrick BS, Chang BS, Manning MC, Randolph TW (1999) Inhibition of stress-induced aggregation of protein therapeutics. Methods Enzymol 309:236–255

Challener CA (2014) Advances in linker technology: improving the safety and efficacy of antibody drug conjugates. http://wwwbiopharminternationalcom, Mar 1, 2014

Chang BS, Kendrick BS, Carpenter JF (1996) Surface-induced denaturation of proteins during freezing and its inhibition by surfactants. J Pharm Sci 85:1325–1330

Chang L, Shepherd DA, Sun J, Tang X, Pikal MJ (2005) Effect of sorbitol and residual moisture on the stability of lyophilized antibodies: implications for the mechanism of protein stabilization in the solid state. J Pharm Sci 94:1445–1455

Chih HW, Gikanga B, Yang Y, Zhang b (2011) Identification of amino acid residues responsible for the release of free drug from an antibody–drug conjugate utilizing lysine–succinimidyl ester chemistry. J Pharm Sci 100(7): 2518–2525

Ducry L (ed) (2013) Antibody-drug conjugates. Methods Mol Biol, p 1045, Humana Press
Flygare JA, Pillow TH, Aristoff P (2013) Antibody-drug conjugates for the treatment of cancer. Chem Biol Drug Des 81:113–121
Gieger T, Clarke S (1987) Deamidation, isomerization and racemization at asparaginyl and aspartyl residues in peptides. J Biol Chem 262:785–794
Goswami S, Wang W, Arakawa T, Ohtake S (2013) Developments and challenges for mAb-based therapeutics. Antibodies 2:452–500
Ha E, Wang W, Wang YJ (2002) Peroxide formation in polysorbate 80 and protein stability. J Pharm Sci 91(10):2252–2264
Harmon PA, Kosuda K, Nelson E, Mowery M, Reed RA (2006) A novel peroxy radical based oxidative stressing system for ranking the oxidizability of drug substances. J Pharm Sci 95(9):014–2028
Hovorka S, Schoneich C (2001) Oxidative degrada- tion of pharmaceuticals: theory, mechanisms and inhibition. J Pharm Sci 90:253–269
Ji AJ (2009) A novel stress model to study protein oxidation. Oral presentation at IBC the 4th Annual bioProcess international analytical & quality summit, San Diego, CA 4–6 May, 2009
Ji JA, Ingham E, Wang YJ (2009a) Effect of EDTA and methionine on preventing loss of viscosity of cellulose-based topical gel. AAPS PharmSciTech 10(2):678–683
Ji JA, Zhang B, Cheng W, Wang YJ (2009b) Methionine, tryptophan, and histidine oxidation in a model protein, PTH: mechanisms and stabilization. J Pharm Sci 98:4485–4500
Ji JA, Liu J, Wnag YJ (2013) Compositions and methods useful for stabilizing protein-containing formulations. WO2011119487 A3 May 2013
Johnson R, Lewis L (2011) Freeze-drying protein formulations above their collapse temperatures: possible issues and concerns. Am Pharm Rev:50–54
Junutula JR, Raab H, Clark S, Leipold DD, Weir S, Chen Y et al (2008) Site-specific comjugation of a cytotoxid drug to an antibody improves the therapeutic index. Nat Biotechnol 26:925–932
Junutula JR, Glagella KM, Graham RA, Parsons KI, Ha E, Raab H, Bhakta S, Nguyen T, Dugger DL, Li G, Mai E, Lewis PGD, Hiraragi H, Fuji RN, Tibbitts J, Vandlen R, Spencer SD, Scheller RH, Polakis P, Sliwkowski MX (2010) Engineered thio-trastuzumab–DM1 conjugate with an improved therapeutic index to target human epidermal growth factor receptor 2-positive beast cancer. Clin Cancer Res 16:4769–4778
Kerwin BA (2008) Polysorbates 20 and 80 used in the formulation of protein biotherapeutics: structure and degradation pathways. J Pharm Sci 97(8):2924–2935
Kim SA (2009) Antibiotic-conjugated antibodies. US Patent 7569677 B2, 4 Aug, 2009
Kishore RS, Kiese S, Fischer S, Pappenberger A, Grauschopf U, Mahler HC (2011) The degradation of polysorbates 20 and 80 and its potential impact on the stability of biotherapeutics. Pharm Res 28(5):1194–2210
Kreilgaard L, Frokjaer S, Flink JM, Randolph TW, Carpenter JF (1998) Effects of additives on the stability of recombinant human factor XIII during freeze-drying and storage in the dried solid. Arch Biochem Biophys 360:121–134
Kumru OS, Liu J, Ji JA, Cheng W, Wang YJ, Wang T, Joshi SB, Middaugh CR, Volkin DB (2012) Compatibility, physical stability, and characterization of an IgG4 monoclonal antibody after dilution into different intravenous administration bags. J Pharm Sci 101(10):3636–3650
Labrenz SR (2014) Ester hydrolysis of polysorbate 80 in mAb drug product: evidence in support of the hypothesized risk after the observation of visible particulate in mAb formulations. J Pharm Sci 103:2268. https://doi.org/10.1002/jps.24054
Lam XM, Yang JY, Cleland JL (1997) Antioxidants for prevention of methionine oxidation in recombinant monoclonal antibody HER2. J Pharm Sci 86(11):1250–1255
Li SH, Schoneich C, Borchardt RT (1995) Chemical instability of protein pharmaceuticals: mechanisms of oxidation and strategies for stabilization. Biotechnol Bioeng 48:490–500
Lin D, Saleh S, Liebler DC (2008) Reversibility of covalent electrophile-protein adducts and chemical toxicity. Chem Res Toxicol 21(12):2361–2369
Liu D, Ren D, Huang H, Dankberg J, Rosenfeld R et al (2008a) Structure and stability changes of human IgG1 Fc as a consequence of methionine oxidation. Biochemist 47:5088–5100

Liu H, Gaza-Bulseco G, Xiang T, Chumsae C (2008b) Structural effect of deglycosylation and methionine oxidation on a recombinant monoclonal antibody. Mol Immunol 45:701–708

Luthra SA, Hodge IM, Utz M, Pikal MJ (2008) Correlation of annealing with chemical stability in lyophilized pharmaceutical glasses. J Pharm Sci 97:5240–5251

Lyon RP, Meyer DL, Setter JR, Senter PD (2012) Conjugation of anticancer drugs through endogenous monoclonal antibody cysteine residues. Methods Enzymol 502:123–138

Lyon RP, Doronina SO, Bovee TD (2013a) Self-stabilizing linker conjugates. WO Patent 2013173337 A2, Nov 21, 2013

Lyon RP, Setter JR, Bovee TD, Doronina SO, Anderson ME, Balasubramanian B, Duniho S, Leiske CL, Senter PD (2013b) Self-stabilizing ADCs: conjugates prepared with maleimido drug-linkers that catalyze their own thiosuccinimide ring hydrolysis. Posted presented at American Association for Cancer Research, Washington, DC. April 2013

Maggio ET (2012) Polysorbates, peroxides, protein aggregation, and immunogenicity—a growing concern. J Excipients Food Chem 3(2):45–53

Mather BD, Viswanathan K, Miller KM, Long TE (2006) Michael addition reactions in macromolecular design for emerging technologies. Prog Polym Sci 31:487–531

Naito K, Takeshita A, Shieno K, Nakamura S, Fujisawa S, Shinjo K, Yoshida H, Ohnishi K, Mori M, Terakawa S, Ohno R (2000) Calicheamicin-conjugated humanized antiCD33 monoclonal antibody (gemtuzumab ozogamicin, CMA-676) shows cytocidal effect on CD33-positive leukemia cell lines, but is inactive on P-glycoprotein-expressing sublines. Leukemia 14:1436–1443

Nakajima Y, Suzuki S (2013) Environmental stresses induce misfolded protein aggregation in plant cells in a microtubule-dependent manner. Int J Mol Sci 14(4):7771–7783

Nema S, Washkuhn RJ, Brendel RJ (1997) Excipeints and their use in injectable products. PDA J Pharm Sci Tech 51:166–171

Nguyen TH (1994) In: Cleland JL, Langer R (eds) Oxidation degradation of protein pharmaceuticals. Chapter 4 in Formulation and delivery of protein and peptides. American Chemical Society, Washington, DC, pp 59–71

Nireesha GR, Divya L, Sowmya C, Venkateshan N, Babu MN, Lavakumar V (2013) Lyophilization/freeze drying - a review. Int J Novel Trends Pharm Sci 3(4):87–98

Nolting B (2013) Linker technologies for antibody-drug conjugates. Methods Mol Biol 1045:71–100

Oeswein OJ, Ongpipattanakul B, Shahrokh Z, Weissburg WS, Wong RP (1998) Stabilized antibody formulation. US Patent 2292730, 1998

Panowksi S, Bhakta S, Raab H, Polakis P, Junutula J (2014) Site-specific antibody drug conjugates for cancer therapy. MAbs 6(1):34–45

Patel AR, Kulkarni S, Nandekar TD, Vavia PR (2008) Evaluation of alkyl polyglucoside as an alternative surfactant in the preparation of peptide-loaded nanoparticles. J Microencapsul 25(8):531–540

Patel SM, Lobo B, Shah A (2013) Practical considerations for freeze-drying process design, development and scale-Up. http://wwwamericanpharmaceuticalreviewcom 25 Oct 2013

Pearlman R, Bewley TA (1993) Stability and characterization of human growth hormone. In: Wang YJ, Pearlman R (eds) Pharmaceutical biotech- nology, Vol. 5: Formulation, characterization, and stability of protein drugs: case histories, Chapter 1. Plenum NY, pp 1–58

Phillip GDL, Li GM, Dugger DL, Crocker LM, Parsons KL, Mai E, Blattelr WA, Lambert JM, Chari RVJ, Lutz RJ, Wong WLT, Jacobson FS, Koeppen H, Schwall RH, Kenkare-Mitra SR, Spencer SD, Sliwkowski MX (2008) Targeting HER2-positive breast cancer with trastuzumab-DM1, an antibody-cytotoxic drug conjugate. Cancer Res 68:9280–9290

Qi P, Volkin DB, Zhao H, Nedved ML, Hughes R, Bass R, Yi SC, Panek ME, Wang D, DalMonte P, Bond MD (2009) Characterization of the photodegradation of a human IgG1 monoclonal antibody formulated as a high-concentration liquid dosage form. J Pharm Sci 98(9):3117–3130

Rey L, May J (eds) (2010) Freeze-drying/lyophilization of pharmaceutical and biological products, (3rd edn, Drugs and the pharmaceutical sciences). CRC Press

Rosati S, Yang Y, Barendregt A, Heck JR (2014) Detailed mass analysis of structural heterogeneity in monoclonal antibodies using native mass spectrometry. Nat Protoc 9:967–976

Roy I, Gupta MN (2004) Freeze-drying of proteins: some emerging concerns. Biotechnol Appl Biochem 39(Pt 2):165–177

Sasaoki K, Hiroshima T, Kusumoto S, Nishi K (1989) Oxidation of methionine residues of recombinant human interleukin 2 in aqueous solutions. Chem Pharm Bull (Tokyo) 37:2160–2164

Schneiderheinze J (2011) Analytical challenges in the analysis of peptide bioconjugates, WCBP, Washington, DC, pp 9–11

Shen BQ et al (2012) Conjugation site modulates the in vivo stability and therapeutic activity of antibody-drug conjugates. Nat Biotechnol 30:184–189

Son K, Kwon C (1995) Stabilization of human epidermal growth factor (hEGF) in aqueous formulation. Pharm Res 12:451–454

Steimann D, Ji AJ, Wang YJ, Schöneich C (2012) Oxidation of human growth hormone by oxygen-centered radicals: formation of Leu-101 hydroperoxide and Tyr-103 oxidation products. Mol Pharm 9(4):803–814

Stroop S, Conca D, Lundgard R, Renz M, Peabody L, Leigh S (2011) Photosensitizers form in histidine buffer and mediate the photodegradation of a monoclonal antibody. J Pharm Sci 100:5142–5155

Teh LC, Murphy LJ, Hug NL, Surus AS, Friesen HG, Lazarus L, Chapman GE (1987) Methionine oxidation in human growth hormone and human chorionic somatomammotropin: effects on receptor binding and biological activities. J Biol Chem 262:6472–6477

Terrett J, Gangwar S, Rao-Naik C, Pan C, Guerlavais V, Huber M, Chong C, Green L, Cardarelli P, King D (2007) Single, low dose treatment of lymphoma and renal cancer xenografts with human anti-CD70 antibody-toxin conjugates, results in long term cures. In proceedings of the 98th annual meeting of the American association for cancer research, Los Angeles, CA, USA, 14–18 April 2007

Tse KF, Jeffers M, Pollack VA, McCabe DA, Shadish ML, Khramtsov NV, Hackett CS, Shenoy SG, Kuang B, Boldog FL, MacDougall JR, Bornstein G, Senter PD, Meyer DL, Lichenstein HS, LaRochelle WJ (2006) CR011, a fully human monoclonal antibody-auristatin E conjugate for the treatment of melanoma. Clin Cancer Res 12:1373–1382

Tyler-Cross R, Schirch V (1991) Effects of amino acid sequence, buffers, and ionic strength on the rate and mechanism of deamidation of asparagine residues in small peptides. J Biol Chem 266:22549–22556

Valliere-Douglass JF, McFee WA, Salas-Solano O (2012) Native intact mass determination of antibodies conjugated with monomethyl Auristatin E and F at interchain cysteine residues. Anal Chem 84:2843–2849

Wagner-Rousset E, Janin-Bussat MC, Colas O, Excoffier M, Ayoub D, Haeuw JF, Rilatt I, Perez M, Corvaïa N, Beck A (2014) Antibody-drug conjugate model fast characterization by LC-MS following IdeS proteolytic digestion. MAbs 6(1):173–184

Wakankar AA, Borchardt RT (2006) Formulation considerations for proteins susceptible to asparagine deamidation and aspartate isomerization. J Pharm Sci 95(11):2321–2336

Wakankar AA, Feeney MB, Rivera J, Chen Y, Kim M, Sharma VK, Wang YJ (2010) Physicochemical stability of the antibody-drug conjugate trastuzumab-DM1: changes due to modification and conjugation processes. Bioconjug Chem 21:1588–1595

Wang W (1999) Instability, stabilization, and formulation of liquid protein pharmaceuticals. Int J Pharm 185:129–188

Wei Z, Feng J, Lin HY, Mullapudi S, Bishop E et al (2007) Identification of a single tryptophan residue as critical for binding activity in a humanized monoclonal antibody against respiratory syncytial virus. Anal Chem 79:2797–2805

Xu K, Liu L, Saad OM, Baudys J, Williams L, Leipold D, Shen B, Raab H, Junutula JR, Kim A, Kaur S (2011) Characterization of intact antibody-drug conjugates from plasma/serum in vivo by affinity capture capillary liquid chromatography-mass spectrometry. Anal Biochem 412(1):56–66

Yu J, Anchordoquy J (2009) Effects of moisture content on the storage stability of dried lipoplex formulations. J Pharm Sci 98(9):3278–3289

Zhao F, Ghezzo-Schoneich E, Aced GI, Hong J, Milby T, Scho¨neich C (1997) Metal-catalyzed oxidation of histidine in human growth hormone: mechanism, isotope effects, and inhibition by a mild denaturing alcohol. J Biol Chem 272:9019–9029
Zheng K, Ren D, Lilyestrom W, Gokarn Y, Bayer R, Scherer T, Wang YJ, Ji JA (2012) AAPH induced mAb oxidation and aggregation. Poster presented at the 243rd ACS national meeting, San Diego, CA, 25–29 march 2012
Zheng K, Chen Y, Wang YJ, Zheng L, Hutchinson M, Gorrell J, Liu J, Ji AJ (2013) Characterization of ring-opening reaction of thio-succinimide linker in ADC, San Antonio, Texas, Presented at AAPS annual meeting and Exposition, 10–14 Nov 2013

Chapter 6
Bioanalytical Assay for Characterization of Antibody–Drug Conjugates

Chen Xie and Zhijun Kevin Wang

Abstract Antibody–drug conjugates (ADCs) combine the characteristics of large-molecule biologics and small-molecule drugs. Although a number of assays have been reported to characterize the ADCs, a new and comprehensive bioanalytical strategy to overcome the limitations of the current methods is required for better characterization of ADCs for clinical use. An appropriate bioanalytical assay is essential for characterizing ADCs, which include physicochemical and pharmacokinetic properties due to their complex and heterogeneous structures. The commonly used bioanalytical methods typically implemented to characterize various ADCs are summarized with a focus on the pharmacokinetic studies and determination of drug-to-antibody ratios. The challenges and perspectives of the assays are discussed as well.

Keywords Antibody–drug conjugates · Bioanalysis

6.1 Introduction

Antibody–drug conjugates (ADCs) are heterogeneous molecules composed of cytotoxic drugs (payload) linked to an antibody (a whole monoclonal antibody, mAb, or its fragment) via a stable chemical linker (Polakis 2005; Teicher and Chari 2011). ADCs are becoming increasingly important for the treatment of cancer. The drug molecules are usually stably associated with the antibody in the systemic

C. Xie
College of Veterinary Medicine, Western University of Health Sciences, Pomona, CA, USA

Z. K. Wang (✉)
Department of Clinical Pharmacy Practice, School of Pharmacy & Pharmaceutical Sciences, University of California, Irvine, CA, USA
e-mail: zhijunw1@uci.edu

J. L. Zaro, J. Wang (eds.), *Antibody-Drug Conjugates*, AAPS Advances in the Pharmaceutical Sciences Series 17, https://doi.org/10.1007/978-3-031-83005-1_6

circulation. The ADC can be internalized and trafficked to the lysosomes when the antibody binds to the antigen-expressing tumor cells. The linker can be hydrolyzed by the low intra-lysosomal pH (pH 4.5 ~ 5.0) or digested by proteases to release the free payload. ADCs have complex molecular structures that can incorporate the features of both large and small molecules. The small-molecule drugs are conjugated to an antibody via various amino acid residue conjugation sites (Carter and Senter 2008; Junutula et al. 2008), which will result in the heterogeneity of ADCs. An example is the conjugation of the linker drug at lysine residues and cysteine residues through the reduction of the interchain disulfide bonds or at engineered cysteine residues (Fig. 6.1). The conjugation reaction can result in heterogeneous mixtures of ADC molecules with various drug-to-antibody ratios (DARs). If the conjugation is at the lysine residue, the distribution of DARs can range from 0 to 9 drugs (Stephan et al. 2011). A range of drug numbers, such as 0, 2, 4, 6, and 8, will be conjugated at different disulfide bonds where the cysteine residues are located. ADCs with mainly DAR 2 have been reported for conjugation at engineered cysteines (Stephan et al. 2011; Kaur et al. 2013).

ADCs can be explored for treating other diseases as well. DSTA4637S, a novel THIOMAB™ antibody–antibiotic conjugate, is under investigation as a new therapy for complicated *Staphylococcus aureus* (*S. aureus*) bloodstream infections. This ADC consists of a monoclonal THIOMAB™ IgG1 targeting *S. aureus*, linked to a rifamycin-class antibiotic (dmDNA31) via a protease-cleavable linker. Preclinical studies in rats and monkeys have demonstrated the complex pharmacokinetics of DSTA4637S, providing valuable insights into its systemic concentrations and informing its further clinical development (Deng et al. 2019).

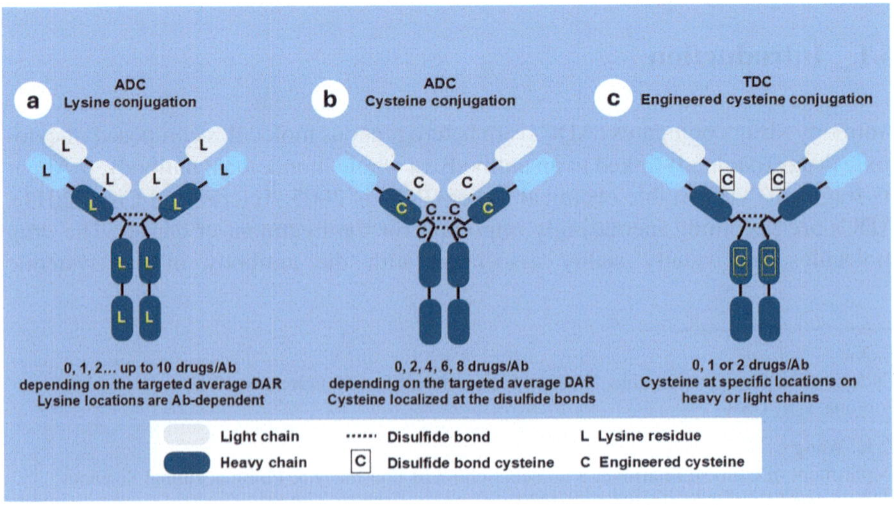

Fig. 6.1 Antibody–drug conjugate conjugation sites and drug load characteristics. (**a**) Conjugation through lysines, (**b**) conjugation through reduced interchain disulfide bonds, (**c**) conjugation through engineered cysteines. *Ab:* Antibody, *ADC:* Antibody–drug conjugate, *DAR*: Drug-to-antibody ratio, *TDC:* ThiomAb–drug conjugate. (Adapted by permission from Future Science Ltd.: Stephan et al. (2011))

The complexity of ADCs may increase in vivo due to the biotransformation resulting from catabolism and metabolism. In addition to drug deconjugation by chemical or enzymatic cleavage, other biotransformations such as adduct formation, generation of peptide fragments, or linker drug fragments can also lead to further complications (Kaur et al. 2013; Shen et al. 2012a). Diverse components exist in vivo upon administration in addition to the original intact or total ADCs.

- Total ADC: the intact antibody–drug conjugates (antibody linked to drug payload(s) via chemical linkers). The number of payloads may vary.
- Total antibody: all forms of antibody, including both unconjugated and conjugated antibodies.
- Intact antibody: fully assembled antibody (both heavy and light chains connected).
- Active ADC: ADC retaining its therapeutic functionality (intact structure, targeting binding ability, and cytotoxic effect).
- Conjugated payload (Antibody-conjugated payload): drug payload conjugated to the antibody.
- Unconjugated payload: drug payload that is not attached to the antibody (some articles use the name of free payload, although the free payload includes both that released from the conjugate or as a standalone substance.)
- Conjugated payload metabolite: metabolites of the payload that remain conjugated to the antibody.
- Payload catabolites: drug payload released together with the linkers or partial linkers.

Active ADC may become inactivated or lose activity caused by payload deconjugation and metabolism, soluble target binding, or neutralizing antibody binding to the active domains.

The total amount of antibodies can be used to assess the protein component of the ADC. The content of conjugated antibodies and antibody-conjugated drugs can provide an assessment of the conjugate efficiency. The free payload could be used to evaluate safety (Gorovits et al. 2013).

The difference between the ADC and total antibody profiles theoretically indicates the degree of drug deconjugation.

$$[\text{Total Ab}] = [\text{Active ADC}] + [\text{Inactive ADC}] + [\text{Unconjugated Ab}]$$

DAR is essential for the quality evaluation of an ADC in terms of potency and toxicity. Thus, reliable in vitro and in vivo methods to measure DAR distribution are highly required. A variety of methods have been utilized to analyze the distribution of drug-linked forms (e.g., fraction of antibodies containing different numbers of drugs), which is an important characteristic of ADC, because these forms may lead to different pharmacokinetic and toxicological properties (Hamblett et al. 2004; Wakankar et al. 2011). The information and typical bioanalytical methods for quantification and qualification of ADC are listed in Table 6.1. A single assay may serve

Table 6.1 Analytes and related parameters commonly assessed for antibody–drug conjugate bioanalysis

Analyte types and parameters	Details	Typical bioanalytical method
Total antibody	Conjugated and unconjugated antibodies (DAR ≥ 0)	LBA
Conjugated antibody	Antibody with drugs (minimum of DAR ≥ 1)	LBA
Antibody-conjugated drug	Total drug conjugated to antibody	MS/LBA
Free drug	Drug fallen off the antibody	MS/LBA
Dar	Average number of conjugated drug	HPLC/UV/Vis/HIC/MS
Drug distribution	Location of drug on the antibody	MS (LC-ESI-MS/MALDI-TOF-MS)/HIC/HPLC/IEC/SEC

LBA ligand-binding assay, *DAR* drug-to-antibody ratio, *MS* mass spectrometry, *HIC* hydrophobic interaction chromatography, *HPLC* High-performance liquid chromatography

multiple purposes and provide different information. However, a set of assay methods is required to comprehensively describe the concentration and composition of heterogeneous ADCs. The commonly used approaches for bioanalytical characterization of ADCs are summarized in this chapter.

6.2 Typical Bioanalytical Methods for ADCs

Bioanalytical assays play an essential role in drug development for both small chemical drugs and complex biologicals. Chemical drugs are usually small molecules with well-defined structures, so the assays primarily focus on the quantification of the parent drugs and/or their metabolites. On the contrary, the complexity of ADCs requires assays that can characterize each component, including antibodies, drug payloads, and linkers, and provide detailed information on their molecular structure, conjugation efficiency, stability, and activity. Accurate and reliable bioanalytical methods are vital for the development, regulatory approval, and clinical application of ADCs.

6.2.1 Assay for Pharmacokinetic Studies

For a typical pharmacokinetic study of ADCs, three assays are necessary to quantify total antibody, total ADC, and unconjugated payload. However, if the modification to the payload/linker is significant, more assays need to be configured to measure more forms of ADCs, including active ADC, conjugated payload, conjugated payload metabolite, conjugated payload catabolites, etc. (Table 6.2).

Table 6.2 Examples of using LBA and mass spectrometry-based methods for quantification of ADC components

ADC	Component	Method	Note	Ref
N/A	Total antibody	Microflow LC-MS/MS	Immunocapture; digestion by papain; signature peptide from heavy chain	Suh et al. (2023)
	Intact antibody	Microflow LC-MS/MS	Immunocapture; digestion by papain; signature peptide from light chain	
	Total ADC	Microflow LC-MS/MS	Surrogate by payload	
Maytansinoids, Auristatins pyrrolo[2,1-c][1,4] benzodiazepine-dimer	Total antibody	ELISA	Biotinylated HER2 target antigen to capture ADC	Zhang et al., (2018) and Zhang et al. (2019)
	Catabolites	LC-MS/MS	Protein precipitation	
	Free payload	LC-MS/MS	Protein precipitation	
5T4 antibody ADC	Total antibody	ELISA	Biotinylated goat anti-human IgG antibody	Giddabasappa et al. (2016)
	Conjugated payload	LC-MS/MS Qtrap	Biotinylated anti-MMAF antibody for the ADC.	
	Catabolites	LC-MS/MS		
Trastuzumab emtansine	Total ADC	LC-HRMS	Immunoaffinity capture	Jin et al. (2018)
PYX-201 PYX201	Total ADC		Immunocapture by protein A beads; digestion by papain; Aur0101 as the surrogate analyte	Yin et al. (2023a)
	Free payload Aur0101	LC/MS/MS	Solid-liquid extraction	Yin et al. (2023b)
Trastuzumab-SG3584	Total Ab Intact ADC	LC-HRMS	IgG heavy and light chain polyclonal antibody was biotinylated and conjugated to streptavidin beads	Huang et al. (2021)
	Total Ab	LC-MS/MS	Immunocapture by Protein A Digestion by trypsin	
	Total ADC		Immunocapture by Protein A Digestion by trypsin surrogate by released payload	

(continued)

Table 6.2 (continued)

ADC	Component	Method	Note	Ref
Trastuzumab Emtansine, Trastuzumab deruxtecan Sacituzumab govitecan	Free payload	LC–MS/MS	Protein precipitation by method	Cheng et al. (2024)
LCB14–0110 (a HER2-targeting ADC based on trastuzumab)	Conjugated payload	LC–TOF MS	Immunocapture Cleavage of acDrug by beta-glucuronidase	Lee et al. (2020)
A microtubule polymerization inhibitor	Conjugated payload	LC-MS/MS	Biotin labeled anti-idiotype mAb Digestion by Cathepsin B	Liu et al. (2015)
	Conjugated payload	LC-MS/MS	Immunocapture by anti-payload mAb	
	Payload with linker	LC-MS/MS	Immunocapture by anti-payload mAb (cleave dipeptide linker to release the payload)	
	Catabolites	LC-MS/MS	Immunocapture by anti-payload mAb	
Emtansine Ansamitocin	Total antibody	LC/MS	Immunocapture by Protein A; Nano-surface and molecular-orientation limited (nSMOL) proteolysis	Iwamoto et al. (2017)
	Conjugated payload	LC/MS	Liquid-liquid extraction	
DSTA4637A	Total ADC	LC-MS/MS	Immunocapture by protein A; Surrogate by conjugated payload	Zhou et al. (2016)
	Unconjugated dmDNA31	LC-MS/MS	Protein precipitation	
	Total antibody	ELISA	ELISA	
Antisense oligonucleotides antibody conjugate	Conjugated payload	LC-MS/MS		Jiang et al. (2024)
Monomethyl auristatin F	Unconjugated payload and metabolites	LC-TOF-MS/MS	Protein precipitation	Park et al. (2019)
Belatacept	Total antibody	Two-dimensional LC-MS	Protein G immunocapture in-silico trypsin digestion	Truffot et al. (2023)

(continued)

Table 6.2 (continued)

ADC	Component	Method	Note	Ref
A microtubule polymerization inhibitor	Total ab	ELISA	Antihuman IgG Fc mAb	Myler et al. (2015)
	Active ADC	ELISA	Anti-payload mAb recognizes only the active payload	
	Total ADC	ELISA	Anti-payload mAb recognizes both active and inactive payload	
	Conjugated payload	ELISA	Anti-idiotype mAb biotinylated; Cleaved by cathepsin B	
	Unconjugated payload	LC-MS/MS	Protein precipitation	

There are a variety of assay methods that have been used to analyze ADCs, including ligand-binding assay (LBA), ultraviolet–visible spectrophotometry (UV/Vis), mass spectrometry (MS), and chromatography-based assays. Typically, the bioanalytic assay for ADCs' pharmacokinetic studies are large-molecule LBA methods and mass spectrometry-based methods.

6.2.1.1 LBAs

Ligand-binding assay (LBA) refers to an assay method based on binding a ligand to its specific receptor (Luckey et al. 1993). Large molecules have well-defined tertiary structures that are suitable for LBAs. LBA has been commonly used to quantify total antibody concentration (free antibody and conjugated antibody or ADCs) and conjugated antibody (ADCs). LBAs are generally performed using enzyme-linked immunosorbent assays (ELISAs), cell-based binding assays, or other binding assays.

ELISA

ELISA is a test using the basic immunology concept of an antigen binding to its specific antibody, which allows the detection of various substances, including both small molecules (such as chemical drugs) and large molecules (such as peptides, proteins, or antibodies). It is commonly used to determine the ADC concentration based on the specific antigen that can capture the antibody. ELISA can measure the total antibody, including fully conjugated antibody, partially deconjugated antibody, and fully deconjugated antibody, using specific reagents that can either bind to the antibody or the payloads. In addition, ELISA can also measure the concentration of the free payloads released from the ADC, although this is not popular.

ELISA is a common assay for the measurement of total antibody. Figure 6.2a shows a typical ELISA binding assay used for large-molecule analysis. The capture reagent (antigen, anti-complementarity-determining region mAb, or anti-drug mAb) is attached to a solid matrix. The target analytes can be recognized and bond to the capture reagent, while other molecules are washed out. The amount of analytes can be quantified by a detection reagent. Figure 6.2b shows an example of a measurement of total antibody using reagents that bind to the antibody (Kaur et al. 2013). This format uses the specific antigen to capture either monoclonal or humanized antibodies, followed by the detection of captured antibodies with enzyme-conjugated anti-murine or human IgG (Stephan et al. 2011). This method has been applied to the quantification of several ADCs, including huC242-DM1 (Tolcher et al. 2003), trastruzumab-DM1 (Lewis Phillips et al. 2008), anti-CD33-Calicheamicin (gemtuzumab ozogamicin) (Dowell et al. 2001), anti-MUC16-vc-MMAE (Junutula et al. 2008), and CR011-vc-MMAE (Pollack et al. 2007). When using the direct antigen coat approach, the purified protein as a capture reagent must be available. The anti-drug antibody can be used either as the capture reagent (Fig. 6.2c) (Lewis Phillips et al. 2008; Advani et al. 2010; Kovtun et al. 2006; Stephan et al. 2008; Xie et al. 2004) or as a detection reagent (Junutula et al. 2008; Tolcher et al. 2003; Stephan et al. 2008; DiJoseph et al. 2004; Sanderson et al. 2005). The major difference is the assay sensitivity to drug load. The anti-drug

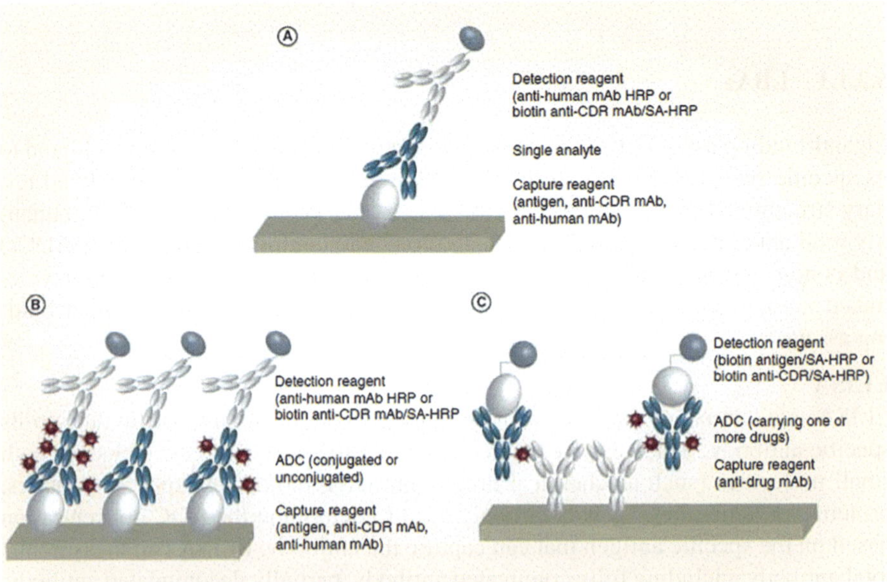

Fig. 6.2 ELISA-based binding assays. (**a**) Typical ELISA, (**b**) ELISA for ADC total antibody measurement, (**c**) ELISA for ADC conjugated antibody measurement. *ADC* Antibody–drug conjugate, *Anti-CDR* Anti-complementarity-determining region, *HRP* Horseradish peroxidase, *mAb* Monoclonal antibody, *SA–HRP* Streptavidin–horseradish peroxidase. (Adapted by permission from Future Science Ltd.: Kaur et al. (2013))

antibody is expected to capture any payloads, so the assay is not sensitive to the drug load. However, the signal intensity will be proportional to the amount of drug conjugated to the antibody if the anti-drug antibody is used as the detection reagent (Stephan et al. 2011). As exemplified by the study performed by Kovtun et al. (2006), the concentration of cantazumab mertansine maytansinoid conjugate (huC242-DM1) is measured by using a murine anti-maytansinoid mAb as the capture reagent. The conjugate from either the standard or the test samples was detected using the horseradish peroxidase-labeled donkey anti-human IgG. The challenge to the assay using anti-drug antibody as the capture antibody is that they might not be able to measure all the ADCs with different drug loads. As discussed by Xie (Xie et al. 2004), such an assay may underestimate the loss of drug molecules from the conjugate when the clearance rate for the huC242-DM1 is identified in a PK study. The anti-drug antibodies have also been used as detection reagents for ADC quantification. For example, Stephan et al. (2008) have used a biotinylated anti-DM1 antibody or a biotinylated anti-MMAF antibody to detect anti-CD22-MCC-DM1 or MC-MMAF ADCs, respectively. The conjugated assays described above can provide signals dependent on the drug load with a signal proportional to the number of DM1 or MMAF. Sanderson et al. (2005) have used an anti-idiotype cAC10 mAb as capture and a biotinylated anti-MMAE mAb antibody to detect the anti-CD30 ADC (cAC10-Val-Cit-MMAE) with 2, 4, and 8 val-cit-MMAE drug linkers per antibody (DAR 2, 4, 8). It has been found that cAC10-Val-Cit-MMAE, with a DAR of 2, can generate lower conjugated antibody assay signal intensity than that with a DAR of 4. This suggests that the ADC conjugate circulating concentration might be higher than what has been measured, considering the potential drug loss in the circulation versus the original ADC used as the standard material. It is noteworthy that ADC ELISA is not capable of measuring the DAR or the overall drug loading (Stephan et al. 2011).

For measuring total antibodies, a specific antibody is coated on a plate to capture the target antibody regardless of whether the drug is attached. A detection antibody usually linked to an enzyme is then added to bind to another site of the target antibody. The substrate for the enzyme is added to react with the enzyme and produce the signal for quantification. The method for detecting ADCs is similar, except for using a detection antibody specific to the payload or another part of the ADC. The signal intensity correlates with the concentration of ADCs in the sample (Chang and Shah 2020).

ELISA can also measure the concentration of the free drugs released from ADCs, although mass spectrometry is considered the most common method (Tolcher et al. 2003; Sanderson et al. 2005). After determining the amount of total mAb in plasma by the cAC10 mAb ELISA, Sanderson has subsequently used an MMAE competition ELISA to detect the amount of MMAE released from ADC (cAC10-Val-Cit-MMAE) following in vitro incubation with cathepsin (Sanderson et al. 2005). For free MMAE competition ELISA, an anti-MMAE mAb (clone SG3.218) is the most sensitive reagent for the detection of free MMAE. To determine the sensitivity of the assay using mAb SG3.218 for capturing the free and HRP-conjugated MMAE, a series of dilutions of free MMAE standards are mixed with the HRP-MMAE conjugate using known concentrations to compete for binding to anti-MMAE mAb SG3.218-coated microtiter plates. The binding of the HRP-MMAE reporter is effectively completed by free drug in a dose-dependent manner.

ELISAs are high-throughput and relatively inexpensive. However, this kind of measurement has some limitations. The reagents are susceptible to various interferences. It is not able to distinguish ADCs with different DARs. Such information is very important because the efficacy and safety of ADCs can potentially be compromised due to the release of conjugated drugs from the ADC in systemic circulation over time, which may result in a change in the DAR. ELISA offers high specificity if well-characterized antibodies that selectively bind to target ADCs are available. However, specific reagents, such as highly selective well-characterized antibodies, might not be available in the early stages of drug discovery and development. As the development of ADCs progresses, mass spectrometry-based methods like LC–MS/MS (liquid chromatography–mass spectrometry/mass spectrometry) become increasingly important, as they can provide greater analytical power and flexibility.

Cell-Based Binding Assay

Although ELISA is very commonly used, it is not applicable when the antigen is not available. In this case, another common approach to characterizing ADC-binding activity is the use of cells expressing either the endogenous or transfected antigen, which could be recognized by the ADCs. In vitro cell-binding experiments have been performed to determine the targeting effectiveness of ADC to antigen-positive cells. The cells need to express a sufficient amount of target protein, while the non-specific binding to the cells has to be as low as possible to provide sufficient assay sensitivity. The cells are generally placed in a 96-well plate and subsequently incubated for 30 min to several hours with serial dilutions of ADC. After incubation, the plates are washed and detected with a specific secondary antibody radiolabeled or conjugated to a fluorescent dye. In order to reduce possible antigen internalization upon ADC binding, the cell-based binding assays are performed at the temperature between 0 and 4 °C. When no purified antigen is available, the cell-based binding assay will have its own advantage, although this approach seems to be a little cumbersome. At present, there are various types of cell-binding assays that can evaluate ADCs, such as direct cell binding assay (Sapra et al. 2005), and competitive cell-bind assay (McDonagh et al. 2006). Flow cytometry-based analysis can also be used in cell binding assays (Junutula et al. 2008; Chen et al. 2007). Cell surface binding is usually analyzed by incubating whole cells with either radio-labeled or fluorescent-labeled ligands, followed by detecting cell-binding labels. To decrease non-specific binding to the cell, extensive washing is imperative with this strategy. Since the cell surface binding assay has insufficient reproducibility due to variable cell loss from the plate during the multiple washing steps, an alternative procedure for measuring antibody binding to cell surface antigens using an immobilized plasma membrane fraction has been developed (Vater et al. 1995). In this method, isolated plasma membrane fraction exhibiting cell surface antigen is bound to a 96-well plate and incubated with antibodies that can recognize a certain cell surface protein. These approaches can provide alternatives to mitigate the limitations of conventional cell-based assay.

Other Types of Binding Assays
Surface plasmon resonance (SPR) binding analysis is another type of binding assay for detecting the interaction of two different molecules (Ramakrishnan et al. 2006; Schuck 1997), in which the antigen is immobilized onto the chip and has been used for various ADCs, including inotuzumab ozogamon (CMC-544) (Boghaert et al. 2008) and gemtuzumab ozogamicin (CMA-676) (DiJoseph et al. 2004). In principle, the detection system relies on the measurement of the changes in refractive index caused by the interaction of macromolecules on the biosensor chips. Because SPR directly detects mass (concentration) with no need for special radio or fluores cent labeling of interacting components before measurement, it presents a great advantage in eliminating possible changes in their molecular properties. This method is very sensitive; however, these instruments need to be dedicated to handling toxic materials, which can limit their use in ADC analysis. To prevent contamination of the instrument, a convenient alternative is the implementation of biolayer interferometry-based platforms. This optical technique analyzes the interference pattern of white light reflected from two surfaces: one is a layer of immobilized protein on the biosensor tip and the other is an internal reference layer (Stephan et al. 2011). The antigen immobilized on the biosensor tip surface can specifically capture ADC in the solution. This type of binding will result in a wavelength shift. Regardless of the format of the assay, ADC binding should be performed along with the control of the unmodified antibody.

6.2.1.2 Mass Spectrometry-Based Method

Mass spectrometry (MS) is an analytical technique for determining the masses of molecules such as peptides and other chemical compounds by their mass-to-charge (m/z) ratios. The MS-based assay has been utilized to characterize DAR fractions, determine the relative ratios of ADCs with different DARs, analyze free drugs and metabolites, and monitor various ADC molecular entities (Kozak and Raab 2013). The process involves ionizing molecules to generate charged species or molecular fragments measured by a mass analyzer.

The common ionization methods include electrospray ionization (ESI), atmospheric pressure chemical ionization (APCI), atmospheric pressure photoionization (APPI), matrix-assisted laser desorption (MALDI), laser ablation electrospray ionization (LAESI), desorption electrospray ionization (DESI), and secondary ion mass spectrometry (SIMS). For ADC quantification, ESI and MALDI are commonly used. ESI produces ionic species from the solution into the gas by electrospray under a high voltage. Matrix-assisted laser desorption/ionization (MALDI) is a soft ionization technique using a laser energy-absorbing matrix to create ionic fragments from large molecules.

The mass analyzers in mass spectrometry commonly used for ADCs are quadrupole mass spectrometry, time-of-flight (TOF), and orbitrap mass spectrometry. Quadrupole mass spectrometry uses a quadrupole mass filter consisting of four parallel metal rods arranged in a square to separate ions based on their mass-to-charge ratio (m/z). TOF analyzers measure the flight time for ions to travel a certain distance, and the flight time is inversely proportional to m/z. Orbitrap traps ions in an electrostatic field to generate a frequency signal that is proportional to their mass-to-charge ratio (m/z). These analyzers can also be combined into a hybrid mode, such as quadrupole-orbitrap and quadrupole-time-of-flight.

The free drug payload or deconjugated payload can be typically quantified by the conventional LC–MS/MS method as small chemical drugs extracted using protein precipitation, liquid–liquid extraction, or solid-phase extraction methods. However, the payload may exist in different forms. The deconjugation of an ADC with a noncleavable linker produces different forms of unconjugated payload compared to the ADC with a cleavable linker (Shen et al. 2012b). For ADCs with cleavable linkers, drug payload can be isolated by linker cleavage using proteases (cathepsin B and papain for peptide linkers) or reducing agents (dithiothreitol) for disulfide bond linkers. For ADCs with noncleavable linkers, the drug payload is typically isolated by more complex methods such as chemical hydrolysis and enzymatic digestion.

Separating the various components of ADCs is usually required before quantification. One common method is the ligand-based method to separate and enrich the target species of ADCs. The captured components of ADCs can be digested to generate small fragments of peptides and release the conjugated payloads. These resultants can then be detected and quantified, typically using tandem mass spectrometry coupled with chromatography. The combination of immunocapture and LC–MS/MS is a hybrid immunocapture LC–MS/MS technique to selectively isolate, separate, and identify or quantify specific biomolecules in complex biological samples, which generally involves three steps.

Step 1: Immunocapture of the ADC of Interest from the Biological Matrix
The purpose is to capture the ADC of interest from the biometrics. For drug discovery or early nonclinical studies, protein A/G or anti-human IgG can be used as the immune capture reagents. Protein A is a 42 kDa surface protein that can bind to the Fc portion of immunoglobulin. For clinical studies, the capture can be achieved using a biotinylated anti-idiotypic antibody or target antigen immobilized using streptavidin-coated magnetic beads. The total antibody, including both ADC and free antibody, can be captured and eluted for further processing. For ADC quantification, the anti-payload antibody can be used as the capture reagent to capture ADC, but not free antibody.

Step 2: Digestion Using Proteases Such as Trypsin and Papain
The immune-enriched ADC and/or free antibody can be analyzed directly using HRAMS (high-resolution and accuracy mass spectrometry) to detect intact biological molecules. Therefore, this step may not be necessary if the HRAMS is employed.

In the absence of HRAMS, conventional mass spectrometry methodologies can also effectively measure the concentrations of ADC conjugates, although they may lack the enhanced mass accuracy and resolving power provided by HRAMS. The captured ADCs (including free antibodies) can be denatured/reduced, alkylated, and digested with enzymes (trypsin or papain). The signature peptides of the antibody and m/z transition for the drug payload can then be measured (Qin and Gong 2022).

Step 3: Detection and Quantification

HRAMS is an advanced analytical technique able to detect the mass-to-charge ratio of ions with high precision. Such technology can be used for the determination and quantification of complex compounds with very similar masses, and therefore, it is essential in fields like proteomics, metabolomics, and pharmaceuticals. Usually, drug payloads are small molecules with molecular weights around 300–1000 Da, while the molecular weight of monoclonal antibody (mAb) is about 130,000 Da. So, the percent difference in mass due to the conjugated linker payloads between unconjugated mAb and ADC is usually between 0.1% and 1%. HRAMS can directly detect the little differences (<1%) in m/z values between conjugated ADC and unconjugated mAb. The quantification can be conducted using the XIC method (extracted ion chromatogram) or deconvoluted mass approach. XIC method plots the intensity of specific ions (or a narrow range of ions) over the retention time from a chromatogram, while the deconvoluted mass approach can separate the overlapping peaks into individual peptide spectra from a complex mass spectrum (Zhu et al. 2020).

Currently, conventional mass spectrometry is more common for detecting and quantifying the peptide fragments and released payloads by assessing the intensity of the detected signals for the signature peptides and payloads. The relative concentrations of these components are determined by comparing them to known standards or calibration curves.

HRAMS can identify new species (new biotransformation of ADC) that cannot be observed with the conventional MS/MS method. The data analysis of HRAMS is more complex and requires specialized software and expertise. In contrast, conventional MS instruments are typically less expensive and more widely available.

Currently, most pharmacokinetic studies include three assays for species of total antibody, total ADC, and conjugated payload. ELISA was the most common method for quantifying total antibodies, but in recent years, mass spectrometry has been increasingly used.

In preclinical studies, the capture of target antibodies can be achieved by using protein A beads, biotinylated target antigen, or biotin-labeled anti-idiotype mAb. The enriched ADCs were digested by papain or trypsin to release small peptide fragments and the payload molecules with cleavable linkers. The signature peptides from heavy chains and light chains serve different purposes. Heavy chain signature peptide was used as the marker for total antibodies, while light chain signature peptides can be the marker for the intact antibody (Table 6.2).

Only a few studies have distinguished between intact and total antibodies, with most research focusing on total antibody levels. A hybrid immunoaffinity capture microflow LC–MS/MS (μLC-MS/MS) assay was developed using the ligand that binds to the human IgG-Fc biotin conjugate. The captured antibody and ADC were digested by papain and purified using solid-phase extraction. The signature peptide from the heavy and light chains was used to quantify the total antibody and intact antibody assays, respectively. A high correlation of total antibody levels between μLC-MS and ELISA was observed (<20% difference) (Suh et al. 2023).

Regarding the total ADC, the conjugated payload was often used as the surrogate marker based on the premise that the amount of conjugated payload directly correlates with the total ADCs present in vivo, assuming a stable DAR. PYX-201 comprises a fully human IgG1 antibody, a cleavable mcValCitPABC linker, and four Auristatin 0101 (Aur0101, PF-06380101) payloads. Protein A beads were used to enrich PYX-201 and digested using papain to release Aur0101. In another study of LCB14-0110, a novel prenyl transferase-mediated, site-specific, antibody–drug conjugate composed of a proprietary beta-glucuronide linker and monomethyl auristatin F (an inhibitor of tubulin polymerization) payload attached to HER2-targeting trastuzumab. The method involved affinity capture and enzymatic cleavage. The released conjugated payloads were then quantified by LC–MS/MS analysis in the positive ion mode (Yin et al. 2023a; Lee et al. 2020).

In a preclinical study on DSTA4637S (a THIOMAB™ antibody–antibiotic conjugate against *Staphylococcus aureus*), systemic concentrations of three analytes, total antibody, antibody-conjugated dmDNA31 (ac-dmDNA31), and unconjugated dmDNA31, were measured. Protein A-coated magnetic beads were used to capture the target antibodies from the serum, followed by enzymatic digestion. A unique representative signature peptide, exclusive to human antibodies, was selected as a surrogate marker for the total antibody concentration determination, which was then analyzed using liquid chromatography–tandem mass spectrometry (LC–MS/MS) through multiple reaction monitoring. The released payload (dmDNA31) was measured to determine the antibody-conjugated dmDNA31. In addition, unconjugated dmDNA31 was extracted using protein precipitation and quantified by LC-MS/MS (Deng et al. 2019).

ADCs can undergo modification in vivo and render them inactive, so it is important to differentiate between active and inactive ADCs. A study on an ADC with a microtubule polymerization inhibitor payload and cathepsin B-cleavable linker detected total antibody, active ADC, total ADC, antibody-conjugated payload, and unconjugated payload levels (Myler et al. 2015). As active ADCs are the most pharmacologically relevant, it is essential to develop bioanalytical assays capable of identifying and quantifying each species to better understand the exposure–response relationship. This understanding is crucial for optimizing dosing and improving overall treatment outcomes.

Some studies only measured the free payloads. In these studies, the ADC molecules were not captured. The biological samples were processed using similar methods for small chemical drugs. In a study, a liquid chromatography–quadrupole-time-of-flight-mass spectrometric (LC–TOF-MS/MS) assay has been developed to evaluate the metabolism and pharmacokinetics (PK) of monomethyl auristatin F antibody–drug conjugates in rats, which is used in (ADCs). The metabolite profiling was also identified. These findings enhance understanding of the PK, catabolism, and metabolism of MMAF, aiding future ADC development (Park et al. 2019).

HRAMS can be a complementary approach to conventional mass spectrometry, providing enhanced precision and detail in the analysis of complex biomolecules. In a study, HRAMS was employed to detect and quantify both Trastuzumab emtansine and Trastuzumab-SG3584, two ADCs (Table 6.2). Following the immunocapture step, where antibodies were selectively captured and purified from a complex biological mixture, HRAMS was used to directly analyze the total antibody content and the intact ADCs. After data acquisition, the mass spectra were deconvoluted, a process that transforms complex raw data into interpretable mass profiles to determine the molecular weights of the intact proteins accurately. This allowed for the characterization of the ADCs, including any modifications or conjugations present in the antibodies. The results were compared with the conventional mass method, and these two appeared consistent.

6.2.2 *Drug-to-Antibody Ratio*

One of the most important attributes of an ADC is the average number of drugs conjugated to a single monoclonal antibody (Drug-to-Antibody Ratio: DAR). Various methods have been used to measure DAR, depending on the properties of the drug and how it is linked to the protein.

6.2.2.1 UV/Vis Spectroscopy

Ultraviolet-visible spectrophotometry (UV/Vis) is the simplest analytical method to determine DAR and ADC concentrations using the extinction coefficients (ε) of the antibody, the drug, and the ADC. ε can be easily measured by the Beer–Lamber law (A = ε·c·l, where A is the absorbance at specific wavelengths, c is the molar concentration, and l is the path length of the light). The absorbance of ADC at drug-specific wavelengths and 280 mm can be measured to get the ratio of these two absorbances (R). DAR can then be measured using the following equation:

$$\mathrm{DAR} = \frac{\epsilon_{Ab,D} - R \times \varepsilon_{Ab,280}}{R \times \varepsilon_{D,280} - \varepsilon_{D,D}}$$

where εA,D and εAb,280 are extinction coefficients of ADC at drug-specific wavelength and 280 nm, respectively, R is the ratio of absorbance of drug-specific wavelength to 280 nm, and $\varepsilon_{D,280}$ and $\varepsilon_{D,D}$ are extinction coefficients of the drug at 280 and its specific wavelength, respectively (Abedi et al. 2019).

Examples include the characterization of calicheamicin analogs (Hinman et al. 1993), cA10-Val-Cit-MMAE (Hamblett et al. 2004) and maytansinoid DM1 (Chari et al. 1992). The absorbance of the ADC and extinction coefficients of the antibody are used to determine the average DAR; however, the calculation based on UV absorption can be complicated by similarities in the antibody and drugs (Chen 2013; Zhao et al. 2011).

6.2.2.2 Mass Spectrometry-Based Assay

MS-based assay has been utilized to characterize DAR fractions and determine the relative ratios of ADCs with different DARs (Kozak and Raab 2013). DARs are typically assessed by electrospray ionization mass spectrometry (ESI-MS). Matrix-assisted laser desorption/ionization (MALDI) is a soft ionization technique to profile and monitor biomolecules. MALDI TOF-MS is another reported new approach for determining drug loading (Quiles et al. 2009; Safavy et al. 2003).

A recent study used this technology to determine the intact mass of trastuzumab and its conjugated forms. Samples were desalted and mixed with a sinapinic acid matrix solution. The samples were then analyzed in a high linear positive model. The molecular weights of each conjugate and mertansine were determined, and the molar ratios were calculated (Abedi et al. 2019).

The observed mass shifts of the peak centroids are used to calculate the average drug loading, and the peak profiles are used to model the drug distribution mathematically. IR MALDI has been used to evaluate the analysis of calicheamicin conjugates (Siegel et al. 1997). MALDI, in combination with UV (UV MALDI-TOF MS), has been used to analyze the average DAR of lysine-linked conjugates prepared using activated paclitaxel (Safavy et al. 2003).

Although the MS-based assay can be used alone to characterize the ADCs without additional steps such as separation and purification, it is usually coupled with chromatography technology, e.g., it is used in combination with reverse-phase high-performance liquid chromatography (RP-HPLC), LC-MS, and LC-MS/MS for determination of the DAR which it is suitable for lysine-linked ADC. First, the ADC samples are desalted using RP-HPLC with a reversed-phase LC column and then the MS spectrum is processed, deconvoluted, and converted to a series of zero charge state masses that corresponds to the increasing number of drugs in the ADC (Beck 2014). The quantification of small molecules such as free drugs or their metabolites is commonly analyzed by LC–MS or LC–MS/MS following extraction from tissues or plasma/serum, while both LBA (Buckwalter et al. 2004) and MS (Kaur et al. 2013; Alley et al. 2008) analytical platforms can be applied for the unconjugated drug assay. The cAC10-vc-MMAE incubated in human, mouse, or dog plasma can be analyzed by LC–MS/MS for determination of the release of free MMAE (Francisco et al. 2003). High-resolution accurate mass systems such as affinity capture capillary LC–MS can characterize drug release by quantifying DAR distributions of the ADC in plasma/serum in a semiquantitative manner (Clark et al. 2013; Xu et al. 2011).

Recently, LC–MS with electrospray ionization coupled to time-of-flight (TOF) or triple–quadrupole mass detectors (Wakankar et al. 2011) has been used to analyze ADC entities such as huN901-DM1 (Wang et al. 2005), T-DM1(trastruzumab-MCC-DM1) (Junutula et al. 2010), and C242-DM4 (Lazar et al., 2005). Wang L et al. studied the drug distribution profile of huN901-DM1 using size-exclusion chromatography (SEC) coupled with ESI-TOF-MS (Wang et al. 2005). Figure 6.3 shows an example of the deconvoluted spectrum of deglycosylated huN901-DM1 using this method. The seven major peaks can be assigned to naked dghuN901 (0D) and dghuN901 with one to six covalently linked DM1 drugs (1D-6D) with the expected mass difference (852 Da) between drug-containing forms. For the reduced conjugate, the light and heavy chains can be separated by RP-HPLC. The number of linker attached to the light and heavy chains can be obtained directly from the deconvoluted MS spectrum. The three prominent peaks in the light-chain spectrum are for species with zero, one and two linkers, while the four prominent peaks in the heavy-chain spectrum correspond to chains with zero, one, two, and three attached linkers (Fig. 6.4). Therefore, both chains of the antibody are modified with an attached linker and conjugated with DM1 drugs.

One of the challenges of LC–MS is that extra consideration is needed when preparing ADC samples. For example, a protease inhibitor should be added to the homogenization buffer to prevent any degradation of antibodies and link-drug when

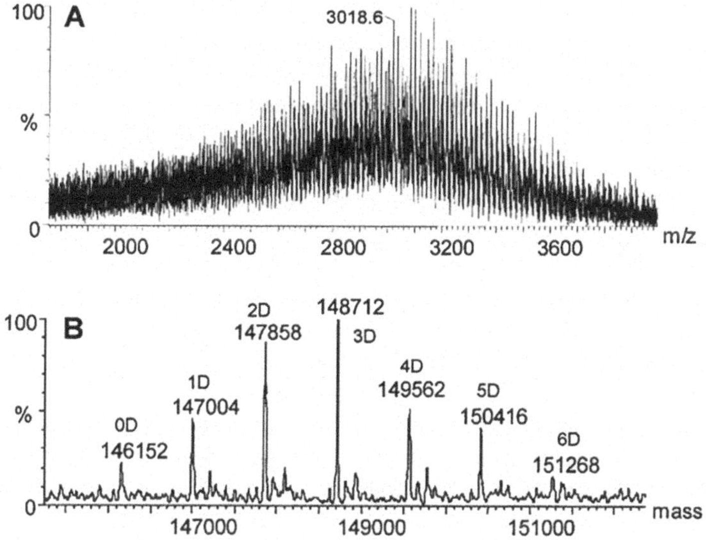

Fig. 6.3 ESI-TOF MS analysis of deglycosylated huN901-DM1. (**a**) Raw ESI-MS spectrum spanned an *m/z* range of 2000–4000. (**b**) Deconvoluted MS spectrum of seven prominent peaks which were assigned to naked dghuN901(0D) and dghuN901 with one, two, three, four, five, and six covalently linked DM1 drugs (1D ~ 6D). “D” in B designates DM1 drug. (Adapted by permission from John Wiley & Sons: Wang et al. (2005))

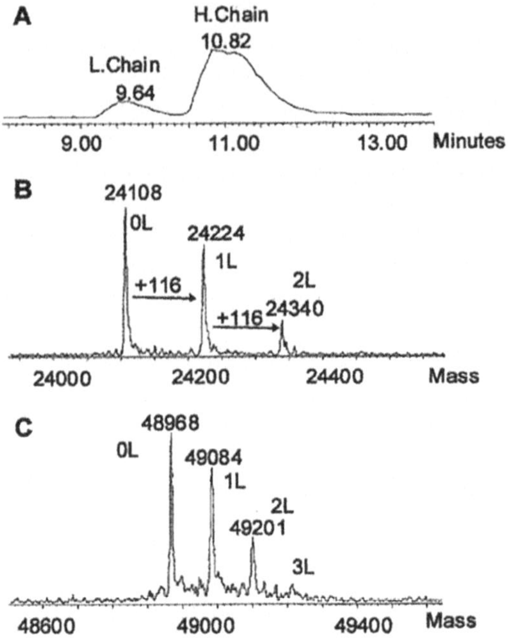

Fig. 6.4 RP-HPLC coupled with ESI-TOF MS analysis of deglycosylated and reduced huN901-DM1. (**a**) RP-HPLC separation of light (L) chain and heavy (H) chain, (**b**) deconvoluted MS spectra of conjugated light chains with zero, one and two linkers (0 L ~ 2 L), (**c**) deconvoluted MS spectra of conjugate heavy chains with zero, one, two, and three attached linkers (0 L ~ 3 L). (Adapted by permission from John Wiley & Sons: Wang et al. (2005))

tissues are homogenized. Any pH alteration of the matrix, such as that used for protein precipitation by organic solvent or Solid-Phase Extraction (SPE) along with LC conditions, may lead to a change in DAR or loss of the drug (Clark et al. 2013). Another issue is that conventional LC–MS for small molecule drug measurement is to quantify priori postulated forms of the drug released by the ADC from catabolism. However, it is theoretically possible that the released drugs may bind to the plasma peptides/protein or contain part of the ADCs linker. Therefore, the putative released drug analyte may not be the major form of the released drug. The acidic conditions during electrospray ionization of the ADC, which may cause the dissociation of drug-derivatized chains, have limited the application of this technique to ADCs produced by conjugation with only cysteine residues (Clark et al. 2013; Wang et al. 2005; Lazar et al. 2005).

LBA-LC–HRMS (High-Resolution Mass Spectrometry) combines affinity capture with intact mass analysis to accurately characterize the Drug-to-Antibody Ratio (DAR) at the intact protein or subunit level. This method involves deconvoluting the mass spectra into a series of "zero-charge" masses and computing the average DAR by integrating and weighting the spectral peak areas or intensities. This approach also allows for determining DAR distribution at various time points. For instance, Kaur et al. demonstrated the capture of all DAR species using biotinylated HER2 immobilized on streptavidin-coated paramagnetic beads. The captured ADCs were deglycosylated and analyzed by LC-MS in positive TOF-MS mode. The raw data were deconvoluted and peak areas for each DAR were obtained to determine directly (Qin and Gong 2022).

6.3 Chromatography

Chromatographic separation is usually necessary to ensure accurate analysis by isolating individual components from complex biometrics. Techniques such as Reverse-Phase High-Performance Liquid Chromatography (RP-HPLC), Hydrophobic Interaction Chromatography (HIC), Ion-Exchange Chromatography (IEC), and Size-Exclusion Chromatography (SEC) are commonly coupled with mass spectrometry (LC-MS or LC-MS/MS).

6.3.1 RP-HPLC

RP-HPLC can separate, identify, quantify, and purify the individual components of a mixture. RP-HPLC is the most widely used analytical technique to separate the free drug released from ADC. The components are then monitored with a UV/Vis detector with the wavelength at the absorption maxima of the drug. The amount of free drug can be calculated based on the HPLC peak area with a standard curve constructed by plotting the peak areas versus the corresponding nominal concentrations. RP-HPLC in combination with mass spectrometry (MS) has also been reported to monitor the free drug species (Francisco et al. 2003).

In addition, reverse-phase chromatography has been used to assess the drug-linker distribution on antibody heavy and light chains for cys-linked ADCs (McDonagh et al. 2006; Sun et al. 2005), which allows calculation of the average drug-load distribution of ADCs. The method involves a reduction reaction to completely dissociate the heavy and light chain of the ADC. Then, the light and heavy chains and their corresponding drug-loaded forms can be separated by an RP column. The weighted average DAR is calculated by the percentage peak area from the integration of the light and heavy chain peaks and the drug number (Beck 2014; Ouyang 2013). For example, the RP-HPLC methods can determine the location of the drug monomethyl auristatin E (MMAE) attachment to the anti-CD30 monoclonal antibody in heavy and light chains (Hamblett et al. 2004; McDonagh et al. 2006). It is essential to mention that direct injection of protein-containing ADC samples onto the RP-HPLC column may result in column deterioration due to the irreversible binding of proteins to the stationary phase. Therefore, guard columns and various sample cleanup procedures are required to separate the free drug from the protein-drug conjugate in the sample prior to the RP-HPLC analysis (Wakankar et al. 2011).

6.3.2 HIC

HIC is another powerful technique used to separate ADC fractions containing various drug loads and determine DAR and drug load distribution for Cys-linked ADCs. Separation is performed using a gradient elution with decreasing ionic strength, and detection is achieved by monitoring the UV absorbance of the eluting species (Alley and Anderson 2013). The drug-loaded species can be resolved based on their hydrophobicity. The hydrophobicity increases with the increasing number of conjugated drugs. For example, an ADC can bind up to eight drug molecules. The unconjugated form is the least hydrophobic, while the eight-drug form is the most hydrophobic. Thus, the unconjugated form will be eluted first, and the eight-drug conjugation will be eluted last. The eluted components can be detected using a UV detector and the peak areas represent the relative percentage distribution of drug-loaded ADCs. Then the weighted average DAR can be calculated using the information of the percentage peak area and the drug number (Beck 2014; Ouyang 2013). As shown in Fig. 6.5, Hamblett et al. used a HIC-HPLC method to purify cAC10-val-cit-MMAE into two, four, and eight drugs per antibody and the corresponding UV spectra of the individual peaks (Hamblett et al. 2004). The five major peaks can be identified because attachment of the drug MMAE results in greater absorbance at ~248 nm (λmax for MMAE) relative to 280 nm (λmax for cAC10). The separation by HIC allows the isolation and purification of chromatographically pure species, which can be further analyzed using ELISA and cell-based assay. In addition to profiling the drug-load distribution, the average drug load for the ADC can be determined by the mean weighted peak area. HIC possesses a great advantage in preserving the integrity of the ADC. However, high salt is usually present in the mobile phase initially,

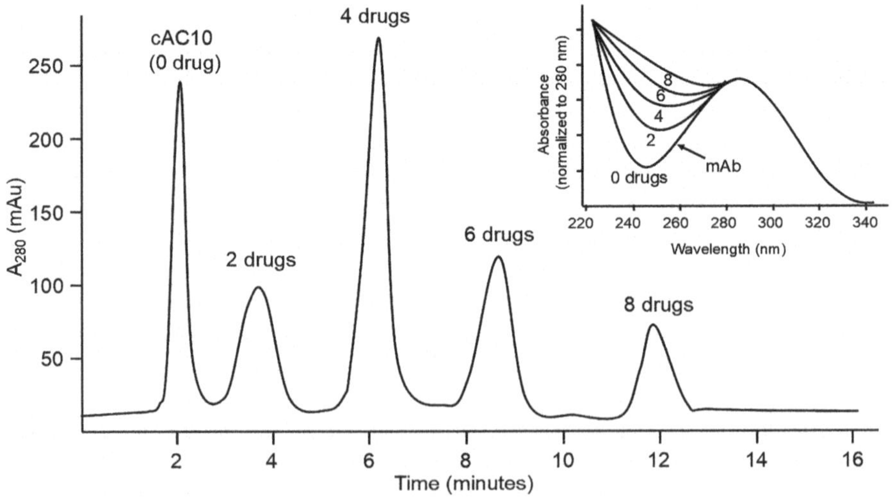

Fig. 6.5 HIC analysis of a cAC10-Val-Cit-MMAE yields five predominant peaks that correspond to mAb (cAC10) conjugated to zero, two, four, six, and eight drugs per antibody. Insert: an overlay of the UV spectra of the starting mAb and the HIC peaks showing the increase in absorbance at 248 nm relative to 280 nm as the level of conjugated drug-linker per mAb increases. (See also Hamblett et al. 2004; Sanderson et al. 2005)

which is necessary for protein binding, and the compounds can be eluted with decreasing salt gradients. Thus, this technique cannot be directly coupled to the mass spectrometer (Wakankar et al. 2011).

6.3.3 Other Types of Chromatography

Charge-based chromatography, such as ion-exchange chromatography (IEC), has been used to determine the distribution of ADCs, while size-based chromatography, such as size-exclusion chromatography (SEC), can also be used to analyze the drug distribution as well as characterize ADCs for protein fractionation, aggregation, and degradation during storage in liquid formulation (Kozak and Raab 2013; Cordoba et al. 2005; King et al. 2002).

6.4 General Assay Validation Considerations

The ADC is a complex molecule composed of a large molecule (antibody) and multiple molecule drugs. Therefore, an appropriate assay should be validated with the capability to detect both molecules precisely and accurately. Large molecules have well-defined tertiary structures and are often more hydrophilic. In contrast, small molecules lack a tertiary structure and are often more hydrophobic (Dere et al. 2013). Due to the significant differences in physicochemical properties between small and large molecules, the combination of different approaches, such as LBAs for large molecules and LC–MS for small molecules, is usually employed to quantify ADCs. Currently, there are no regulatory guidelines and best standard industry practices for bioanalytical methods of ADCs. However, the current widely accepted guidelines for large and small molecules can be a good reference. For example, the FDA guidance on bioanalytical method validation (US FDA 2018) has described the basic elements of assay validation. Some critical items are listed in Table 6.3. Based on these guidelines, additional requirements unique to ADCs can be added.

Current guidelines used to validate chromatographic assays can be applied when LC–MS methodology is used to measure unconjugated cytotoxic drug concentrations. As mentioned previously, ELISA is designed to measure the total antibody analyte, including conjugated, partially unconjugated, and fully unconjugated antibodies. Therefore, additional experiments should be performed during validation to demonstrate the ability of the assay to quantify both conjugated and completely unconjugated antibodies with acceptable accuracy and precision. Another aspect unique to ADCs includes an assessment of the stability of free drugs in the presence of ADC because free drug concentration may increase due to the release of additional drugs from the ADC under storage conditions. Chemical stability assessment is required for small molecules, but stability assessment for large molecules is more complicated and needs to evaluate the physicochemical properties and biological integrity.

Table 6.3 Assay validation parameters for ligand-binding and chromatographic assay

Validation process	Ligand-binding assays (LBA)	Chromatographic assays
Selectivity	Matrix from ≥10 individual sources Spiked samples should be ±20% LLOQ.	Matrix from ≥6 individual sources For 80% of sources, unspiked matrix should be BQL, and spiked samples should be ±25% at LLOQ, and ±20% at HQC.
Specificity	QCs should meet ±20%, or 25% at the LLOQ and ULOQ	Same as selectivity.
Carryover	Not applicable	Carryover should not exceed 20% of LLOQ
Sensitivity	The accuracy should be ±25% of the nominal concentration The precision should be ±25% CV The total error should be ≤40%.	LLOQ response should be ≥5 times the analyte response of the zero calibrator The accuracy should be ±20% of nominal concentration The precision should be ±20% CV
Accuracy	±20% of nominal concentrations; except for LLOQ and ULOQ (±25%)	±15% of nominal concentrations, except for LLOQ (± 20%)
Precision	±20% CV, except for LLOQ and ULOQ (±25%)	±15% CV, except for LLOQ (±20%)
Calibration curve	A blank and at least six, non-zero calibrator levels covering the quantitation range, including LLOQ on-zero calibrators should be ±20% of nominal concentration (except for LLOQ and ULOQ: ± 25%) 75% and a minimum of six non-zero calibrator levels should meet the above criteria in each validation run. Anchor points should not be included in the curve fit	A blank (no analyte, no IS), a zero calibrator (blank plus IS), and at least six, non-zero calibrator levels covering the quantitation range, including LLOQ Non-zero calibrators should be ±15% of nominal concentration (except for LLOQ: ±20%) 75% and a minimum of six non-zero calibrator levels should meet the above criteria in each validation run.

HQC high-quality control, *QC* quality control, *LLOQ* lower limit of quantification, *ULOQ* upper limit of quantification

When the assay is applied to analyze the real sample (such as samples from a pharmacokinetic study), it should be noted that the calibration curve for LBA assays such as ELISA is usually prepared using the product reference standard that has a fixed DAR distribution. Such standards are only similar to the samples of initial time points. Moreover, most LBA calibration curves are inherently nonlinear over the nominal concentration range, and a nonlinear regression is needed to achieve a good fitting. Thus it is important to verify that the calibration curve is appropriate for quantifying the dynamically changing mixture of ADC for PK evaluation (Dere et al. 2013).

The current FDA guidance for bioanalytical method validation contains an additional section on endogenous compounds. Small-molecule assays often include a pre-assay extraction to alleviate problems from individual matrix variability. In contrast, assays to quantify large molecules are often developed to measure an analyte in complex matrices without pre-extraction where endogenous protein may be present. Therefore, special considerations must be taken regarding the matrix effect (DeSilva et al. 2003).

6.5 Challenges and Future Perspectives

Bioanalysis of ADCs is more complex compared to conventional large- or small-molecule bioanalysis. The multicomponent nature of ADCs, which arises from the heterogeneity of the conjugation and the biotransformation in vivo represents a significant challenge in developing reliable and accurate bioanalytical analysis. The DAR composition continuously changes in circulation due to drug deconjugation and antibody degradation. Therefore, the ADC reference standard used for assay quantification may not be identical to the ADC mixture from in vivo samples (Gorovits et al. 2013). Additional attention to assay reagents is required to ensure that they can detect all DAR species formed in vivo without sacrificing selectivity. Given the inherent limitations of current bioanalytical assays for characterizing ADCs, developing a comprehensive bioanalytical strategy is highly recommended for ADCs.

The choice of proper forms of ADCs for analysis should align with the goals of each individual study, and bioanalytical strategies should be adapted to the specific molecular properties of these forms. For the bioanalytical assay of a particular ADC, a tailored bioanalytical approach based on available assay platforms may be necessary. As technologies advance, including the development of bispecific antibodies, dual payloads, and immune-stimulating payloads, new analytical strategies are necessary to address the challenges posed by these innovations (Qin and Gong 2022).

6.6 Conclusions

We have summarized the assays that are commonly used to study the pharmacokinetics and determine DAR for ADCs. Multiple methods are required to fully characterize the physiochemical and pharmacokinetic properties of ADCs. The complexity of ADCs presents challenges in the development of bioanalytical methods, so new bioanalytical techniques will need to be developed to ensure assay accuracy.

References

Abedi M, Ahangari Cohan R, Mahboudi F, Shafiee Ardestani M, Davami F (2019) MALDI-MS: a rapid and reliable method for drug-to-antibody ratio determination of antibody-drug conjugates. Iran Biomed J 23(6):395–403. https://doi.org/10.29252/ibj.23.6.395

Advani A, Coiffier B, Czuczman MS, Dreyling M, Foran J, Gine E et al (2010) Safety, pharmacokinetics, and preliminary clinical activity of inotuzumab ozogamicin, a novel immunoconjugate for the treatment of B-cell non-Hodgkin's lymphoma: results of a phase I study. J Clin Oncol 28(12):2085–2093. https://doi.org/10.1200/JCO.2009.25.1900

Alley SC, Anderson KE (2013) Analytical and bioanalytical technologies for characterizing antibody–drug conjugates. Curr Opin Chem Biol 17(3):406–411. https://doi.org/10.1016/j.cbpa.2013.03.022

Alley SC, Benjamin DR, Jeffrey SC, Okeley NM, Meyer DL, Sanderson RJ et al (2008) Contribution of linker stability to the activities of anticancer Immunoconjugates. Bioconjug Chem 19(3):759–765. https://doi.org/10.1021/bc7004329

Beck A (2014) Review of antibody-drug conjugates, methods in molecular biology series: a book edited by Laurent Ducry. MAbs 6(1):30–33

Boghaert ER, Khandke KM, Sridharan L, Dougher M, DiJoseph JF, Kunz A et al (2008) Determination of pharmacokinetic values of calicheamicin-antibody conjugates in mice by plasmon resonance analysis of small (5 microl) blood samples. Cancer Chemother Pharmacol 61(6):1027–1035. https://doi.org/10.1007/s00280-007-0560-2

Buckwalter M, Dowell JA, Korth-Bradley J, Gorovits B, Mayer PR (2004) Pharmacokinetics of Gemtuzumab Ozogamicin as a single-agent treatment of pediatric patients with refractory or relapsed acute myeloid leukemia. J Clin Pharmacol 44(8):873–880. https://doi.org/10.1177/0091270004267595

Carter PJ, Senter PD (2008) Antibody-drug conjugates for cancer therapy. Cancer J 14(3):154–169. https://doi.org/10.1097/PPO.0b013e318172d704

Chang HP, Shah DK (2020) Determination of ADC concentration by ligand-binding assays. Methods Mol Biol 2078:361–369. https://doi.org/10.1007/978-1-4939-9929-3_26

Chari RV, Martell BA, Gross JL, Cook SB, Shah SA, Blattler WA et al (1992) Immunoconjugates containing novel maytansinoids: promising anticancer drugs. Cancer Res 52(1):127–131

Chen Y (2013) Drug-to-antibody ratio (DAR) by UV/Vis spectroscopy. In: Ducry L (ed) Antibody-Drug Conjugates. Humana Press, pp 267–273

Chen Y, Clark S, Wong T, Dennis MS, Luis E, Zhong F et al (2007) Armed antibodies targeting the mucin repeats of the ovarian cancer antigen, MUC16, are highly efficacious in animal tumor models. Cancer Res 67(10):4924–4932. https://doi.org/10.1158/0008-5472.CAN-06-4512

Cheng CN, Liao HW, Lin CH, Chang WC, Chen IC, Lu YS et al (2024) Quantifying payloads of antibody–drug conjugates using a postcolumn infused-internal standard strategy with LC–MS. Anal Chim Acta 1303:342537. https://doi.org/10.1016/j.aca.2024.342537

Clark T, Han X, King L, Barletta F (2013) Insights into antibody–drug conjugates: bioanalysis and biomeasures in discovery. Bioanalysis 5(9):985–987. https://doi.org/10.4155/bio.13.65

Cordoba AJ, Shyong BJ, Breen D, Harris RJ (2005) Non-enzymatic hinge region fragmentation of antibodies in solution. J Chromatogr B Anal Technol Biomed Life Sci 818(2):115–121. https://doi.org/10.1016/j.jchromb.2004.12.033

Deng R, Zhou C, Li D, Cai H, Sukumaran S, Carrasco-Triguero M et al (2019) Preclinical and translational pharmacokinetics of a novel THIOMAB™ antibody-antibiotic conjugate against Staphylococcus aureus. MAbs 11(6):1162–1174. https://doi.org/10.1080/19420862.2019.1627152

Dere R, Yi JH, Lei C, Saad OM, Huang C, Li Y et al (2013) PK assays for antibody-drug conjugates: case study with ado-trastuzumab emtansine. Bioanalysis 5(9):1025–1040. https://doi.org/10.4155/bio.13.72

DeSilva B, Smith W, Weiner R, Kelley M, Smolec J, Lee B et al (2003) Recommendations for the bioanalytical method validation of ligand-binding assays to support pharmacokinetic assessments of macromolecules. Pharm Res 20(11):1885–1900

DiJoseph JF, Armellino DC, Boghaert ER, Khandke K, Dougher MM, Sridharan L et al (2004) Antibody-targeted chemotherapy with CMC-544: a CD22-targeted immunoconjugate of calicheamicin for the treatment of B-lymphoid malignancies. Blood 103(5):1807–1814. https://doi.org/10.1182/blood-2003-07-2466

Dowell JA, Korth-Bradley J, Liu H, King SP, Berger MS (2001) Pharmacokinetics of gemtuzumab ozogamicin, an antibody-targeted chemotherapy agent for the treatment of patients with acute myeloid leukemia in first relapse. J Clin Pharmacol 41(11):1206–1214

Francisco JA, Cerveny CG, Meyer DL, Mixan BJ, Klussman K, Chace DF et al (2003) cAC10-vcMMAE, an anti-CD30–monomethyl auristatin E conjugate with potent and selective antitumor activity. Blood 102(4):1458–1465. https://doi.org/10.1182/blood-2003-01-0039

Giddabasappa A, Gupta VR, Norberg R, Gupta P, Spilker ME, Wentland J et al (2016) Biodistribution and targeting of anti-5T4 antibody-drug conjugate using fluorescence molecular tomography. Mol Cancer Ther 15(10):2530–2540. https://doi.org/10.1158/1535-7163.Mct-15-1012

Gorovits B, Alley SC, Bilic S, Booth B, Kaur S, Oldfield P et al (2013) Bioanalysis of antibody-drug conjugates: American Association of Pharmaceutical Scientists antibody-drug conjugate working group position paper. Bioanalysis 5(9):997–1006. https://doi.org/10.4155/bio.13.38

Hamblett KJ, Senter PD, Chace DF, Sun MM, Lenox J, Cerveny CG et al (2004) Effects of drug loading on the antitumor activity of a monoclonal antibody drug conjugate. Clin Cancer Res Off J Am Assoc Cancer Res 10(20):7063–7070. https://doi.org/10.1158/1078-0432.CCR-04-0789

Hinman LM, Hamann PR, Wallace R, Menendez AT, Durr FE, Upeslacis J (1993) Preparation and characterization of monoclonal antibody conjugates of the calicheamicins: a novel and potent family of antitumor antibiotics. Cancer Res 53(14):3336–3342

Huang Y, Mou S, Wang Y, Mu R, Liang M, Rosenbaum AI (2021) Characterization of antibody-drug conjugate pharmacokinetics and in vivo biotransformation using quantitative intact LC-HRMS and surrogate Analyte LC-MRM. Anal Chem 93(15):6135–6144. https://doi.org/10.1021/acs.analchem.0c05376

Iwamoto N, Shimomura A, Tamura K, Hamada A, Shimada T (2017) LC-MS bioanalysis of Trastuzumab and released emtansine using nano-surface and molecular-orientation limited (nSMOL) proteolysis and liquid-liquid partition in plasma of Trastuzumab emtansine-treated breast cancer patients. J Pharm Biomed Anal 145:33–39. https://doi.org/10.1016/j.jpba.2017.06.032

Jiang D, Li P, Yuan L (2024) Bioanalysis of free antisense oligonucleotide payload from antibody-oligonucleotide conjugate by hybridization LC-MS/MS. Bioanalysis 16:1–10. https://doi.org/10.1080/17576180.2024.2368339

Jin W, Burton L, Moore I (2018) LC-HRMS quantitation of intact antibody drug conjugate trastuzumab emtansine from rat plasma. Bioanalysis 10(11):851–862. https://doi.org/10.4155/bio-2018-0003

Junutula JR, Raab H, Clark S, Bhakta S, Leipold DD, Weir S et al (2008) Site-specific conjugation of a cytotoxic drug to an antibody improves the therapeutic index. Nat Biotechnol 26(8):925–932. https://doi.org/10.1038/nbt.1480

Junutula JR, Flagella KM, Graham RA, Parsons KL, Ha E, Raab H et al (2010) Engineered Thio-Trastuzumab-DM1 conjugate with an improved therapeutic index to target human epidermal growth factor receptor 2–positive breast cancer. Clin Cancer Res 16(19):4769–4778. https://doi.org/10.1158/1078-0432.ccr-10-0987

Kaur S, Xu K, Saad OM, Dere RC, Carrasco-Triguero M (2013) Bioanalytical assay strategies for the development of antibody-drug conjugate biotherapeutics. Bioanalysis 5(2):201–226. https://doi.org/10.4155/bio.12.299

King HD, Dubowchik GM, Mastalerz H, Willner D, Hofstead SJ, Firestone RA et al (2002) Monoclonal antibody conjugates of doxorubicin prepared with branched peptide linkers: inhibition of aggregation by methoxytriethyleneglycol chains. J Med Chem 45(19):4336–4343

Kovtun YV, Audette CA, Ye Y, Xie H, Ruberti MF, Phinney SJ et al (2006) Antibody-drug conjugates designed to eradicate tumors with homogeneous and heterogeneous expression of the target antigen. Cancer Res 66(6):3214–3221. https://doi.org/10.1158/0008-5472.CAN-05-3973

Kozak K, Raab H (2013) Assay methodologies and challenges. In: Phillips GL (ed) Antibody-drug conjugates and immunotoxins. Springer, New York, pp 41–56

Lazar AC, Wang L, Blättler WA, Amphlett G, Lambert JM, Zhang W (2005) Analysis of the composition of immunoconjugates using size-exclusion chromatography coupled to mass spectrometry. Rapid Commun Mass Spectrom 19(13):1806–1814. https://doi.org/10.1002/rcm.1987

Lee BI, Park MH, Byeon JJ, Shin SH, Choi J, Park Y et al (2020) Quantification of an antibody-conjugated drug in fat plasma by an affinity capture LC-MS/MS method for a novel Prenyl transferase-mediated site-specific antibody-drug conjugate. Molecules 25(7). https://doi.org/10.3390/molecules25071515

Lewis Phillips GD, Li G, Dugger DL, Crocker LM, Parsons KL, Mai E et al (2008) Targeting HER2-positive breast cancer with trastuzumab-DM1, an antibody-cytotoxic drug conjugate. Cancer Res 68(22):9280–9290. https://doi.org/10.1158/0008-5472.CAN-08-1776

Liu A, Kozhich A, Passmore D, Gu H, Wong R, Zambito F et al (2015) Quantitative bioanalysis of antibody-conjugated payload in monkey plasma using a hybrid immuno-capture LC-MS/MS approach: assay development, validation, and a case study. J Chromatogr B Anal Technol Biomed Life Sci 1002:54–62. https://doi.org/10.1016/j.jchromb.2015.08.007

Luckey JA, Drossman H, Kostichka T, smith LM. (1993) High-speed DNA sequencing by capillary gel electrophoresis. In: Ray W (ed) Methods in enzymology. Academic Press, pp 154–172

McDonagh CF, Turcott E, Westendorf L, Webster JB, Alley SC, Kim K et al (2006) Engineered antibody–drug conjugates with defined sites and stoichiometries of drug attachment. Protein Eng Des Sel 19(7):299–307. https://doi.org/10.1093/protein/gzl013

Myler H, Rangan VS, Wang J, Kozhich A, Cummings JA, Neely R et al (2015) An integrated multiplatform bioanalytical strategy for antibody-drug conjugates: a novel case study. Bioanalysis 7(13):1569–1582. https://doi.org/10.4155/bio.15.80

Ouyang J (2013) Drug-to-antibody ratio (DAR) and drug load distribution by hydrophobic interaction chromatography and reversed phase high-performance liquid chromatography. In: Ducry L (ed) Antibody-drug conjugates. Humana Press, pp 275–283

Park MH, Lee BI, Byeon JJ, Shin SH, Choi J, Park Y et al (2019) Pharmacokinetic and metabolism studies of monomethyl Auristatin F via liquid chromatography-quadrupole-time-of-flight mass spectrometry. Molecules 24(15). https://doi.org/10.3390/molecules24152754

Polakis P (2005) Arming antibodies for cancer therapy. Curr Opin Pharmacol 5(4):382–387. https://doi.org/10.1016/j.coph.2005.04.008

Pollack VA, Alvarez E, Tse KF, Torgov MY, Xie S, Shenoy SG et al (2007) Treatment parameters modulating regression of human melanoma xenografts by an antibody-drug conjugate (CR011-vcMMAE) targeting GPNMB. Cancer Chemother Pharmacol 60(3):423–435. https://doi.org/10.1007/s00280-007-0490-z

Qin Q, Gong L (2022) Current analytical strategies for antibody-drug conjugates in biomatrices. Molecules 27(19). https://doi.org/10.3390/molecules27196299

Quiles S, Raisch KP, Sanford LL, Bonner JA, Safavy A (2009) Synthesis and preliminary biological evaluation of high-drug-load paclitaxel-antibody conjugates for tumor-targeted chemotherapy. J Med Chem 53(2):586–594. https://doi.org/10.1021/jm900899g

Ramakrishnan NA, Drescher MJ, Sheikhali SA, Khan KM, Hatfield JS, Dickson MJ et al (2006) Molecular identification of an N-type Ca2+ channel in saccular hair cells. Neuroscience 139(4):1417–1434. https://doi.org/10.1016/j.neuroscience.2006.01.064

Safavy A, Bonner JA, Waksal HW, Buchsbaum DJ, Gillespie GY, Khazaeli MB et al (2003) Synthesis and biological evaluation of paclitaxel-C225 conjugate as a model for targeted drug delivery. Bioconjug Chem 14(2):302–310. https://doi.org/10.1021/bc020033z

Sanderson RJ, Hering MA, James SF, Sun MM, Doronina SO, Siadak AW et al (2005) In vivo drug-linker stability of an anti-CD30 dipeptide-linked auristatin immunoconjugate. Clin Cancer Res Off J Am Assoc Cancer Res 11(2 Pt 1):843–852

Sapra P, Stein R, Pickett J, Qu Z, Govindan SV, Cardillo TM et al (2005) Anti-CD74 antibody-doxorubicin conjugate, IMMU-110, in a human multiple myeloma xenograft and in monkeys. Clin Cancer Res Off J Am Assoc Cancer Res 11(14):5257–5264. https://doi.org/10.1158/1078-0432.CCR-05-0204

Schuck P (1997) Use of surface plasmon resonance to probe the equilibrium and dynamic aspects of interactions between biological macromolecules1. Annu Rev Biophys Biomol Struct 26(1):541–566. https://doi.org/10.1146/annurev.biophys.26.1.541

Shen BQ, Xu K, Liu L, Raab H, Bhakta S, Kenrick M et al (2012a) Conjugation site modulates the in vivo stability and therapeutic activity of antibody-drug conjugates. Nat Biotechnol 30(2):184–189. https://doi.org/10.1038/nbt.2108

Shen BQ, Bumbaca D, Saad O, Yue Q, Pastuskovas CV, Khojasteh SC et al (2012b) Catabolic fate and pharmacokinetic characterization of trastuzumab emtansine (T-DM1): an emphasis on preclinical and clinical catabolism. Curr Drug Metab 13(7):901–910. https://doi.org/10.2174/138920012802138598

Siegel MM, Tabei K, Kunz A, Hollander IJ, Hamann RR, Bell DH et al (1997) Calicheamicin derivatives conjugated to monoclonal antibodies: determination of loading values and distributions by infrared and UV matrix-assisted laser desorption/ionization mass spectrometry and electrospray ionization mass spectrometry. Anal Chem 69(14):2716–2726

Stephan JP, Chan P, Lee C, Nelson C, Elliott JM, Bechtel C et al (2008) Anti-CD22-MCC-DM1 and MC-MMAF conjugates: impact of assay format on pharmacokinetic parameters determination. Bioconjug Chem 19(8):1673–1683. https://doi.org/10.1021/bc800059t

Stephan JP, Kozak KR, Wong WL (2011) Challenges in developing bioanalytical assays for characterization of antibody-drug conjugates. Bioanalysis 3(6):677–700. https://doi.org/10.4155/bio.11.30

Suh MJ, Powers JB, Daniels CM, Wu Y (2023) Enhanced pharmacokinetic bioanalysis of antibody-drug conjugates using hybrid Immunoaffinity capture and microflow LC-MS/MS. AAPS J 25(4):68. https://doi.org/10.1208/s12248-023-00835-0

Sun MMC, Beam KS, Cerveny CG, Hamblett KJ, Blackmore RS, Torgov MY et al (2005) Reduction–alkylation strategies for the modification of specific monoclonal antibody disulfides. Bioconjug Chem 16(5):1282–1290. https://doi.org/10.1021/bc050201y

Teicher BA, Chari RV (2011) Antibody conjugate therapeutics: challenges and potential. Clin Cancer Res Off J Am Assoc Cancer Res 17(20):6389–6397. https://doi.org/10.1158/1078-0432.CCR-11-1417

Tolcher AW, Ochoa L, Hammond LA, Patnaik A, Edwards T, Takimoto C et al (2003) Cantuzumab mertansine, a maytansinoid immunoconjugate directed to the CanAg antigen: a phase I, pharmacokinetic, and biologic correlative study. J Clin Oncol 21(2):211–222

Truffot A, Jourdil JF, Veyret-Gautier E, Noble J, Jouve T, Malvezzi P et al (2023) Quantification of belatacept by liquid chromatography-tandem mass spectrometry in human plasma: application to a pharmacokinetic study in renal transplant recipients. Clin Biochem 113:17–20. https://doi.org/10.1016/j.clinbiochem.2022.12.014

US FDA (2018) Guidance for Industry: Bioanalytical method validation. US FDA, Washington, DC

Vater CA, Reid K, Bartle LM, Goldmacher VS (1995) Characterization of antibody binding to cell surface antigens using a plasma membrane-bound plate assay. Anal Biochem 224(1):39–50. https://doi.org/10.1006/abio.1995.1006

Wakankar A, Chen Y, Gokarn Y, Jacobson FS (2011) Analytical methods for physicochemical characterization of antibody drug conjugates. MAbs 3(2):161–172

Wang L, Amphlett G, Blättler WA, Lambert JM, Zhang W (2005) Structural characterization of the maytansinoid–monoclonal antibody immunoconjugate, huN901–DM1, by mass spectrometry. Protein Sci 14(9):2436–2446. https://doi.org/10.1110/ps.051478705

Xie H, Audette C, Hoffee M, Lambert JM, Blattler WA (2004) Pharmacokinetics and biodistribution of the antitumor immunoconjugate, cantuzumab mertansine (huC242-DM1), and its two components in mice. J Pharmacol Exp Ther 308(3):1073–1082. https://doi.org/10.1124/jpet.103.060533

Xu K, Liu L, Saad OM, Baudys J, Williams L, Leipold D et al (2011) Characterization of intact antibody–drug conjugates from plasma/serum in vivo by affinity capture capillary liquid chromatography–mass spectrometry. Anal Biochem 412(1):56–66. https://doi.org/10.1016/j.ab.2011.01.004

Yin F, Adhikari D, Sun M, Shane Woolf M, Ma E, Mylott W et al (2023a) Bioanalysis of an antibody drug conjugate (ADC) PYX-201 in human plasma using a hybrid immunoaffinity LC-MS/MS approach. J Chromatogr B Anal Technol Biomed Life Sci 1223:123715. https://doi.org/10.1016/j.jchromb.2023.123715

Yin F, Adhikari D, Li Y, Turner D, Shane Woolf M, Lebarbenchon D et al (2023b) A sensitive and rapid LC-MS/MS assay for quantitation of free payload Aur0101 from antibody drug conjugate (ADC) PYX-201 in human plasma. J Chromatogr B Anal Technol Biomed Life Sci 1226:123786. https://doi.org/10.1016/j.jchromb.2023.123786

Zhang D, Yu SF, Khojasteh SC, Ma Y, Pillow TH, Sadowsky JD et al (2018) Intratumoral payload concentration correlates with the activity of antibody-drug conjugates. Mol Cancer Ther 17(3):677–685. https://doi.org/10.1158/1535-7163.Mct-17-0697

Zhang D, Dragovich PS, Yu SF, Ma Y, Pillow TH, Sadowsky JD et al (2019) Exposure-efficacy analysis of antibody-drug conjugates delivering an excessive level of payload to tissues. Drug Metab Dispos 47(10):1146–1155. https://doi.org/10.1124/dmd.119.087023

Zhao RY, Wilhelm SD, Audette C, Jones G, Leece BA, Lazar AC et al (2011) Synthesis and evaluation of hydrophilic linkers for antibody-maytansinoid conjugates. J Med Chem 54(10):3606–3623. https://doi.org/10.1021/jm2002958

Zhou C, Lehar S, Gutierrez J, Rosenberger CM, Ljumanovic N, Dinoso J et al (2016) Pharmacokinetics and pharmacodynamics of DSTA4637A: a novel THIOMAB™ antibody antibiotic conjugate against Staphylococcus aureus in mice. MAbs 8(8):1612–1619. https://doi.org/10.1080/19420862.2016.1229722

Zhu X, Huo S, Xue C, An B, Qu J (2020) Current LC-MS-based strategies for characterization and quantification of antibody-drug conjugates. J Pharm Anal 10(3):209–220. https://doi.org/10.1016/j.jpha.2020.05.008

Chapter 7
Pharmacokinetics/Pharmacodynamics and Disposition of Antibody–Drug Conjugates

Siddharth Sukumaran and Kedan Lin

Abstract Antibody–drug conjugates (ADCs) combine the high target specificity and favorable pharmacokinetics of monoclonal antibodies with the potent tumor-killing properties of cytotoxic agents and have demonstrated convincing antitumor effects in both animal models and patients. However, the inherent complexity of ADCs with their multiple components often makes their development challenging. Pharmacokinetic and ADME (absorption, distribution, metabolism, and excretion) characterization of ADCs reflects the dynamic interactions between the biological system and ADC and provides critical assessments in lead selection, optimization, and clinical development. A rational strategy integrating the mechanistic understanding of pharmacokinetics/pharmacodynamics and ADC disposition helps to inform target selection, drug selection, and linker design and ultimately to maximize the therapeutic window. In this chapter, we give an overview of ADC PKPD and disposition and discuss our current understanding of the major determinants, unique challenges, and lessons learned from the current ADC landscape. The utility of PKPD modeling is also discussed in the context of providing guidance to assist in the successful development of these complex molecules.

Keywords Pharmacokinetics (PK) · Pharmacodynamics (PD) · Disposition · Monoclonal antibody (mAb) · Antibody drug conjugates (ADCs) · Targeted therapy · Optimization

S. Sukumaran · K. Lin (✉)
Preclinical and Translational Pharmacokinetics, Genentech Research and Early Development, South San Francisco, CA, USA
e-mail: lin.kedan@gene.com

J. L. Zaro, J. Wang (eds.), *Antibody-Drug Conjugates*, AAPS Advances in the Pharmaceutical Sciences Series 17, https://doi.org/10.1007/978-3-031-83005-1_7

7.1 Introduction

There is a growing need for evaluating and developing new therapeutic modalities in our battle against cancer. In the past few decades, development of targeted therapies resulted in drugs with improved efficacy and safety (Gerber 2008). Monoclonal antibodies, as a class of highly targeted therapeutics, have been used for the treatment of a variety of cancers. Recently, antibody drug conjugates (ADCs) have shown impressive potential in advancing cancer treatment to the next level (Sieversand and Senter 2013). Conjugation of potent cytotoxic drugs to an antibody can increase the potency of the antibody itself, which usually acts by blocking/activating signal transduction, Antibody-Dependent Cellular Cytotoxicity (ADCC), and/or Complement-Dependent Cytotoxicity (CDC) (Waldmann 2003). By combining the high target specificity of monoclonal antibodies with the potent tumor-killing properties of cytotoxic agents, ADCs have demonstrated convincing antitumor effects in both animal models and patients (Sieversand and Senter 2013).

ADCs consist of a potent cytotoxic agent conjugated through a linker molecule to an antibody that can bind with a high specificity to a target antigen. The expression of the target antigen is selected for its overexpression in tumor compared to normal cells with the exception of B cell targets, such as CD22 and CD79 (Bander 2013). The antibody part of the ADC is designed to bind with high affinity and selectivity to its target antigen. Cytotoxic drugs that are attached to the antibody are usually potent antimitotic or DNA-modifying agents (Ducryand and Stump 2010). Administration of these drugs by themselves may have minimal therapeutic index, but conjugation to an antibody renders an acceptable window through enhanced delivery to tumor, reduction of systemic exposure, and minimizing distribution and uptake of drug to non-target tissues. The third component of an ADC is the linker that conjugates the cytotoxic agent with that of the antibody, usually through the cysteine or lysine amino acid residues (Nolting 2013).

There are two ADCs that were currently approved for marketing. Adcetris (Brentuximab vedotin) is an anti-CD30 antibody conjugated to an antimitotic agent monomethyl auristatin E (MMAE) through a cathepsin-cleavable linker and was approved for the treatment of patients with Hodgkin's lymphoma after failure of autologous stem cell transplant (Cao et al. 2013). Kadcyla (ado-trastuzumab emtansine) consists of transtuzumab (targeted against her2 antigen) conjugated to mertansine (DM1) and was approved for the treatment of HER2-positive metastatic breast cancer (Ballantyneand and Dhillon 2013). In addition, more than 30 ADCs with different cytotoxic drug and linker combinations are in various stages of clinical development against a variety of targets (Mullard 2013). Furthermore, there is tremendous breakthrough in developing next-generation ADCs through antibody engineering, site-specific conjugation, and novel linker/cytotoxic drug combination (Flemming 2014).

However, the inherent complexity of ADCs with their multiple components often makes their development challenging. Pharmacokinetic and ADME (absorption, distribution, metabolism, and excretion) characterization of ADCs reflects the dynamic interactions between the biological system and ADC and provides critical assessments in lead selection, optimization, and clinical development. A rational strategy integrating the mechanistic understanding of pharmacokinetics/pharmacod ynamics and ADC disposition helps to inform target selection, drug selection, linker design, and ultimately to maximize the therapeutic window.

Target, antibody, linker, cytotoxic drug, and their interactions make distinct contributions to the mechanism of action for a given ADC, and the disposition of an ADC defines its therapeutic window (Sieversand and Senter 2013). Two pharmacologically distinct components, the antibody and the cytotoxic small-molecule drug, necessitate the understanding of the behavior and fate of both components in vivo, and monitoring of their PK/disposition requires a comprehensive analytical strategy.

7.2 Analytical Strategy and Its Application in ADC Selection and Optimization

A typical ADC assay strategy evolves with the stage of development with the goal of selecting the right analyte to inform the safety and efficacy and describe the exposure–response relationship (Kaur et al. 2013). Multiple analytes help to capture the many facets of the behavior of these complex molecules, such as the rate of drug loss from an ADC (i.e. linker stability), the effect of conjugation on ADC clearance, and ultimately the exposure–response relationship. However, the desire to be comprehensive must be balanced by the practicality, the availability of the technology and reagents, and ultimately by the purpose of each study. At the discovery stage, many optimization factors are being explored, such as different linkers, drug, drug-to-antibody ratio, and site of conjugation, and a diverse array of assays are often used to understand the stability and disposition of ADCs and its initial interactions with biological systems. At the clinical stage, with the confirmation of preclinical prediction and availability of human data, the assay strategy may be streamlined to increase efficiency and patient compliance (Dere et al. 2013). In general, a panel of assays is required to measure these disparate components: total antibodies (conjugated and unconjugated antibodies), conjugated antibodies, conjugated drugs, unconjugated antibodies, and unconjugated (free) drugs (Kaur et al. 2013). For example, in the development of T-DM1, a comprehensive assay strategy consisting of three validated assays was developed. These validated assays included an ELISA designed to measure total trastuzumab, an ELISA to measure conjugated trastuzumab, and a small-molecule LC–MS/MS assay to measure the amount of free DM1 catabolite. These assays are capable of quantifying DAR analytes in circulation to characterize the PK and stability of T-DM1 in nonclinical and clinical studies (Dere et al. 2013).

The total antibody (Tab) concentration captures both the conjugated and unconjugated forms of an ADC and is usually determined using an enzyme-linked immunosorbent assay (ELISA)-based format. It provides the best assessment of the in vivo stability and integrity of the antibody over time and serves a key role in ADC optimization, particularly in evaluating the impact of conjugation and selecting a drug load.

Conjugated antibody concentrations in systemic circulation are usually determined using an ELISA assay format that measures a mixture of ADC species bearing at least one conjugated cytotoxic drug. Since the detection in this assay requires the presence of both intact antibody and cytotoxic drug components of the ADC, conjugated antibody concentration is commonly used as an estimate of the active ADC concentration and is the basis for most ADC PK analyses. The limitation of this assay is its inability to differentiate ADCs with varying numbers of conjugated cytotoxic drugs (DARs). Given ADC species with different DARs may have different potencies, the measured concentrations may not accurately reflect the associated pharmacological activity (Stephan et al. 2008).

On the other hand, measuring a conjugated drug provides a measure of the total amount of cytotoxic drug covalently bound to the antibody (Sanderson et al. 2005; Xu et al. 2013), and changes in the conjugated drug concentration could reflect both the elimination of ADC from the systemic circulation and the loss of cytotoxic drug from the antibody. Affinity capture LC–MS is a method of choice by which ADC is specifically extracted from plasma and then analyzed using LC–MS/MS (Xu et al. 2013). The power of this method lies in its ability to provide direct measurement of average concentrations of a drug associated with the antibody. Xu et al. used affinity capture LC–MS to assess the site-specific loss of cytotoxic drug from a thiomab ADC, providing valuable insights into the impact of conjugation site on linker stability, a critical factor in ADC safety and efficacy (Xu et al. 2011). These examples illustrate the value of evolving analytical technologies in exploring the behavior of ADCs, which in turn can lead to improvements in ADC design and development.

Unconjugated cytotoxic drug concentrations are used to infer the systemic exposure to the cytotoxic drug released from the ADC and are often associated with the loss of efficacy and increased toxicity. Assays for drug-containing products usually employ LC-MS or ELISA methods (Wang et al. 2005; Xie et al. 2004). LC-MS methods are highly specific for the measured analyte, while ELISA methods may be less specific and able to quantitate multiple analytes of similar structure. A critical consideration of assay selection ensures that the analyte(s) selected for measurement are relevant for efficacy or toxicity and able to provide meaningful PKPD relationship.

7.3 Disposition of an ADC and Its Implication on PKPD

The systemic PK profiles of an ADC provide only partial narratives of its fate. By design, an ADC is a prodrug. Its activity depends on the interaction with the target, subsequent internalization of the antigen–ADC complex, and the final release of active drug inside the cells. The full absorption, distribution, metabolism, and elimination (ADME) properties of ADCs are crucial for the therapeutic window rendered by ADCs. Biologically, the disposition of ADCs is strongly influenced by the underlying antibody backbone conferring properties such as target-specific binding, neonatal Fc receptor (FcRn)-dependent recycling, and Fc (fragment, crystallizable) effector functions. Similarly, the absorption, distribution, metabolism, and elimination (ADME) properties of ADCs possess similar attributes associated with unconjugated antibodies (Lobo et al. 2004; Deng et al. 2012; Boswell et al. 2011a; Linand and Tibbitts 2012).

7.3.1 *Absorption*

All ADCs that are currently on the market or in clinical development are dosed intravenously, and hence absorption is not considered when assessing the PK properties of ADCs. However, for convenience, intraperitoneal dosing of ADCs is sometimes used in early preclinical studies, and in the future, there is always a possibility for the development of subcutaneous dosing strategies for ADC similar to that of biologics treatment (Sharkey et al. 2012; Sharkey et al. 2011). In both cases, the absorption properties of ADCs in general will be similar to that of any monoclonal antibody. Although the mechanism is still unclear, it is generally accepted that lymphatic drainage is the major route for absorption from the site of subcutaneous administration to systemic circulation (Zhao et al. 2013; Wang et al. 2008). In addition, it is also acknowledged that diffusion of the molecule across blood vessels can contribute to the absorption kinetics.

7.3.2 *Distribution*

The distribution characteristics of ADC generally mimic that of monoclonal antibodies. Because of the cytotoxicity of the conjugated drug, distribution, and accumulation of an ADC to nontarget tissues, via either antigen-specific or nonspecific processes, may have profound pharmacologic/toxic effects. Owing to the higher molecular weight, the initial distribution is limited to the vascular space, with the volume of distribution in the central compartment similar to the total plasma volume in any species (~40 mL/kg) (Tabrizi et al. 2010; Mouldand and Green 2010). However, with time, distribution extends to the tissue interstitial space. The extent of distribution into tissues is determined by a variety of factors including blood flow, tissue porosity, structure and heterogeneity, and target properties. The movement of molecules into the tissues occurs by convection, transcytosis, and diffusion across the capillary, and studies show that the partitioning of antibodies into tissues ranges between 5% and 15% of what is observed in circulation, with brain being an exception, having a very low penetration of antibody because of the tough blood–brain barrier (Shahand and Betts 2013).

Beyond these similarities, ADCs bring in a different level of complexity as a result of interactions between target antigen and ADCs. Factors such as binding affinity, tissue expression profile, target turnover/internalization rates, target density, and conjugation impact on ADC distribution and influence the therapeutic index of an ADC (Tabrizi et al. 2010). Unlike the targets for monoclonal antibodies which are developed for blocking specific function/pathway in tumors leading to tumor growth inhibition or killing, targets for ADCs do not need to be causal or implicated in tumor progression. The role of the target antigen for an ADC is to provide high specificity, either expressed only by the tumor or at levels significantly higher in tumor (Bander 2013; Silver et al. 1997). In some cases, the low level of target antigen expression in normal tissues and their subsequent uptake of ADC may lead to decreased ADC delivery to tumor and/or increased delivery of cytotoxic drug to normal tissues, the phenomenon that could affect the therapeutic index of the ADC (Boswell et al. 2013). Employing a "predosing" or capping nontumor expression site with an inactive antibody may be helpful to mitigate the toxicity (Boswell et al. 2013).

The level of target antigen expression and binding affinity of the target to the antibody are critical in determining the amount of ADC delivered to the target. Irrespective of how specific the target expression is, if the target expression levels and/or antibody-binding affinity are low, ADCs delivered to tumor compared to normal tissues (nonspecific uptake) will be low, resulting in a poor therapeutic ratio. Hence, in addition to being specific, the level of target expression (and binding affinity) should enable adequate accumulation of the drug at the site of interest (Bander 2013).

Most of the cytotoxic agents in the ADC act intracellularly, and hence internalization and rate of internalization of the target play an important role in determining the drug accumulation inside the tumor cells and their subsequent killing (Sieversand and Senter 2013). The linker molecules are designed to be stable in circulation but are cleaved in endosome/lysosome compartments to effectively release the cytotoxic agents inside the cell. In addition to a rapid internalization process, the recycling or replenishment of the target at the cell surface is important to sustain the delivery of ADCs into the tumor cells. Furthermore, shedding of target antigen from the cell surface can play an important role in determining the efficacy and toxicity of ADCs (Tolcher et al. 2003). If shed target antigens are present, ADC binding to these shed antigens will affect ADC distribution and can result in a higher liver uptake (as these complexes are cleared by liver) and hence liver-related toxicities (Lovdal et al. 2000).

A few studies have assessed the effects of conjugation on the distribution of ADCs compared to unconjugated antibodies (Ericksonand and Lambert 2012). Distribution of DM1- and DM4-conjugated ADCs (lysine conjugation) in mice were similar to that of typical unconjugated IgG molecules (Xie et al. 2004; Ericksonand and Lambert 2012). However, auristatin-conjugated ADCs (cysteine conjugation) show modest differences in tissue distribution compared to unconjugated antibodies (Boswell et al. 2011b). For example, MMAE-conjugated STEAP1 ADC showed a general trend toward increased hepatic uptake and reduced levels in other highly vascularized organs in rats. Similarly, CMD-193 antibody conjugated with calicheamicin showed increased hepatic uptake compared to the parental huS193 unconjugated antibody, in a phase I biodistribution and PK study in patients with advanced epithelial cancer (Scott et al. 2007). Conjugation likely increases the hydrophobicity of the antibody, which leads to enhanced uptake by liver (Boswell et al. 2011b). Efforts on antibody engineering fronts have also shown promise in altering the distribution and mitigating the toxicity. Mutated antibody with deficiency in FcR binding with attenuated nonspecific uptake led to a longer circulation time and enhanced efficacy (Sussman et al. 2011).

7.3.3 Metabolism and Elimination

Therapeutic antibodies are eliminated from the body predominantly via target-mediated and nonspecific uptake into cells, followed by proteolytic degradation into inert small peptides and amino acids (Lobo et al. 2004). ADCs, bearing cytotoxic drugs, become active following release from or degradation of their associated antibodies. Understanding the mechanism of ADC metabolism and elucidating the identity of its active metabolites have practical implications on ADC optimization, clinical monitoring, and PKPD relationship. Metabolism of ADCs includes at least three processes: deconjugation as cytotoxic drug release from the ADC via enzymatic or chemical processes; proteolytic metabolism of the antibody, which can result in small peptides or amino acids that are still conjugated to the cytotoxic small-molecule drug (Tabrizi et al. 2006); and the metabolism of the cytotoxic drug (primarily the free drug) by the typical small-molecule phase I and phase II metabolizing enzymes (Linand and Lu 1997).

Nonspecific cleavage of the linker could happen in circulation, resulting in the deconjugation of one or more molecule of the cytotoxic drug from the antibody. Once internalized into the cell (by target-mediated cellular uptake in target cells or nonspecific pinocytosis), depending on the type of linker, the linker can be cleaved by specific proteases (like cathepsin B involved in the cleavage of valine–citrulline (vc) dipeptide linkage used for MMAE or MMAF conjugates) or the proteolytic processing in endosomes/lysosomes (Sutherland et al. 2006; Chuand and Polson 2013). For maleimide-containing drug linkers which are conjugated to the antibody cysteine residues to form thio-succinimide bonds, there are two competing reactions: exchange with sulfhydryl-containing molecules with the direct consequence of drug loss (Alley et al. 2009) versus hydrolysis of the thiosuccinimide ring resulting in a stable bond (Xu et al. 2013; Lyon et al. 2012). Site of conjugation and solvent accessibility have a direct influence on drug loss and subsequent efficacy (Shen et al. 2012a), which led to the selection of L and HC over FC variants; meanwhile, taking advantage of chemically induced hydrolysis, Lyon et al. showed the improved therapeutic window by making self-stabilizing maleimido-DPR (Lyon et al. 2012).

The other dominant mechanism that results in cytotoxic drug-containing catabolites is ADC catabolism, a process driven by either receptor-mediated endocytosis or fluid-phase pinocytosis with subsequent trafficking to the lysosome, followed by enzymatic degradation (Alley et al. 2009; Erickson et al. 2006; Okeley et al. 2010). Both deconjugation and proteolytic degradation of the antibody can occur simultaneously, and the contribution of each process to the release of the cytotoxic drug (or drug-containing products) depends on multiple factors including target properties, linker stability, drug load per antibody, and conjugation site. For example, apart from the free DM1 (cytotoxic drug), MCC-DM1 (linker conjugated to the drug) and lysine-MCC-DM1 were detected in circulation in both preclinical studies and in patients treated with T-DM1 (Shen et al. 2012b). Furthermore, for some ADCs, studies have shown other mechanisms of cytotoxic drug release or deconjugation like the transfer of linker drug from the antibody to circulating albumin in the case of MC-MMAF and VC-MMAE (Shen et al. 2012a). Nonspecific uptake and processing of ADCs may be the main source of ADC toxicity, and modulation of this process has a potential benefit in expanding TI. For example, by mutating Fc residues that are in contact with FcγRIII, Sussman et al. have shown that the S239C variant does not bind FcγRIII, blocks localization to nontarget tissues, and decreases off-target toxicity (Sussman et al. 2011; Jeffrey et al. 2013).

Once the cytotoxic drug is released from the antibody, they are subject to the metabolism and elimination processes associated with small-molecule drugs, including CYPs and drug transporters, and possess the theoretical potential for drug–drug interactions (DDIs) (Lu et al. 2013). In vitro studies in human liver microsome preparations suggest that DM1 is metabolized mainly by CYP3A4 and to a lesser extent by CYP3A5 (Shen et al. 2012b; Davis et al. 2012). The elimination of DM1-containing products and their metabolites was primarily through the biliary/fecal route in rats, with 80% recovery of the radio-labeled drug. Similar observations were made with VC–MMAE conjugated ADCs, with ~80% of the radio-labeled MMAE observed in feces and only ~6% of MMAE recovered from urine (Pastuskovas et al. 2005).

It has been recognized that one of the key factors that modulates ADC disposition is heterogeneity of an ADC, the consequence of both its conjugation process during manufacturing and its deconjugation in vivo. Traditional conjugation through either cysteine or lysine residues on the monoclonal antibody results in a mixture of ADC species differing not only in the number of drugs attached to the antibody but also in the sites of drug linkage (Hamblett et al. 2004; Singhand and Erickson 2009). For instance, ADCs conjugated by cysteines are mixtures of conjugated antibodies, with the DAR ranging from 0 to 8, representing drugs conjugated to some or all of the cysteines in unconjugated antibodies that form the interchain disulfide bonds (Singhand and Erickson 2009). ADCs conjugated via lysine residues have the potential for even greater variability in the number of conjugated drugs and their locations (Wang et al. 2005; Hamblett et al. 2004). Heterogeneity from ADC mixtures potentially could diminish the therapeutic window with higher DAR species contributing to significant toxicity while lower DAR species compromising on efficacy. In addition, it presents a major challenge in describing the collective ADC PKPD profile with multiple species with different pharmacological potency and PK behavior. One particular concern has been the presence of higher DAR species. For the reasons that have been touched upon before, multiple-site conjugation tends to increase hydrophobicity and decrease the overall stability on two fronts: faster loss of drug or deconjugation and accelerated antibody clearance (Sun et al. 2011).

Several studies helped assess the heterogeneity on PK, efficacy, and toxicityand guided the field towards more homogeneous ADC production. McDonagh et al. engineered the cysteines associated with the interchain disulfides, resulting in an antibody with fewer cysteines for conjugation and potentially resulting in less heterogeneity (McDonagh et al. 2008). ThioMab–drug conjugates (TDC) with site-specifically engineered cysteines provided a defined DAR of 2 per mAb with minimized heterogeneity and increased therapeutic index (Junutula et al. 2008). Alternatively, biosynthetically incorporating non-native amino acids into a given Ab scaffold provides homogeneous antibody drug conjugates with precise control over the site and stoichiometry of drug conjugation (Tian et al. 2014). Minimizing heterogeneity and optimizing DARs also help to establish a more defined PKPD relationship.

7.4 Application of PK in ADC Optimization

Integrating information from multiple assays is critical in characterizing the in vivo behavior of these complex molecules, interpreting ADC pharmacologic effects, and ultimately building exposure–response relationship. Accumulated knowledge on ADC PK helps to shed light on the mechanism of its stability, such as the effect of conjugation on ADC clearance and the rate of drug loss from an ADC (i.e., linker stability), and has profound impact on ADC optimization.

It has been observed that conjugation causes an increase in the clearance of the antibody for several ADCs (Boswell et al. 2011b; Hamblett et al. 2004; Herbertson et al. 2009), which may result in more rapid delivery of cytotoxic drug-bearing ADC to the normal organs or tissues with potentially toxic consequences (Hamblett et al. 2004). Comparison of Table PK of the ADC with the Table PK of unconjugated antibody (administered as unconjugated antibody) provides information regarding the effect of conjugation on antibody clearance (Xie et al. 2004; Sapra et al. 2005). There are several mechanisms being postulated ranging from the disruption of tertiary structure, weakening of disulfide bonds, to increase in hydrophobicity from conjugation (Boswell et al. 2011b; Hamblett et al. 2004). Among them, the hydrophobicity charge seems to be plausible. Conjugation with higher drug antibody ratio (DAR) leading to faster clearance seems to corroborate this hypothesis. For example, cAC10-vc MMAE ADCs with high DARs were observed to have a faster Tab clearance than lower DAR ADCs (Hamblett et al. 2004). Adem et al. showed that high-drug-load species leads to thermal unfolding and aggregation and affects the physical stability of ADCs (Jeffrey et al. 2013).

Another key parameter in ADC optimization is the rate of drug loss from ADC. Loss of drug from the ADC can result in decreased efficacy and changes in the toxicity associated with ADC administration. Comparative assessment of Table

PK with either conjugated antibody PK or conjugated drug PK can provide qualitative guidance on the rate of drug loss from the ADC (Linand and Tibbitts 2012). When comparing Table PK with conjugated antibody PK, it is typically observed that conjugated antibody concentrations decline more rapidly than Tab concentrations, and the degree of divergence of the curves is indicative of the rate of complete drug loss from the ADC (Tolcher et al. 2003; Tijink et al. 2006; Burris 3rd et al. 2011; Kantarjian et al. 2012). A better understanding of the mechanism of drug loss from a conjugated antibody helps assess the stability of linkers, conjugation sites, and impact of DARs. Illustrative cases for differences in linker stability between disulfide and thioether linkers, and the effect of conjugation site on thiomab-ADC linker stability have also been reported (Shen et al. 2012a; Lewis Phillips et al. 2008). One example of different efficacy among anti-Her2 ThioMab Drug Conjugate (TDC) variants (light chain, heavy chain, and Fc site-specific) was evidenced by the differential rate of drug loss or deconjugation despite similar total antibody clearance (Shen et al. 2012a).

Conjugated drug measurements have provided insights when compared with total antibody profiles. Antibody-conjugated drug concentrations decline more rapidly than Tab concentrations because two processes drive the decrease in conjugated drug concentrations: loss of drug from the ADC and elimination of ADC, while Tab concentration changes are driven solely by the elimination of ADC and unconjugated antibody. As such, the difference in the concentration decrease can be used to infer the rate of drug loss from the ADC (Kaur et al. 2013; Sanderson et al. 2005; Xie et al. 2004). Assessment and integration of ADC PK can be valuable not only in understanding a single ADC but also in evaluating multiple ADCs with different structural and pharmacologic characteristics, allowing improved design and development of these complex molecules.

7.5 PKPD Modeling for ADCs

As described earlier, ADC is a heterogeneous mixture of multiple species described by a variety of analytes. Because of this heterogeneity and dynamics between different species, quantitative characterization of ADC pharmacokinetics and pharmacodynamics becomes challenging. Mathematical modeling approaches can provide help in gaining a better understanding of the complex PK and PD behaviors of ADCs. In addition, it can help in the integration of the dynamic changes in different ADC species and provide a way to assess the drivers of efficacy and toxicities.

A variety of PKPD mathematical models ranging from simple data-driven empirical models to semi-mechanistic and physiology-based models have been developed to characterize and describe ADC disposition and actions. For example, for T-DM1, Jumbe et al. described the mouse drug-conjugated antibody pharmacokinetics using a simple two-compartment model with linear drug clearance (Jumbe et al. 2010). In this study, drug-conjugated antibody was assumed to be the driving force for efficacy, and the simple model could describe reasonably well the disposition profile. A first-order growth model with tumor transit compartments was used for describing the tumor progression, and tumor killing was modeled as nonlinear killing driven by the drug-conjugated antibody (T-DM1). In contrast, Bender et al. developed a mechanism-based multicompartment model that described the pharmacokinetic profile of different heterogeneous species representing T-DM1 ADC (Bender et al. 2012). Preclinical data for individual drug-to-antibody ratio (DAR) species were utilized, and each DAR species was explicitly modeled with a central and peripheral compartment for distribution. Deconjugation of the cytotoxic drug DM1 from the antibody was modeled with variable rate constants (differentiating deconjugation between higher and lower DAR species). Both in vivo and in vitro measurements were utilized to develop the model. Gibiansky et al. adapted the target-mediated drug disposition model to describe ADC PK, and with a number of approximations including rapid binding, quasi-steady-state, and Michaelis-Menten approximations derived a reduced model that can still describe ADC PK with load-independent properties (Gibianskyand and Gibiansky 2014).

Although most of the modeling work has been done with T-DM1, some of the recent efforts have been in modeling other ADCs like that of brentuximab–vedotin which consist of a cleavable linker conjugating the cytotoxic drug with the cysteine residues of the antibody. Shah et al. developed a multiscale mechanism-based model incorporating a variety of physiological processes including ADC and drug payload PK at the cellular level, in circulation and tumor tissue to characterize brentuximab–vedotin disposition and action and to predict the clinical response to the drug in cancer patients including progression-free survival rates and complete response rates (Fig. 7.1) (Shah et al. 2012). The known mechanisms of ADC and payload

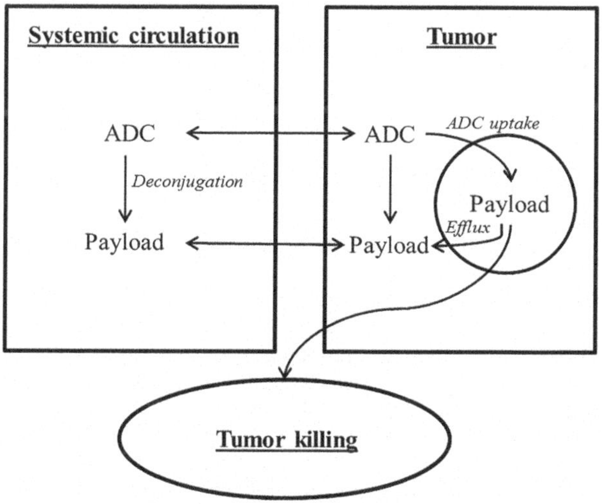

Fig. 7.1 Simplified model representation of disposition and action of brentuximab–vedotin ADC and its payload described by Shah et al. (2012). The integrated model takes into account the disposition of ADC and its payload in systemic circulation, tumor tissue, and inside cells and link intracellular payload concentration to tumor killing

disposition and tumor growth kinetics were used for the model development, and the model utilized in vitro, preclinical, and clinical measurements of ADC and payload and tumor growth inhibition data in xenograft mouse models to develop the model.

7.6 Conclusions

As a hybrid between antibody therapeutics and small-molecule cytotoxic drugs, ADCs exhibit unique pharmacological and PK properties. Among them, the heterogeneity from ADC production and in vivo processing, the necessity to monitor multiple active ADC analytes, and less understood catabolic and metabolic species, all of which demand meaningful PKPD relationships. As with other therapeutics, PKPD can aid in understanding exposure–response relationships, determining the optimal dose and dose regimen, predicting human PK, facilitating the translation of nonclinical data to clinical outcome, and allowing quantitative understanding of mechanistic pharmacology (Morgan et al. 2012). The growth of interest in ADCs, the evolvement of powerful analytical tools, and generation of crucial mechanistic data indicate a promising future for ADC development.

References

Alley SC, Zhang X, Okeley NM, Anderson M, Law CL, Senter PD, Benjamin DR (2009) The pharmacologic basis for antibody-auristatin conjugate activity. J Pharmacol Exp Ther 330:932–938

Ballantyneand A, Dhillon S (2013) Trastuzumab emtansine: first global approval. Drugs 73:755–765

Bander NH (2013) Antibody-drug conjugate target selection: critical factors. Methods Mol Biol 1045:29–40

Bender B, Leipold D, Liu L, Xu K, Shen BQ, Friberg LE, Tibbitts J (2012) A multicompartmental population PK model elucidating the complex disposition of trastuzumab emtansine (T-DM1): an antibody-drug conjugate for the treatment of HER2-positive cancer. In: Population Approach Group in Europe

Boswell CA, Ferl GZ, Mundo EE, Bumbaca D, Schweiger MG, Theil FP, Fielder PJ, Khawli LA (2011a) Effects of anti-VEGF on predicted antibody biodistribution: roles of vascular volume, interstitial volume, and blood flow. PLoS One 6:e17874

Boswell CA, Mundo EE, Zhang C, Bumbaca D, Valle NR, Kozak KR, Fourie A, Chuh J, Koppada N, Saad O, Gill H, Shen BQ, Rubinfeld B, Tibbitts J, Kaur S, Theil FP, Fielder PJ, Khawli LA, Lin K (2011b) Impact of drug conjugation on pharmacokinetics and tissue distribution of anti-STEAP1 antibody-drug conjugates in rats. Bioconjug Chem 22:1994–2004

Boswell CA, Mundo EE, Firestein R, Zhang C, Mao W, Gill H, Young C, Ljumanovic N, Stainton S, Ulufatu S, Fourie A, Kozak KR, Fuji R, Polakis P, Khawli LA, Lin K (2013) An integrated approach to identify normal tissue expression of targets for antibody-drug conjugates: case study of TENB2. Br J Pharmacol 168:445–457

Burris HA 3rd, Tibbitts J, Holden SN, Sliwkowski MX, Lewis Phillips GD (2011) Trastuzumab emtansine (T-DM1): a novel agent for targeting HER2+ breast cancer. Clin Breast Cancer 11:275–282

Cao H, Yamamoto K, Yang LX, Weber R (2013) Brentuximab vedotin: first-line agent for advanced Hodgkin lymphoma. Anticancer Res 33:3879–3885
Chuand YW, Polson A (2013) Antibody-drug conjugates for the treatment of B-cell non-Hodgkin's lymphoma and leukemia. Future Oncol 9:355–368
Davis JA, Rock DA, Wienkers LC, Pearson JT (2012) In vitro characterization of the drug-drug interaction potential of catabolites of antibody-maytansinoid conjugates. Drug Metab Dispos 40:1927–1934
Deng R, Jin F, Prabhu S, Iyer S (2012) Monoclonal antibodies: what are the pharmacokinetic and pharmacodynamic considerations for drug development? Expert Opin Drug Metab Toxicol 8:141–160
Dere R, Yi JH, Lei C, Saad OM, Huang C, Li Y, Baudys J, Kaur S (2013) PK assays for antibody-drug conjugates: case study with ado-trastuzumab emtansine. Bioanalysis 5:1025–1040
Ducryand L, Stump B (2010) Antibody-drug conjugates: linking cytotoxic payloads to monoclonal antibodies. Bioconjug Chem 21:5–13
Erickson HK, Park PU, Widdison WC, Kovtun YV, Garrett LM, Hoffman K, Lutz RJ, Goldmacher VS, Blattler WA (2006) Antibody-maytansinoid conjugates are activated in targeted cancer cells by lysosomal degradation and linker-dependent intracellular processing. Cancer Res 66:4426–4433
Ericksonand HK, Lambert JM (2012) ADME of antibody-maytansinoid conjugates. AAPS J 14:799–805
Flemming A (2014) Antibody engineering: fine-tuning antibody-drug conjugates. Nat Rev Drug Discov 13:178
Gerber DE (2008) Targeted therapies: a new generation of cancer treatments. Am Fam Physician 77:311–319
Gibianskyand L, Gibiansky E (2014) Target-mediated drug disposition model and its approximations for antibody-drug conjugates. J Pharmacokinet Pharmacodyn 41:35–47
Hamblett KJ, Senter PD, Chace DF, Sun MM, Lenox J, Cerveny CG, Kissler KM, Bernhardt SX, Kopcha AK, Zabinski RF, Meyer DL, Francisco JA (2004) Effects of drug loading on the antitumor activity of a monoclonal antibody drug conjugate. Clin Cancer Res 10:7063–7070
Herbertson RA, Tebbutt NC, Lee FT, MacFarlane DJ, Chappell B, Micallef N, Lee ST, Saunder T, Hopkins W, Smyth FE, Wyld DK, Bellen J, Sonnichsen DS, Brechbiel MW, Murone C, Scott AM (2009) Phase I biodistribution and pharmacokinetic study of Lewis Y-targeting immunoconjugate CMD-193 in patients with advanced epithelial cancers. Clin Cancer Res 15:6709–6715
Jeffrey SC, Burke PJ, Lyon RP, Meyer DW, Sussman D, Anderson M, Hunter JH, Leiske CI, Miyamoto JB, Nicholas ND, Okeley NM, Sanderson RJ, Stone IJ, Zeng W, Gregson SJ, Masterson L, Tiberghien AC, Howard PW, Thurston DE, Law CL, Senter PD (2013) A potent anti-CD70 antibody-drug conjugate combining a dimeric pyrrolobenzodiazepine drug with site-specific conjugation technology. Bioconjug Chem 24:1256–1263
Jumbe NL, Xin Y, Leipold DD, Crocker L, Dugger D, Mai E, Sliwkowski MX, Fielder PJ, Tibbitts J (2010) Modeling the efficacy of trastuzumab-DM1, an antibody drug conjugate, in mice. J Pharmacokinet Pharmacodyn 37:221–242
Junutula JR, Raab H, Clark S, Bhakta S, Leipold DD, Weir S, Chen Y, Simpson M, Tsai SP, Dennis MS, Lu Y, Meng YG, Ng C, Yang J, Lee CC, Duenas E, Gorrell J, Katta V, Kim A, McDorman K, Flagella K, Venook R, Ross S, Spencer SD, Lee Wong W, Lowman HB, Vandlen R, Sliwkowski MX, Scheller RH, Polakis P, Mallet W (2008) Site-specific conjugation of a cytotoxic drug to an antibody improves the therapeutic index. Nat Biotechnol 26:925–932
Kantarjian H, Thomas D, Jorgensen J, Jabbour E, Kebriaei P, Rytting M, York S, Ravandi F, Kwari M, Faderl S, Rios MB, Cortes J, Fayad L, Tarnai R, Wang SA, Champlin R, Advani A, O'Brien S (2012) Inotuzumab ozogamicin, an anti-CD22-calecheamicin conjugate, for refractory and relapsed acute lymphocytic leukaemia: a phase 2 study. Lancet Oncol 13:403–411
Kaur S, Xu K, Saad OM, Dere RC, Carrasco-Triguero M (2013) Bioanalytical assay strategies for the development of antibody-drug conjugate biotherapeutics. Bioanalysis 5:201–226

Lewis Phillips GD, Li G, Dugger DL, Crocker LM, Parsons KL, Mai E, Blattler WA, Lambert JM, Chari RV, Lutz RJ, Wong WL, Jacobson FS, Koeppen H, Schwall RH, Kenkare-Mitra SR, Spencer SD, Sliwkowski MX (2008) Targeting HER2-positive breast cancer with trastuzumab-DM1, an antibody-cytotoxic drug conjugate. Cancer Res 68:9280–9290

Linand JH, Lu AY (1997) Role of pharmacokinetics and metabolism in drug discovery and development. Pharmacol Rev 49:403–449

Linand K, Tibbitts J (2012) Pharmacokinetic considerations for antibody drug conjugates. Pharm Res 29:2354–2366

Lobo ED, Hansen RJ, Balthasar JP (2004) Antibody pharmacokinetics and pharmacodynamics. J Pharm Sci 93:2645–2668

Lovdal T, Andersen E, Brech A, Berg T (2000) Fc receptor mediated endocytosis of small soluble immunoglobulin G immune complexes in Kupffer and endothelial cells from rat liver. J Cell Sci 113(Pt 18):3255–3266

Lu D, Sahasranaman S, Zhang Y, Girish S (2013) Strategies to address drug interaction potential for antibody-drug conjugates in clinical development. Bioanalysis 5:1115–1130

Lyon RP, Meyer DL, Setter JR, Senter PD (2012) Conjugation of anticancer drugs through endogenous monoclonal antibody cysteine residues. Methods Enzymol 502:123–138

McDonagh CF, Kim KM, Turcott E, Brown LL, Westendorf L, Feist T, Sussman D, Stone I, Anderson M, Miyamoto J, Lyon R, Alley SC, Gerber HP, Carter PJ (2008) Engineered anti-CD70 antibody-drug conjugate with increased therapeutic index. Mol Cancer Ther 7:2913–2923

Morgan P, Van Der Graaf PH, Arrowsmith J, Feltner DE, Drummond KS, Wegner CD, Street SD (2012) Can the flow of medicines be improved? Fundamental pharmacokinetic and pharmacological principles toward improving Phase II survival. Drug Discov Today 17:419–424

Mouldand DR, Green B (2010) Pharmacokinetics and pharmacodynamics of monoclonal antibodies: concepts and lessons for drug development. BioDrugs 24:23–39

Mullard A (2013) Maturing antibody-drug conjugate pipeline hits 30. Nat Rev Drug Discov 12:329–332

Nolting B (2013) Linker technologies for antibody-drug conjugates. Methods Mol Biol 1045:71–100

Okeley NM, Miyamoto JB, Zhang X, Sanderson RJ, Benjamin DR, Sievers EL, Senter PD, Alley SC (2010) Intracellular activation of SGN-35, a potent anti-CD30 antibody-drug conjugate. Clin Cancer Res 16:888–897

Pastuskovas CV, Maruoka EM, Shen BQ, Koeppen H, Doronina SO, Senter PD, Zioncheck TF (2005) Tissue distribution, metabolism, and excretion of the antibody-drug conjugate Herceptin-monomethyl auristatin E in rats. AACR Meeting Abstracts. 2005:1195-d-1196

Sanderson RJ, Hering MA, James SF, Sun MM, Doronina SO, Siadak AW, Senter PD, Wahl AF (2005) In vivo drug-linker stability of an anti CD30 dipeptide linked auristatin immunoconjugate. Clin Cancer Res 11:843–852

Sapra P, Stein R, Pickett J, Qu Z, Govindan SV, Cardillo TM, Hansen HJ, Horak ID, Griffiths GL, Goldenberg DM (2005) Anti-CD74 antibody-doxorubicin conjugate, IMMU-110, in a human multiple myeloma xenograft and in monkeys. Clin Cancer Res 11:5257–5264

Scott AM, Tebbutt N, Lee FT, Cavicchiolo T, Liu Z, Gill S, Poon AM, Hopkins W, Smyth FE, Murone C, MacGregor D, Papenfuss AT, Chappell B, Saunder TH, Brechbiel MW, Davis ID, Murphy R, Chong G, Hoffman EW, Old LJ (2007) A phase I biodistribution and pharmacokinetic trial of humanized monoclonal antibody Hu3s193 in patients with advanced epithelial cancers that express the Lewis-Y antigen. Clin Cancer Res 13:3286–3292

Shah DK, Haddish-Berhane N, Betts A (2012) Bench to bedside translation of antibody drug conjugates using a multiscale mechanistic PK/PD model: a case study with brentuximab-vedotin. J Pharmacokinet Pharmacodyn 39:643–659

Shahand DK, Betts AM (2013) Antibody biodistribution coefficients: inferring tissue concentrations of monoclonal antibodies based on the plasma concentrations in several preclinical species and human. MAbs 5:297–305

Sharkey RM, Karacay H, Govindan SV, Goldenberg DM (2011) Combination radioimmunotherapy and chemoimmunotherapy involving different or the same targets improves therapy of human pancreatic carcinoma xenograft models. Mol Cancer Ther 10:1072–1081
Sharkey RM, Govindan SV, Cardillo TM, Goldenberg DM (2012) Epratuzumab-SN-38: a new antibody-drug conjugate for the therapy of hematologic malignancies. Mol Cancer Ther 11:224–234
Shen BQ, Xu K, Liu L, Raab H, Bhakta S, Kenrick M, Parsons-Reponte KL, Tien J, Yu SF, Mai E, Li D, Tibbitts J, Baudys J, Saad OM, Scales SJ, McDonald PJ, Hass PE, Eigenbrot C, Nguyen T, Solis WA, Fuji RN, Flagella KM, Patel D, Spencer SD, Khawli LA, Ebens A, Wong WL, Vandlen R, Kaur S, Sliwkowski MX, Scheller RH, Polakis P, Junutula JR (2012a) Conjugation site modulates the in vivo stability and therapeutic activity of antibody-drug conjugates. Nat Biotechnol 30:184–189
Shen BQ, Bumbaca D, Saad O, Yue Q, Pastuskovas CV, Khojasteh SC, Tibbitts J, Kaur S, Wang B, Chu YW, Lorusso PM, Girish S (2012b) Catabolic fate and pharmacokinetic characterization of Trastuzumab Emtansine (T-DM1): an emphasis on preclinical and clinical catabolism. Curr Drug Metab 13:901–910
Sieversand EL, Senter PD (2013) Antibody-drug conjugates in cancer therapy. Annu Rev Med 64:15–29
Silver DA, Pellicer I, Fair WR, Heston WD, Cordon-Cardo C (1997) Prostate-specific membrane antigen expression in normal and malignant human tissues. Clin Cancer Res 3:81–85
Singhand R, Erickson HK (2009) Antibody-cytotoxic agent conjugates: preparation and characterization. Methods Mol Biol 525:445–467, xiv
Stephan JP, Chan P, Lee C, Nelson C, Elliott JM, Bechtel C, Raab H, Xie D, Akutagawa J, Baudys J, Saad O, Prabhu S, Wong WL, Vandlen R, Jacobson F, Ebens A (2008) Anti-CD22-MCC-DM1 and MC-MMAF conjugates: impact of assay format on pharmacokinetic parameters determination. Bioconjug Chem 19:1673–1683
Sun X, Widdison W, Mayo M, Wilhelm S, Leece B, Chari R, Singh R, Erickson H (2011) Design of antibody-maytansinoid conjugates allows for efficient detoxification via liver metabolism. Bioconjug Chem 22:728–735
Sussman D, Torrey L, Westendorf L, Zhang X, Okeley NM, Alley SC, Lyon R, Meyer D, Miyamoto JB, Benjam DR (2011) Engineered cysteine antibodies: improved antibody-drug conjugate vehicles. In: Proceedings of the AACR-NCI-EORTC international conference: molecular targets and cancer therapeutics
Sutherland MS, Sanderson RJ, Gordon KA, Andreyka J, Cerveny CG, Yu C, Lewis TS, Meyer DL, Zabinski RF, Doronina SO, Senter PD, Law CL, Wahl AF (2006) Lysosomal trafficking and cysteine protease metabolism confer target-specific cytotoxicity by peptide-linked anti-CD30-auristatin conjugates. J Biol Chem 281:10540–10547
Tabrizi MA, Tseng CM, Roskos LK (2006) Elimination mechanisms of therapeutic monoclonal antibodies. Drug Discov Today 11:81–88
Tabrizi M, Bornstein GG, Suria H (2010) Biodistribution mechanisms of therapeutic monoclonal antibodies in health and disease. AAPS J 12:33–43
Tian F, Lu Y, Manibusan A, Sellers A, Tran H, Sun Y, Phuong T, Barnett R, Hehli B, Song F, DeGuzman MJ, Ensari S, Pinkstaff JK, Sullivan LM, Biroc SL, Cho H, Schultz PG, DiJoseph J, Dougher M, Ma D, Dushin R, Leal M, Tchistiakova L, Feyfant E, Gerber HP, Sapra P (2014) A general approach to site-specific antibody drug conjugates. Proc Natl Acad Sci USA 111:1766–1771
Tijink BM, Buter J, de Bree R, Giaccone G, Lang MS, Staab A, Leemans CR, van Dongen GA (2006) A phase I dose escalation study with anti-CD44v6 bivatuzumab mertansine in patients with incurable squamous cell carcinoma of the head and neck or esophagus. Clin Cancer Res 12:6064–6072
Tolcher AW, Ochoa L, Hammond LA, Patnaik A, Edwards T, Takimoto C, Smith L, de Bono J, Schwartz G, Mays T, Jonak ZL, Johnson R, DeWitte M, Martino H, Audette C, Maes K, Chari RV, Lambert JM, Rowinsky EK (2003) Cantuzumab mertansine, a maytansinoid immunocon-

jugate directed to the CanAg antigen: a phase I, pharmacokinetic, and biologic correlative study. J Clin Oncol 21:211–222
Waldmann TA (2003) Immunotherapy: past, present and future. Nat Med 9:269–277
Wang L, Amphlett G, Blattler WA, Lambert JM, Zhang W (2005) Structural characterization of the maytansinoid-monoclonal antibody immunoconjugate, huN901-DM1, by mass spectrometry. Protein Sci 14:2436–2446
Wang W, Wang EQ, Balthasar JP (2008) Monoclonal antibody pharmacokinetics and pharmacodynamics. Clin Pharmacol Ther 84:548–558
Xie H, Audette C, Hoffee M, Lambert JM, Blattler WA (2004) Pharmacokinetics and biodistribution of the antitumor immunoconjugate, cantuzumab mertansine (huC242-DM1), and its two components in mice. J Pharmacol Exp Ther 308:1073–1082
Xu K, Liu L, Saad OM, Baudys J, Williams L, Leipold D, Shen B, Raab H, Junutula JR, Kim A, Kaur S (2011) Characterization of intact antibody-drug conjugates from plasma/serum in vivo by affinity capture capillary liquid chromatography-mass spectrometry. Anal Biochem 412:56–66
Xu K, Liu L, Dere R, Mai E, Erickson R, Hendricks A, Lin K, Junutula JR, Kaur S (2013) Characterization of the drug-to-antibody ratio distribution for antibody-drug conjugates in plasma/serum. Bioanalysis 5:1057–1071
Zhao L, Ji P, Li Z, Roy P, Sahajwalla CG (2013) The antibody drug absorption following subcutaneous or intramuscular administration and its mathematical description by coupling physiologically based absorption process with the conventional compartment pharmacokinetic model. J Clin Pharmacol 53:314–325

Chapter 8
Regulatory Considerations

Karen Jieun Cha

Abstract This chapter on regulatory considerations for antibody-drug conjugates (ADCs) in oncology highlights the unique and complex regulatory landscape required for their development, given their dual nature as biologics and small-molecule drugs. It provides an overview of key regulatory processes, including chemistry, manufacturing, and controls (CMC), where standards for antibodies and cytotoxic linkers must meet distinct regulatory expectations. The chapter also covers the organization of applications in eCTD format, meeting types with Food and Drug Administration (FDA), and key considerations for a successful regulatory application. Regulatory pathways, such as fast track and breakthrough therapy, are discussed as options to expedite ADC development, and detailed CMC requirements for various product components are emphasized to ensure patient safety and treatment efficacy through each development phase.

Keywords Regulatory · Development · Antibody CMC · Investigational new drug application

8.1 Introduction

Antibody-drug conjugates (ADCs) are monoclonal antibodies (mAbs) that are designed to selectively deliver cytotoxic agents to antigen-bearing cells. Since the first edition of this book, ADCs have become well established as an important class of treatment in oncology, with 11 Food and Drug Administration (FDA)-approved ADCs for various cancer types and hundreds more in clinical development.

K. J. Cha (✉)
Regulatory Affairs Consultant, Sapience Regulatory Consulting LLC, Los Angeles, CA, USA
e-mail: kcha@sapienceregulatory.com

J. L. Zaro, J. Wang (eds.), *Antibody-Drug Conjugates*, AAPS Advances in the Pharmaceutical Sciences Series 17, https://doi.org/10.1007/978-3-031-83005-1_8

The regulatory expectations of ADC products combine the set of requirements applicable to both biologic and chemical drugs, and given the focus on oncology treatment, the development of ADCs are obviously subject to the regulatory requirements and expectations of developing cancer therapeutics. Given the dual nature of the conjugates which incorporate both a targeting antibody and a highly potent chemical cytotoxic agent and linker, the developing sponsor is obligated to consider an extremely wide set of requirements and expectations. This substantially expands the effort required in ADC development, and these challenges are reviewed below. In addition, those areas that are seen as unique to ADC products are discussed where possible.

8.2 Regulatory Review

While not specific to ADC products, it is worth mentioning that in the United States, a reorganization was announced in 2011 in the Food and Drug Administration (FDA)s office responsible for reviewing all drug and biologic applications for cancer therapies. Antibody-drug conjugate products for cancer are reviewed by the Office of Oncologic Disease, which comprises five separate review divisions based on indication. The FDA website provides the information necessary to determine which Division will review a given ADC application based on target indication. Given this format, the FDA has often requested sponsors to submit separate investigational new drugs (INDs) to the relevant Divisions for the same product being evaluated in different cancer indications presenting some challenges with maintaining multiple INDs for a given compound, including the cross-filing of safety information.

8.2.1 Organization of the Application

As of May 5, 2017, the FDA began requiring the electronic Common Technical Document (eCTD) format for Biologics License Applications (BLAs) and 1 year later on May 5, 2018 for commercial Investigational New Drug (INDs) Applications. The agreement among the regions participating in the International Conference on Harmonization (ICH) to assemble all the quality, safety, and efficacy information in a common format has significantly improved the regulatory review processes and has led to harmonized electronic submissions in these regions. For sponsors, it has eliminated or minimized the need to reformat the information for submission to the different regulatory authorities in the ICH regions and others countries following the harmonized guidance.

The CTD is organized into five modules in which Module 1 is region-specific and Modules 2–5 are intended to be common for all regions (ICH M4 2004).

Guidance for the content of each module of the CTD can be found on the official ICH website.

The granularity of the CTD format is well suited for ADCs, given the multiple key intermediates in the manufacturing process and life-cycle management of an application that grows and changes over time as a product advances in development.

Antibody-drug conjugates comprise of two key intermediates—the monoclonal antibody (mAb) and the drug-linker (DL)—which are conjugated to generate the bulk drug substance (BDS). Therefore, regulatory applications for ADC products should have three separate drug substance quality [or chemistry, manufacturing and controls (CMC)] sections (Module 3.2.S in the CTD format), one each for the mAb, the DL, and the conjugated BDS, as well as the required drug product (DP) section (Module 3.2.P in the CTD format). The separate sections for each intermediate and the drug substance allow both the sponsor and reviewer to update/review information for specific intermediates and manage more efficiently the life cycle of the application. An example of this Module 3 structure used in the United States is shown in Fig. 8.1. If a drug master file (DMF) exists for the DL as is sometimes the case, the DMF can be cross-referenced within the application with a Letter of Authorization from the DMF holder and the drug substance section for the DL omitted or partially omitted.

In some countries outside the United States, Module 2.3 Quality Overall Summary (QOS) is required during development to provide the overview of the quality information and Module 3 is used only to place supplemental data or those sections containing a significant amount of information. Similarly, the information should be provided in three separate drug substance documents—one QOS each for the mAb, DL, and BDS.

From a review perspective, this structure also supports the fact that regulatory review of CMC information involves multidisciplinary divisions within the US FDA. The mAb will be reviewed by the Division of Monoclonal Antibodies (DMA) in the Office of Biotechnology Products (OBP), the DL will be reviewed by the Office of New Drug Quality Assessment (ONDQA), and the BDS and DP information by both groups. Each of these intermediates and the drug substance and their corresponding sections have particular features—different manufacturing sites, different manufacturing processes, different analytical methods and method validations, different stability protocols, etc., further supporting the separate drug substance sections.

The OBP perspective includes characterization information for the mAb, BDS, and drug product (DP); comparability of the mAb, BDS, and DP; impurities; and testing/specifications. The ONDQA perspective considers starting materials and intermediates of the DL, characterization of the DL, BDS, and DP, and testing/specifications (Miksinski and Shapiro 2012). This high-level list alone affirms that the tasks at hand for fulfilling the expectations of regulatory review for CMC information is much greater for an ADC product because the data requirements for the mAb and DL intermediates are the same as those for drug substance and drug product. Figure 8.2 depicts the various components/intermediates in ADC product

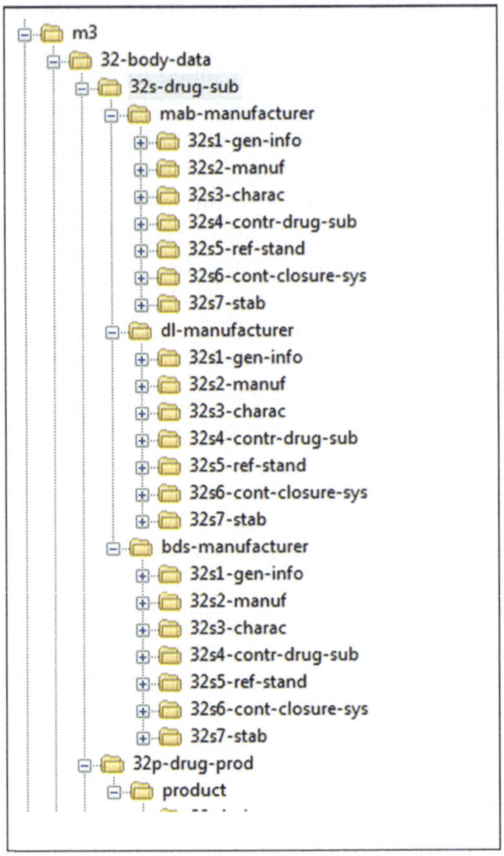

mab – monoclonal antibody; dl = drug-linker; bds = bulk drug substance

Fig. 8.1 Example of Module 3 quality eCTD backbone (Partial) for an ADC product application. Note: mab monoclonal antibody, dl drug-linker, bds bulk drug substance

manufacture and the collaborative review responsibility by both biologic and chemistry disciplines common within regulatory agencies.

8.2.2 Meetings with FDA

It is highly recommended that sponsors entering into any new therapeutic class meet with the FDA review division in a pre-IND (Type B) meeting and ADCs are no exception. The FDA continues to support its position of promoting communication with sponsors under the reauthorization of the Prescription Drug User Fee Act (PDUFA VII: Fiscal Years 2023 through 2027), which includes meeting management goals as well as a general goal for "Promoting Innovation Through Enhanced

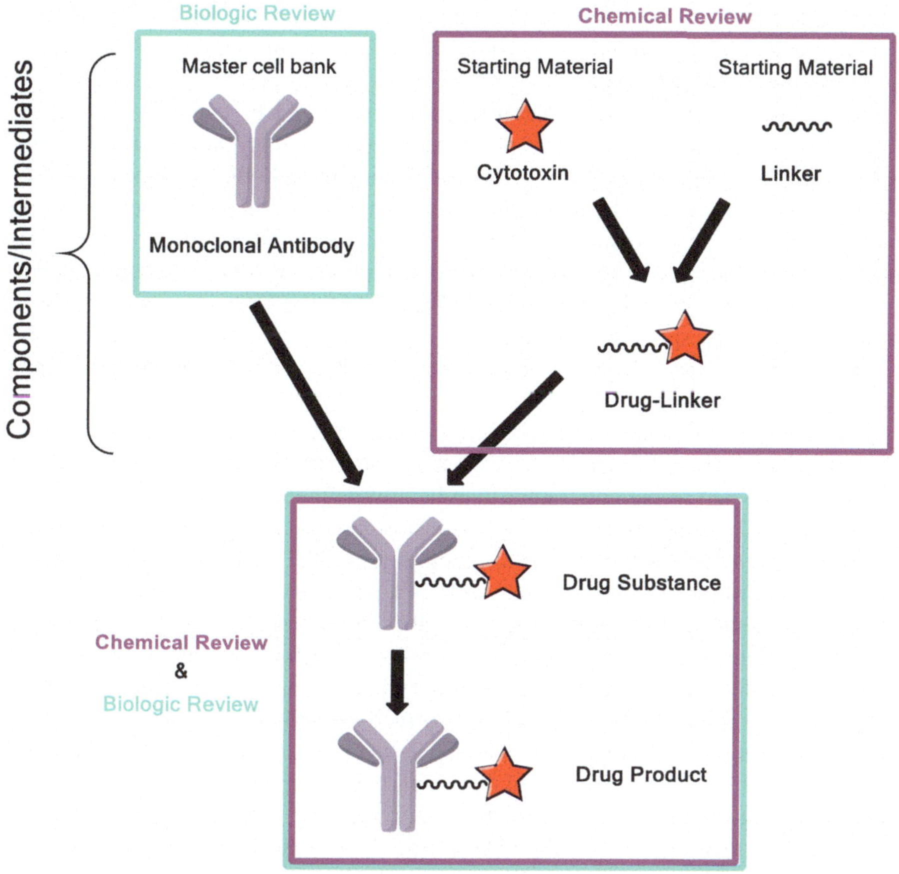

Fig. 8.2 ADC review responsibility

Communication Between FDA and Sponsors During Drug Development" among many other enhancements (PDUFA Reauthorization Performance Goals and Procedures Fiscal Years 2023 Through 2027, www.fda.gov).

The timing of the pre-IND meeting is often a question for new sponsors and the standard regulatory advice relates to the questions to be discussed with the Agency. Some sponsors prefer to meet (or obtain written feedback as is now often the case for pre-IND meetings) with the FDA prior to performing the Good Laboratory Practice (GLP)-compliant toxicology studies to be sure the plans are acceptable. In any case, the questions should be clear, concise, and the relevant background for the Agency to make an assessment and provide meaningful feedback on the question must be available in time for inclusion in the Meeting Briefing Package due at FDA at least 4 weeks prior to the meeting date for a Type B pre-IND meeting (FDA Guidance for Industry, Formal Meetings Between the FDA and Sponsors or Applicants of PDUFA Products, Draft September 2023). Therefore, working backwards from the time the relevant data or information needed for the Briefing Package is available will aid in determining the appropriate timing of the request and feedback or meeting date.

According to US guidance, meeting questions should be categorized into the three main disciplines: nonclinical (pharmacology, toxicology, and pharmacokinetics [PK]), CMC, and clinical. Typical CMC topics include the appropriateness of specifications, test methods, stability protocol, etc. A useful nonclinical question may involve sharing a tabular listing of all studies planned for inclusion in the original IND for Agency feedback. In the clinical discipline, it is highly recommended to include a protocol synopsis in the Briefing Package for Agency review and feedback in addition to any specific clinical questions. The meeting types, content of the request, and briefing package expectations are clearly outlined in the FDA guidance document cited above.

Later in development, it is strongly recommended by the FDA to plan for an End-of-Phase 1 (EOP1) meeting in oncology indications, given that the Phase 1 study is conducted in patients rather than healthy volunteers (typically an End-of-Phase 2 [EOP2]/pre-Phase 3 meeting in most other indications). The EOP1/EOP2 meeting is recommended prior to initiating any pivotal trial(s) that will form the primary basis of an efficacy claim as well as a pre-submission meeting to discuss the planned content of the marketing application with the FDA review division not less than 2 months prior to the planned submission. Such issues can be discussed at both EOP2 (planned) and pre-BLA meeting (when the package is already defined) supporting the concept of keeping FDA informed throughout development.

New meeting types have also been added in recent years that allow for additional interactions with FDA to obtain guidance in early development including the Initial Targeted Engagement for Regulatory Advice on CBER/CDER Products (INTERACT) meeting and the Type D meeting. INTERACT meetings are intended for novel questions and unique challenges in early development typically relating to IND requirements and prior to the typical timing of a pre-IND meeting. The second new type of meeting is the Type D meeting intended for a narrow set of issues, often one or two issues and associated questions requiring no more than three disciplines from FDA. These new meeting types provide sponsors with more opportunities to meet with and obtain guidance from the FDA.

8.2.3 Expedited Pathways

The US FDA Guidance on Expedited Programs for Serious Conditions—Drugs and Biologics (2014) outlines several programs aimed at speeding up the development and review of new drugs and biologics addressing serious or life-threatening conditions with unmet medical needs such as cancer. To summarize the main expedited pathways:

1. Fast Track Designation:
    - This program facilitates drug development and expedites review by enabling frequent FDA-sponsor interactions and a rolling review process, where the sponsor can submit parts of the application before the full application is ready.
    - To qualify, a drug must treat a serious condition and address an unmet medical need. Early stage data, including preclinical or clinical, are required to show the drug's potential.
2. Breakthrough Therapy Designation:
    - For drugs that may represent substantial improvement over available therapies, the breakthrough designation involves intensive FDA guidance to streamline clinical trials, potentially reducing trial size or duration.
    - The designation requires preliminary clinical evidence demonstrating substantial improvement on a significant endpoint, such as survival or symptom relief, compared to current therapies.
3. Accelerated Approval:
    - This pathway allows approval based on surrogate or intermediate clinical endpoints that are "reasonably likely to predict" clinical benefit, rather than direct evidence of clinical improvement.
    - Accelerated approval requires post-marketing studies (already underway at the time of accelerated approval) to confirm clinical benefits, and drugs may have conditional approval withdrawn if they fail to meet these confirmatory standards.
4. Priority Review Designation:
    - Drugs that demonstrate significant improvement in safety or effectiveness can qualify for priority review, shortening the FDA's review target from the standard 10 to 6 months.

The guidance emphasizes the importance of early and ongoing communication with the FDA to align the clinical and manufacturing development processes with regulatory expectations. Each program includes specific criteria and requires adherence to stringent data standards to meet safety and efficacy requirements. These expedited pathways offer different types of support and are often used together to optimize drug development for severe or life-threatening diseases.

Many countries have implemented expedited regulatory pathways similar to the FDA's programs to accelerate the approval of drugs and biologics, especially those addressing serious or life-threatening conditions. Some examples from other key regulatory agencies include:

1. European Union (European Medicines Agency—EMA):
    - Priority Medicines (PRIME): Launched in 2016, PRIME supports the development of drugs addressing unmet medical needs by offering early and enhanced interaction with the EMA. PRIME focuses on drugs that demon-

strate the potential for a significant therapeutic advantage (EMA Guidance for applicants seeking access to PRIME scheme, October 3 2024).
- Accelerated Assessment: For drugs with major public health benefits, the review timeline is shortened from 210 to 150 days.
- Conditional Marketing Authorization: Similar to the FDA's accelerated approval, this pathway allows drugs to be approved based on less comprehensive clinical data, with requirements for post-authorization studies to confirm efficacy and safety.

2. Japan (Pharmaceuticals and Medical Devices Agency—PMDA):

   - Sakigake Designation: This designation offers early consultation, priority review, and post-approval safety measures for innovative drugs targeting severe or life-threatening conditions.
   - Conditional Early Approval System: This allows the approval of drugs based on limited clinical data when traditional trials are challenging, with the requirement for additional studies post-approval.

3. Canada (Health Canada):

   - Priority Review: This provides a 180-day review timeline (reduced from 300 days) for drugs that show substantial clinical benefits over existing therapies.
   - Notice of Compliance with Conditions (NOC/c): Similar to the FDA's accelerated approval, NOC/c grants early market access for drugs treating serious conditions based on promising evidence, with the requirement of confirmatory trials after approval.

4. Australia (Therapeutic Goods Administration—TGA):

   - Priority Review: This offers faster processing for drugs addressing unmet medical needs, reducing the review time frame from 255 to 150 days.
   - Provisional Approval Pathway: Allows temporary approval based on preliminary data for drugs with a favorable risk-benefit profile, with the expectation of ongoing data collection.

These programs share common goals with the FDA's expedited pathways, focusing on early access to promising therapies for patients with limited options, while balancing this with safety through post-market requirements.

8.2.3.1 ADCs with Accelerated Approval

In recent years, an impressive number of ADCs have received accelerated approval from the FDA, given their potential to treat cancers with limited options. The ADCs and the basis of their accelerated approvals are summarized below:

1. Padcev (Enfortumab Vedotin-ejfv): Approved in December 2019 for treating patients with locally advanced or metastatic urothelial carcinoma who had previ-

ously been treated with both a PD-1/PD-L1 inhibitor and platinum-containing chemotherapy. The approval was based on data from trials showing a promising objective response rate, which led the FDA to expedite its availability, given the urgent need for treatments in this population.

2. Loncastuximab Tesirine-lpyl (Zynlonta): Approved in 2021, this CD19-targeting ADC is used for adults with relapsed or refractory large B-cell lymphoma after at least two prior systemic treatments. The approval was based on overall response rates in a single-arm trial, a common criterion for accelerated approval in oncology.
3. Mirvetuximab Soravtansine (Elahere): This ADC targets folate receptor alpha and received accelerated approval in 2022 for treating platinum-resistant ovarian cancer. It showed significant improvements in tumor response and progression-free survival compared to standard chemotherapy, and further trials are underway to confirm its clinical benefits.
4. Polatuzumab Vedotin (Polivy): This ADC was approved in 2019 for use in combination with other chemotherapy agents to treat diffuse large B-cell lymphoma (DLBCL) in patients who have not responded to other treatments. It was approved based on response rates and duration of response, meeting the FDA's accelerated approval criteria.
5. Fam-trastuzumab deruxtecan-nxki (Enhertu): Initially approved for HER2-positive breast cancer, Enhertu has received additional accelerated approvals for HER2-positive gastric and other solid tumors, showcasing its flexibility in targeting different cancers that express HER2.

These approvals reflect the FDA's use of surrogate endpoints (like response rates) under the accelerated pathway, aiming to quickly bring promising therapies to patients with critical needs while continuing confirmatory studies post-approval.

8.3 Chemistry, Manufacturing, and Controls

Because ADCs have both small and large molecule components, the chemistry, manufacturing, and controls (CMC) requirements are multifaceted since they are defined by a set of regulatory requirements covering both areas with no specific CMC guidance for ADC products at this time. Sponsors of various product types understand that CMC issues or roadblocks can very abruptly stop or delay drug development or approval in its tracks. By the same token, getting the CMC foundation right early in development will result in an overall faster development time to the market.

From a regulatory perspective, given that the mAb and DL are considered to be intermediates in the manufacture of the conjugated BDS, this calls for a high level of CMC information for the intermediates, although the CMC requirements for each are quite separate and distinct. Furthermore, when considering the key factors in the control of the drug substance and drug product, it is important to keep in mind that the antibody is directed to a cell surface target and the linker systems are designed to be stable in the bloodstream to release the potent cell-killing agent once inside targeted cancer cells.

Based on the biologic and chemical intermediates described above, ADC manufacture often involves at least three separate and unique facilities, adding complexity in addressing regulatory compliance needs. This will include facilities and processes for the manufacture of the mAb, DL, the conjugated BDS, and the site and process for producing drug product. Furthermore, the facilities for manufacturing DL, conjugation, and drug product manufacture must be able to properly handle highly potent compounds requiring experienced operators, sophisticated systems, and equipment with robust procedures, training, and risk reduction strategies. For those companies not able to handle these highly toxic substances in-house, there are limited options for outsourcing the conjugation of ADC products although the number of contract development and manufacturing sites for ADCs has clearly increased since the first edition.

The following discussion of CMC issues focuses on early development of a new ADC product (pre-IND through Phase 2) during which time the main objective of the regulatory review of CMC information is to assure the safety of patients. During Phase 2 and Phase 3 clinical trials, CMC review continues to evaluate safety but also the bridge of the clinical test material to the intended product to be marketed. Manufacturing development of the product gradually advances to consistent quality through process control and product testing. With the number of regulatory approvals and more advanced understanding of ADC products, the scientific understanding of the various processes can be better balanced with risk-based approaches. Product quality can be demonstrated at logical points where control is required and redundant testing, specifications, or product characterization can be avoided (Bechtold-Peters et al. 2023).

8.3.1 Monoclonal Antibody

Biologics have three major differences from chemical drugs which are reflected by both FDA and EMA in various regulatory guidance documents specific for biologic products: (1) the use of living source materials to produce the biologic, (2) increased complexity of the manufacturing processes, and (3) increased complexity of the products themselves (Geigert 2013). Fortunately, for industry, the available guidance and regulations for biologics are specific and extensive, and the open access to guidance on the websites of the FDA, EMA, as well as ICH provides a great deal of information. Several key areas of focus in the regulatory review of the biologic portion of the ADC application are discussed below.

Given the use of living source materials to produce the mAb intermediate, it is important to demonstrate the biologic product is as free as possible of impurities, contaminants, and adventitious agents. All materials of human or animal origin utilized in the cell line generation and manufacturing process or coming into contact with the mAb intermediate or subsequently conjugated drug substance or drug product must be identified with information evaluating the risk of potential contamination (ICH Q5A(R1) 1999). For example, studies for viral clearance and removal of host cell protein impurities, and the DNA of animal-derived materials should be supported by study reports and documents of certification and testing. Additional detailed guidance is provided in the FDA's Points to Consider in the Manufacture and Testing of Monoclonal Antibody Products for Human Use (FDA 1997). In particular, scientifically sound viral clearance studies to provide a quantitative estimate of the level of virus reduction provided by the removal/inactivation procedures are of great importance. Also, demonstration of genetic stability is a key requirement in both the United States and European Union (ICH Q5B 1996; ICH Q5D 1998).

The mAb component of the ADC must be a well-characterized product. General guidance for biotechnology protein product specifications can be found in ICH Q6B "Specifications: Test Procedures and Acceptance Criteria for Biotechnological/Biological Products" (1999). In addition, the sponsor should consider general aspects of monoclonal antibody characterization with respect to glycosylation profile and other post-translational modifications such as deamidation and lysine variability.

The ADC products currently in clinical development include mAbs of the IgG1, IgG2 and IgG4 isotypes; however, the majority of current ADC products include an IgG1 antibody (Trail 2013). The isotype of the mAb will influence whether the ADC has the potential to kill cells via immune-mediated effector functions such as antibody-dependent cellular cytotoxicity (ADCC) and/or complement-dependent cytotoxicity (CDC). IgG1 mAbs have the potential to mediate CDC, via binding the complement component C1q, and ADCC, via binding to Fcγ receptors expressed on various effector cells, whereas IgG2 and IgG4 mAbs are inefficient in these effector functions (Jiang et al. 2011). The mAb isotype may also influence conjugation. Whether or not the ADCC and/or CDC activity of a given IgG1 mAb is retained following conjugation of the DL is a regulatory consideration in that the sponsor should characterize the Fc-mediated effector function(s) as part of early studies and whether any such activity contributes to the mechanism of action of the ADC product. If effector function(s) contributes to any degree to the mechanism of action (albeit secondary), the sponsor must, as development proceeds, establish a control strategy through control of glycan profile and/or using cell-based potency assay(s) as development proceeds. If effector function activities are found to be pertinent to the mechanism of action, these should be specifically controlled as part of lot release testing.

8.3.2 Drug-Linker

The cytotoxic drug and linker (or DL) is a key intermediate of the ADC as it imparts the cytotoxic effect or payload of the product. Furthermore, the linker plays a central part of the ADC construct as it ideally imparts stability in circulation, favorable pharmacokinetic properties, and efficient release of the toxic agent at the tumor site (Ducry and Stump 2010) (see also Chap. 4). At the time of the first edition, many products in clinical trials were employing a handful of drug-linkers such as the auristatins (Seattle Genetics' monomethyl auristatin E (MMAE) and MMAF)) and maytansinoids (Immunogen's DM1 and DM4) (Trail 2013), which meant that cross-reference to a Drug Master File (DMF) filed with FDA (or Active Substance Master File; ASMF) in Europe may partly or completely cover the information required in the IND or Clinical Trial Authorization (CTA). A DMF or ASMF is a submission made directly to the regulatory authority at the discretion of the master file holder to provide confidential information about a facility, processes, or materials used in the manufacturing of one or more human drugs. The master file is not approved or disapproved but may be cross-referenced by a sponsor in a product application and is reviewed only in connection with the review of the regulatory application.

More recently, for numerous novel drug-linkers not already described in a master file, the Sponsor must fully describe the DL intermediate similar to the level of a drug substance in the regulatory application. Because of the importance placed on the linker for ADC product stability in circulation, favorable pharmacokinetic properties, and efficient release of cytotoxic agent at the tumor sites, some sponsors are evaluating small differences in linkers which may have regulatory impact calling for new studies and additional testing. Again, the phase-specific nature of CMC information required during development is emphasized; however, the minimum CMC information for a novel DL means an added layer to the already complex regulatory dossier for an ADC product.

With regard to testing and specifications, the DL is typically a less heterogeneous product in comparison to the mAb, and specific guidance for specifications can be found in ICH Q6A "Specifications: Test Procedures and Acceptance Criteria for New Drug Substances and New Drug Products: Chemical Substances." Product quality attributes for a new chemical entity focuses on impurities such as heavy metals, solvents, catalysts, and/or genotoxic compounds, with chemistry and safety aspects and reporting criteria listed in ICH Q3A Impurities in New Drug Substances (ICH Q3A(R2) 2006). The following considerations are noted for DL testing:

- Structural characterization.
- Impurity profile.
  - Drug-linker-related impurities.
  - Process-related impurities.
  - Stereochemistry-diastereomers.
- Phase 1 IND/CTA must consider the following:
  - Free drug-related impurities in clinical lot should be toxicologically qualified (no new impurities not found in the material used for toxicology testing).
  - Comparable drug-to-antibody ratios (DARs) should be maintained between the toxicology lot and the clinical lot.

- For Phase 3 trial(s): Characterization of the impurity profile of drug/linker intermediates, including structure determination of individual impurities at levels >0.1% is recommended prior to pivotal clinical trials.

Typically, the impurity profile (quality and quantity) for subsequent lots of DL should be equal to or better than the Phase 1 material. The goal is to have no new impurities and demonstration of process clearance of any impurities introduced into the manufacture of the ADC.

8.3.3 Conjugated Bulk Drug Substance and ADC Drug Product

Finally, the conjugated BDS and the drug product must be described, characterized, and test results presented in the regulatory application. Antibody-drug conjugate products pose challenges from an analytical perspective during early development based on the need to develop multiple assays to support process and formulation development, release, and stability testing as well as having orthogonal methods for product characterization. See also Chap. 6.

Due to multiple conjugation site options, lysines, and carbohydrates (via disulfide reduction, via cysteines and site-specific conjugation), control and characterization can be challenging for the conjugated BDS. At minimum, the following areas should be given special attention in BDS testing and characterization for preparing the necessary regulatory documentation.

- Impact of conjugation on antibody binding (affinity), stability of the antibody backbone and aggregation.
- Control and characterization of drug load (or drug-to-antibody ratio), including residual unconjugated antibody.
- Non-proteinaceous impurities.
- Residual conjugation solvent.
- Cytotoxicity assays in BDS release and stability (including understanding of overall potency of BDS and DP as demonstrated by binding, cytotoxicity assay, drug-to-antibody ratio, etc.)

With regard to impurities, the conjugated BDS requires special attention to evaluate the multiple areas with potential to introduce harmful impurities including:

- The antibody plus associated process and product-related impurities.
- The small molecule plus associated process and product-related impurities.
- Conjugation reaction-associated impurities.

One often challenging area for sponsors developing a new product type can be the setting of early acceptance criteria for release and stability studies. For ADC products, in addition to the control parameters noted above, the sponsor must consider certain acceptance criteria for the mAb and BDS together since the subsequent conjugation step may require tighter specifications for the mAb intermediate (i.e., levels of aggregates and fragments may increase upon conjugation). The effect of the conjugation process must be understood to effectively control specific parameters that may be impacted by conjugation. In addition, certain test parameters of the conjugated drug substance are interdependent and must be considered in concert when setting acceptance criteria for both release and stability studies.

Other CMC topics key to ADCs involve ensuring that the product remains safe and effective throughout the production and delivery process. Chapter 14—summarizes many of these aspects including antibody considerations, chemical conjugation strategies, stability of linker technology, the importance of consistency in drug-to-antibody ratio (DAR) and drug load distribution.

Because first-generation ADCs use naturally occurring amino acids (cysteines, lysines) in the antibody as anchor points to attach a cytotoxic agent at nonsaturating stoichiometries, this results in a heterogeneous mixture of conjugated drug substance including some percentage of unconjugated mAb (no cytotoxic drugs attached). Determination of the average number of drugs that are conjugated to an antibody molecule (average drug-to-antibody molar ratio or DAR) is one of the most important quality attributes of an ADC. The DAR determines the amount of cytotoxic drug or "payload" that can be delivered to the tumor cell and can directly affect both safety and efficacy. A variety of methods have been used to measure this key quality attribute, depending on the properties of the drug and how it is linked to the protein (i.e., site of conjugation and structure of the linker) (Wakankar et al. 2011). Different methods utilized to determine DAR may yield slightly different results, requiring careful consideration and additional studies to determine which assay will provide results closest to the true value.

The level of unconjugated antibody in the final ADC product is another key ADC control parameter noted in regulatory review. Depending on the isotype of the mAb intermediate and various parameters influencing conjugation, the amount of unconjugated mAb in the BDS may be at levels that, according to the regulatory authorities, do not demonstrate control of the process. The regulatory expectation is for control of this "impurity" as the unconjugated mAb, although intended as a sophisticated delivery system for the antitumor cytotoxic agent, may have a pharmacologic affect. For instance, unconjugated antibodies can compete with the ADC for tumor-specific binding sites, which may affect efficacy or pharmacokinetics. From a regulatory perspective, it is not the intended product described in the regulatory dossier (Investigator Brochure during development) and ultimately the product label upon approval.

Moreover, the presence of high DAR species (conjugated antibody with 8, 10, 12, etc. drugs attached) is as much a concern from a safety and product consistency/control point of view. To this end, many firms are pursuing ways to more specifically target and control conjugation at predetermined sites to achieve a specific DAR value with next-generation ADC technologies to produce a more homogeneous ADC drug product.

Finally, it has been presented and discussed informally at industry conferences that the DAR, the percentage of unconjugated antibody, and cytotoxicity results have been shown to be interrelated, calling for studies to understand the relationship between these attributes for a given ADC product. Moreover, the assays employed will have inherent variability, calling for careful consideration to determine even early specifications to assure that these interrelated parameters do not adversely impact one another.

8.3.4 Comparability of ADCs

Contributing to the CMC challenges of ADC development, the multifaceted nature of an ADC further influences and adds complexity to the approach needed to demonstrate comparability, especially during the development stages when changes are common and likely at various intermediates and points involved in manufacturing the ADC product. Many types of changes may be made for improved manufacturability such as change in sites, manufacturing scale, antibody cell line change, and process changes, and the sponsor must assess comparability using a phase-appropriate and justified testing plan to determine the degree of similarity or comparability as a result of such changes. Careful consideration must be given to the comparability testing strategy and analytical methods depending on the specifics of the change and its potential impact on subsequent materials and ultimately the quality, safety, and efficacy of the final drug product. A major point is that the ability to detect changes is directly related to the quality and capability of the analytical methods used, and as a result, ADC products may in fact require the most advanced analytics possible.

The goal of the comparability exercise is to collect the relevant technical information to serve as evidence that the change will not have an adverse impact on the quality, safety, and efficacy of the ADC drug product. Harmonized guidance in this area from ICH (Q5E 2004) greatly aids the sponsor in having a common scientific approach accepted in the ICH regions.

For instance, changes to the mAb intermediate may propagate downstream to the conjugated BDS and final drug product requiring studies be conducted on each material to demonstrate comparability and the lack of an adverse impact on the quality, safety, and efficacy of the ADC. A step-by-step data-driven exercise comparing the pre- and post-change products must be carried out consisting of analytical studies including routine batch analysis and characterization and stability studies. In addition, given the biological component of the ADC product, biological characterization is important to include in the testing. Depending on the differences observed in these set of data that may adversely impact safety and efficacy, the sponsor may need to conduct additional nonclinical (animal pharmacokinetic/pharmacodynamics [PD] studies) and clinical bridging studies.

Sponsors should carefully plan the CMC development path to minimize late-stage changes, which may significantly extend timelines.

8.3.5 *In-Use Compatibility and Stability*

In-use compatibility and stability studies assess the potential impact on product quality by mimicking drug handling throughout the period from drug preparation to administration to the patient including hold times. For many years, the FDA has been carefully scrutinizing the in-use compatibility and stability study conducted for initial IND and in many cases requesting sponsors to submit the Pharmacy Manual for Agency review—a document typically not submitted to the IND, due to an observed disconnect between the technical team that conducts such studies and the clinical team interacting with study pharmacists and staff who administer the drug to patients. The intended drug product usage may involve the dilution of drug formulation into diluents such as saline or dextrose for intravenous (IV) infusion and exposure to new interfaces of administration components like IV bags, syringes, and tubing. Considering the wide range of in-use conditions and administration components available, only limited guidance is available from regulators on expected compatibility and in-use stability data (Blümel et al. 2023; Kamen et al. 2022).

Furthermore, because the payload of ADCs are often hazardous materials, the compatibility study should also consider testing of closed system transfer devices (CSTDs) during dose preparation and administration in the clinical setting as required by USP<800> and the National Institute for Occupational Safety and Health (USP 2019; NIOSH 2020). CSTDs are designed to mechanically prevent both the ingress of contaminants into the system and the escape of drug solution or vapor from the system, for handling hazardous drugs, including ADCs throughout the process from compounding through to patient administration (Besheer et al. 2021).

This study should be carefully designed and reported in the compatibility section of the drug product folder of the initial IND for Agency review. In particular, the conclusions from the study should include clear and easy-to-follow instructions for drug preparation and administration at the clinical sites. Alternatively, the sponsor may submit the Pharmacy Manual to provide the dose preparation and administration instructions.

8.4 Nonclinical Study Requirements

In comparison to the requirements of other indications, the entry of new anticancer agents into Phase 1 clinical trials conducted in patients whose disease is resistant and refractory to available therapy is typically based on a more modest preclinical package as supported by ICH S9 Nonclinical Evaluation for Anticancer Pharmaceuticals, October 29, 2009. In addition, a Questions and Answers document for ICH S9 (June 2018) provides additional useful guidance specific to or applicable to ADCs for various points, including the utility of tissue cross-reactivity studies for biopharmaceuticals, the need for safety studies of unconjugated material such as the antibody, small molecule payload, or linker alone, toxicokinetic analysis, and plasma stability data for a first-in-human study (FDA Guidance for Industry, S9 Questions and Answers, 2018). Specific chapters of this book are dedicated to the development issues related to pharmacokinetic/pharmacodynamic and toxicology studies for ADCs, but this section more generally outlines the regulatory requirements in early development.

In the United States, the original IND for any new drug product should include data from nonclinical studies to demonstrate or determine the below listed items to support the clinical study proposed in the application.

- Proof-of-principle and pharmacological activity of the investigational drug.
- Mechanism of action.
- Dose–response relationship.
- Toxicological effects predictive of the human response.

Rationale for the following:

- Safe starting dose for clinical trials.
- Clinical route of administration.
- Administration schedule.
- Duration of exposure.
- Duration of the follow-up time to detect adverse reactions qualifying as "dose-limiting toxicities".
- Duration of the follow-up time to detect all drug-related adverse reactions.
- The identity of target organs for toxicity.

Some exceptions to the standard nonclinical study requirements do exist for ADC products. For instance, the Harmonized Tripartite Guideline, ICH S6 (R1), recommends that safety of biotherapeutics should be evaluated in two relevant species (2011) and a relevant species is defined in the guidance as one in which the drug candidate is pharmacologically active due to the expression of the receptor or epitope. However, when a second relevant species cannot be identified, the requirement for two species may be altered. The use of one species is also supported in the ICH S9 guideline for Nonclinical Evaluation of Anticancer Pharmaceuticals (2009).

Other aspects of nonclinical studies specific to ADC products are listed below:

- GLP toxicology studies in two biologically relevant species, of which one is usually the cynomolgus monkey (unless a second species cannot be identified); toxicity of the drug alone.
- Measurement of concentrations of all ADC components in circulation from nonclinical and clinical studies to characterize the pharmacokinetics of these molecules in totality (conjugated antibody, total antibody, total drug, free drug).
- Immunogenicity assessments should include characterization to determine if antidrug antibodies are directed toward the antibody or directed toward the drug-linker.

- An additional short-term study or arm in a short-term study should be conducted in at least one species with the unconjugated toxin (ICH S6 Addendum 2011).
- Evaluation of effector function (of both the mAb intermediate and conjugated DS) and other biological activity and whether or not these may contribute to the overall mechanism of action of the ADC drug product.

Next-generation ADCs with a novel DL will require objective testing of the DL and/or its main metabolite, obviating the need to identify the main metabolite(s) early in nonclinical studies. Because ADCs use highly potent cytotoxic drugs that can lead to severe off-target toxicities, the FDA typically demands thorough toxicological evaluations, including studies on how the ADC's cytotoxic drug behaves in the body and how effectively it targets only cancer cells. However, in accordance with ICH S9, pilot studies and the nature of the payload will determine the need for additional studies of the cytotoxic drug or drug with linker.

Finally, one question raised with regulatory reviewers by ADC development community in open discussion has been the amount of information that can be borrowed from prior submissions with respect to use of the same drug-linker since many ADCs employ payloads also used in other ADCs. Since regulatory agencies consider molecule-specific data and information with regard to efficacy, safety, and overall benefit to patients, the situations in which previously submitted information is allowed may be supplemental only and may be considered on a case-by-case basis.

8.5 Clinical Development

Accelerated titration or 3 + 3 dose escalation models are commonly used in early clinical trials for ADCs. These models allow for the cautious increase of doses in subsequent patient cohorts, while monitoring closely for dose-limiting toxicities (DLTs).

Regulatory bodies emphasize the importance of incorporating adaptive designs that allow dose adjustments based on early trial results, particularly when unexpected toxicities arise due to the ADC's payload.

8.5.1 Starting Dose Selection

The selection of the clinical starting dose for an ADC is guided by several regulatory standards, with an emphasis on ensuring patient safety while maximizing therapeutic potential. Both US and European regulatory agencies, provide frameworks for selecting the starting dose in clinical trials. The starting dose is typically based on nonclinical data, pharmacokinetics, and toxicology studies, along with considerations unique to ADCs. Of note, ICH S9 provides guidance on clinical starting dose for cytotoxic payloads in cancer patients using either one-tenth severely toxic dose

(STD) in 10 percent of animals in rodents (STD10) or one-sixth highest non-severely toxic dose in non-rodents (HNSTD) based on body surface area, depending on the most appropriate and/or sensitive species.

Due to their cytotoxic payloads, ADCs can sometimes display more pronounced toxicities, and these toxicological studies also assess the pharmacokinetics and off-target effects of the payload in animal models. The human equivalent dose (HED) is typically calculated from the NOAEL or MABEL. Regulatory guidelines recommend a conservative safety margin to reduce the risk of severe adverse events, particularly because the toxic payload of ADCs can cause significant off-target effects.

For ADCs, the FDA recommends starting doses that incorporate a tenfold safety margin from the HED to account for uncertainties in translating animal toxicology findings to humans.

The pharmacokinetics and pharmacodynamics of both the antibody and the cytotoxic payload must be considered. Since the payload can have distinct toxicity profiles, the FDA expects PK data to demonstrate the clearance rates, bioavailability, and biodistribution of both components.

Specific PK/PD modeling may also be required to understand how the ADC behaves in the body, ensuring that the starting dose allows sufficient drug exposure to observe therapeutic effects without crossing the toxicity threshold.

Regulatory agencies like the FDA recommend that sponsors discuss their dose selection strategy during pre-IND meetings and consider feedback from toxicology and clinical pharmacology experts to ensure that the dose escalation design is robust.

In conclusion, the selection of a starting dose for an ADC in clinical trials is a complex process that integrates nonclinical safety data, pharmacokinetics, and pharmacodynamics of both the antibody and the cytotoxic drug, with careful attention to minimizing toxicity while enabling therapeutic efficacy.

8.5.2 *Dose Optimization*

Most cancer drugs including ADCs enter clinical trials focusing on evaluating the safety, tolerability, pharmacokinetics (PK), and pharmacodynamics (PD) of the drug in humans for the first time. The main goal historically has been to identify the maximum tolerated dose (MTD) and establish a recommended dose for Phase 2 trials. However, such studies have a limited number of patients exposed to an investigational drug for a relatively short duration and frequently do not sufficiently evaluate the safety, tolerability, exposure, and antitumor activity over a wide dosage range.

The FDA has been emphasizing in various types of correspondence with sponsors the need for dose optimization early in development. For example, in pre-IND responses and in Study May Proceed letters, the FDA has been providing early and detailed comments to sponsors to support the proposed dosage(s) to be investigated in clinical trials by investigating multiple dosages early in clinical development and to conduct further dosage exploration beyond the initial dose-escalation phase in a sufficient number of patients with adequate follow up. This may include evaluating additional dose cohorts, investigating multiple dosages in an expansion phase, and conducting a randomized, parallel dose–response trial.

In August 2024, FDA issued a final guidance document, titled "Optimizing the Dosage of Human Prescription Drugs and Biological Products for the Treatment of Oncologic Diseases," which outlines a new approach to identifying optimized dosages for oncology drugs. This guidance emphasizes shifting away from the traditional MTD paradigm, which was common in cytotoxic chemotherapy, toward a more data-driven approach for modern oncology drugs such as ADCs. Key recommendations from the FDA guidance include:

- Clinical Pharmacokinetics and Pharmacodynamics: Trials should focus on collecting relevant PK and PD data early to support population PK and dose–response studies. This allows for a better understanding of how drug exposure correlates with safety and efficacy outcomes.
- Trial Designs: Sponsors are encouraged to compare multiple dosages in randomized, parallel-group trials. Adaptive trial designs may also be employed to assess safety and activity.
- Safety and Tolerability: Persistent toxicities and patient-reported outcomes should be assessed when selecting optimal dosages. This also includes careful monitoring of dose reductions and interruptions.
- Drug Formulation: Sponsors are advised to plan for the manufacturing of multiple dosage strengths during drug development to ensure adequate evaluation in clinical trials.
- Subsequent Indications and Usages: Dose– and exposure–response relationships must be evaluated for subsequent indications, such as in combination therapies or new disease settings.

The FDA guidance encourages early engagement with the agency, as well as the use of model-informed drug development (MIDD) tools to inform dosage selection. This new approach aligns with the overall goal of optimizing the benefit–risk profile of oncology therapies, including ADCs.

In addition to the guidance document on dose optimization summarized above, a related 2024 FDA guidance document, titled "Clinical Pharmacology Considerations for Antibody-Drug Conjugates (March 1, 2024)," should be considered together. This document for industry specific to ADCs reiterates the Agency's recommendations on evaluating a broad dosage range in early drug development to fully characterize the relationship between the exposure of both the intact ADC and its active constituent parts to the safety and activity of the ADC product. It also provides detailed guidance regarding measurement of the ADC and its constituent parts in various clinical pharmacology studies.

8.6 Discussion and Conclusion

With 11 FDA-approved ADCs for various cancer types and hundreds more in clinical development, this class of biologic has experienced great clinical success and validation since the printing of the first edition of this book. Big pharmaceutical companies are heavily investing in ADCs, viewing them as a highly promising therapeutic modality, especially in oncology. Major companies like Pfizer, AstraZeneca, Johnson & Johnson, AbbVie, and Merck have all made significant moves into the ADC space. For example, Pfizer's acquisition of Seagen, a leading ADC developer, signals the industry's confidence in this technology. Additionally, Gilead and Merck have been actively making billion-dollar deals to secure ADC assets.

The components making up the ADC product leads to development complexities and unique challenges and can be more resource-intensive for industry sponsors; therefore, it is important to consider the development-stage-specific nature of CMC and nonclinical information required during the various phases. A proactive approach is to consider the information/data needed at each phase of the clinical study to ensure that CMC development, for example, is keeping pace with clinical development and to effectively utilize formal Agency interaction to share development plans and de-risking strategies at the appropriate times. For example, the End-of-Phase 2 (EOP2) meeting with the FDA may sometimes call for a discussion dedicated to CMC issues separate from clinical topics. Fortunately for this audience, regulatory agencies are considering ways to allow for more rapid access of promising new therapies to patients when a product is shown to have a favorable benefit:risk profile. However, an accelerated approval is only possible if the CMC expectations are also met and the manufacturing is in compliance with current Good Manufacturing Practice standards. As next-generation ADC products advance into clinical development, the CMC challenges are likely to change requiring new insights into this technically and financially demanding area.

As the research and development of ADC products matures, the lessons learned are leading sponsors to develop new technologies by employing synthetic cytotoxic agents and stable linker systems that attach the cytotoxic agents to the antibody at specific sites eliminating the heterogeneous group of molecules in which the linkers can attach in a number of different places on the antibody. The promise of targeted therapy for cancer is attracting new players each year and new sponsors to the area of ADC research and development are encouraged to maintain open communications with the FDA and other regulatory authorities.

Most commonly, these have been developed toward tumor-associated antigens, but a number of groups have demonstrated interest in using ADCs to treat immune and neurological disorders.

References

Besheer A, Burton L, Galas RJ et al (2021) An industry perspective on compatibility assessment of closed system drug-transfer devices for biologics. J Pharm Sci 110(2):610–614

Blümel M, Liu J, de Jong I et al (2023) Current industry best practice on in-use stability and compatibility studies for biological products. J Pharm Sci 112(9):2332–2346

Bechtold-Peters K, Ruggiero A, Vriezen N et al (2023) CMC regulatory considerations for antibody-drug conjugates. J Pharm Sci 112:2965–2980

Ducry L, Stump B (2010) Antibody-drug conjugates: linking cytotoxic payloads to monoclonal antibodies. Bioconjug Chem 21(1):5

European Medicines Agency Guidance for applicants seeking access to PRIME scheme. 03 October 2024

FDA CDER Chemistry Review Summary for brentuximab vedotin, August 2011

FDA Guidance for Industry. Clinical pharmacology considerations for antibody-drug conjugates, March 1, 2024

FDA Guidance for Industry: Formal meetings between the FDA and sponsors or applicants of PDUFA products, September 2023

FDA points to consider in the manufacture and testing of monoclonal antibody products for human use (PTC 1997)

Geigert J (2013) The challenge of CMC regulatory compliance for biopharmaceuticals and other biologics, 2nd edn. Springer

ICH Harmonized Tripartite Guideline M4: Organization of the common technical document for the registration of pharmaceuticals for human use. Step 4, 13 January 2004

ICH Harmonized Tripartite Guideline Q3A(R2): Impurities in new drug substances. Step 4, 25 October 2006

ICH Harmonized Tripartite Guideline Q5A(R1): Viral safety evaluation of biotechnology products derived from cell lines of human or animal origin. Step 4, 23 September 1999

ICH Harmonized Tripartite Guideline Q5B: Quality of biotechnological products: analysis of the expression construct in cell lines used for production of r-DNA derived protein products, July 1996

ICH Harmonized Tripartite Guideline Q5D: Quality of biotechnological products: derivation and characterisation of cell substrates used for production of biotechnological/biological products, March 1998

ICH Harmonized Tripartite Guideline Q5E: Comparability of biotechnological/biological products subject to changes in their manufacturing process. Step 4, 18 November 2004

ICH Harmonized Tripartite Guideline Q6B: Specifications: Test procedures and acceptance criteria for biotechnology/biological products. Step 4, 10 March 1999

ICH Harmonized Tripartite Guideline S6(R1): Preclinical safety evaluation of biotechnology-derived pharmaceuticals and addendum, June 2011

ICH Harmonized Tripartite Guideline S9: Nonclinical evaluation for anticancer pharmaceuticals, 29 October 2009

ICH Harmonized Tripartite Guideline S9: Nonclinical evaluation for anticancer pharmaceuticals, questions and answers, June 2018

Jiang XR, Song A, Bergelson S, Arroll T, Parekh B, May K, Chung S, Strouse R, Mire-Sluis A, Schenerman M (2011) Advances in the assessment and control of the effector functions of therapeutic antibodies. Nat Rev Drug Discov 10:101–111

Kamen DE, Crotts G, Narasimhan C et al (2022) An Intercompany perspective on compatibility and in-use stability studies to enable administration of biopharmaceutical drug products. J Pharm Sci 111(4):1092–1103

Miksinski SP, Shapiro M (2012) Regulatory considerations for antibody-drug conjugates, OPS, CDER, FDA presentation at AAPS

PDUFA VII: Fiscal years 2023–2027, www.fda.gov.Sutter, Sue. Lung cancer "master protocol" eyed as paradigm-changing approach to drug development, The Pink Sheet, November 18 2013, Article 00131118001

The National Institute for Occupational Safety and Health (NIOSH) (2020) NIOSH DRAFT list of hazardous drugs in healthcare settings

Trail P (2013) Antibody drug conjugates as cancer therapeutics. Antibodies 2:113–129

USP<800> General Chapter: Hazardous Drugs—Handling in Healthcare Settings

Wakankar A, Chen Y, Gokarn Y et al (2011) Analytical methods for physicochemical characterization of antibody drug conjugates. MAbs 3(2):161–172

Part IV
Landscape, IP and Pharmacoeconomics

Chapter 9
Major ADC Companies, Current Late-Stage Clinical Trials, Recent Patents Issued and Patent Applications Published, and Cost Analysis of Drug Therapy

Jeffrey Wang, Aryana Sepassi, and Jonathan H. Watanabe

Abstract Antibody-drug conjugates (ADCs) have come to the age of fruition following a long period of trials and errors. Eleven ADCs have been approved by the Food and Drug Administration (FDA) thus far and many more are in the development pipelines. This chapter provides the current landscape of ADC drug discovery and development. It summarizes the major players in ADC drug development worldwide, recently completed or currently running late-stage clinical trials registered in the United States, recently US patents issued, and patent applications published. In addition, the cost and outcomes aspects of ADC drug therapy are discussed.

Keywords ADC landscape · ADC companies · CDMOs · Clinical trials · Patents · Patent applications · Cost and outcomes analysis

9.1 Introduction

Prototype antibody-drug conjugates (ADCs) or immunoconjugates were explored as therapeutics in the 1960s and 1970s in experimental animals with promising results (Arnon and Sela 1982; Hurwitz et al. 1975; Rowland et al. 1975). The first

J. Wang (✉)
Department of Biotechnology and Pharmaceutical Sciences, College of Pharmacy, Western University of Health Sciences, Pomona, CA, USA
e-mail: jwang@westernu.edu

A. Sepassi
Division of Clinical Pharmacy, School of Pharmacy & Pharmaceutical Sciences, University of California San Diego, La Jolla, CA, USA

J. H. Watanabe
Department of Clinical Pharmacy, School of Pharmacy, University of California San Francisco, San Francisco, CA, USA

J. L. Zaro, J. Wang (eds.), *Antibody-Drug Conjugates*, AAPS Advances in the Pharmaceutical Sciences Series 17, https://doi.org/10.1007/978-3-031-83005-1_9

generation of ADCs using monoclonal antibodies and conventional chemotherapeutic drugs such as methotrexate and doxorubicin were subsequently tested in animal models and clinical trials (Hurwitz et al. 1983). The clinical results were not satisfactory, mainly due to the inefficient internalization of the conjugate and intracellular release of the active drug species to the drug target. To solve these problems, novel anticancer drugs with 100–1000 times more potency have been used for the construction and testing of the second generation of ADCs. Most of the payloads in Food and Drug Administration (FDA)-approved ADCs belong to this category. Because of their excellent efficacy and desirable side-effect profiles, the pharmaceutical industry and the medical field have accepted ADCs as a new class of anticancer medications. Many pharmaceutical and biotechnology companies have recognized the trend and dived into the discovery and development of ADC anticancer drugs. This has enriched the research and the pipelines, which benefit millions of cancer patients. This chapter draws a sketch of the global ADC landscape and provides a brief cost and outcomes analysis of the available ADC therapeutics.

9.2 Major ADC Developing and Contract Manufacturing Companies

Wyeth was a pioneer in ADC drug discovery and development with the first FDA-approved product in 2000. Pfizer became involved in ADC space significantly after the 2009 acquisition of Wyeth. Besides, there were two industry heavyweight players specializing in ADC drug discovery and clinical development, Seattle Genetics (Seagen), Inc. and ImmunoGen, Inc., when the first edition of this book was published in 2015. Each company developed unique payloads and linker technologies, leading to FDA approval of Adcetris® of Seattle Genetics and Kadcyla® of Roche/Genentech in 2011 and 2013, respectively. Nearly every major pharma and many biotech companies established collaborations with these two leaders in developing ADCs. The ADC landscape has greatly changed in the past decade. Here we describe some of the companies with active clinical ADC programs and a list of contract development and manufacturing organizations (CDMOs).

9.2.1 Pfizer, Inc./Seagen, Inc.

Pfizer (www.pfizer.com) initially inherited its ADC technology with the acquisition of Wyeth with $68 billion in 2009 (Pfizer 2009). Wyeth was developing ADCs using the payload of calicheamicins and acid-sensitive cleavable linkage. Mylotarg® (gemtuzumab ozogamicin) is an ADC comprising a monoclonal antibody to CD33 linked to a calicheamicin. It was first approved in 2000 by the FDA through the accelerated-approval process to treat acute myelogenous leukemia. But it was voluntarily withdrawn from the market in 2010 because of safety and benefit concerns resulting from a randomized Phase 3 comparative controlled trial (Federal Register 2000). Pfizer conducted additional clinical studies which led to the resurrection of the product in September 2017 (Gamis et al. 2014; Hills et al. 2014; Borthakur et al. 2014; Kharfan-Dabaja 2014; O'Hear and Rubnitz 2014; FDA 2017a). The company also developed another ADC targeting CD22 cells, Besponsa® (inotuzumab ozogamicin), which received FDA approval in August 2017 for the treatment of adults with relapsed or refractory B-cell precursor acute lymphoblastic leukemia (FDA 2017b).

To further enrich its pipelines in oncology and especially in ADC products, Pfizer completed its acquisition of Seagen (formally Seattle Genetics) for $43 billion on December 14, 2023 (Pfizer 2023). Seagen was based in Bothell, Washington, and was founded by Clay B. Siegall in 1998. Peter Senter led research in antibody-drug conjugate technologies, including payloads development, novel linker design, conjugation methodology, and studies on mechanism of action.

Seattle Genetics initially licensed SGN-15 (BMS-182248, BR96-DOX), a first-generation ADC using doxorubicin as the payload, from Bristol-Myers Squibb in 1999 and conducted clinical trials (Smith 2001). It terminated this ADC program due to limited efficacy (Ross et al. 2006). Seattle Genetics licensed from Arizona State University, in 2000, the right to use microtubule-disrupting auristatin E discovered by George Pettit's research laboratory as the potent payload for their ADCs. They developed a cysteine-reactive novel protease-cleavable linker to attached monomethyl auristatins to the antibody. The resulting ADCs are stable in the bloodstream and effective at releasing the cell-killing agent once inside targeted cancer cells (Senter and Sievers 2012).

Seattle Genetics collaborated with Takeda Pharmaceutical Company in developing Adcetris (brentuximab vedotin), which was approved by the FDA in 2011, a decade after the first ADC, Mylotarg, was approved. Adcetris is stable in the bloodstream but it can release MMAE, a monomethyl auristatin, upon internalization into CD30-expressing tumor cells. Adcetris was approved by the FDA in 2011 for two indications: 1) the treatment of patients with Hodgkin lymphoma after failure of autologous stem cell transplant (ASCT) or after failure of at least two prior multi-agent chemotherapy regimens in patients who are not ASCT candidates; and 2) the treatment of patients with systemic anaplastic large-cell lymphoma (ALCL) after failure of at least one prior multi-agent chemotherapy regimen (FDA 2011). According to the co-development agreement, Seattle Genetics has full commercialization rights to Adcetris in the United States and Canada, while Takeda has exclusive rights to commercialize the product in all other countries. Seagen also co-developed Padcev (enfortumab vedotin) with Agensys/Astellas and Tivdak (tisotumab vedotin) with Genmab. These ADCs received FDA approval in 2019 and 2021, respectively (FDA 2019a, 2021a).

Pfizer now has about a dozen of ADCs in its pipeline. These ADCs include vedotin ADCs, ADCs with TOPO1 inhibitor payloads, next-generation auristatin ADCs with potentially improved tolerability, and ADCs with novel payload mechanisms of action. Pfizer's oncology focus includes breast cancer, genitourinary cancers, hematology-oncology, and thoracic cancers.

9.2.2 AbbVie, Inc./ImmunoGen, Inc.

AbbVie (www.abbvie.com) is a pharmaceutical company headquartered in North Chicago, Illinois. AbbVie was in the ADC space before its acquisition of ImmunoGen. They had a promising, but eventually failed, clinical candidate rovalpituzumab tesirine, which they acquired through the purchase of Stemcentrx (Uprety et al. 2021). Their next ADC, telisotuzumab vedotin, was made up of a c-Met-directed monoclonal antibody telisotuzumab and monomethyl auristatin E. Currently, it has received breakthrough therapy designation from the FDA and is in Phase 3 clinical trial as a monotherapy for patients with previously treated non-small-cell lung cancer (NSCLC) with c-Met overexpression (AbbVie 2022).

To elevate itself as a heavyweight player in the ADC space, AbbVie completed the acquisition of ImmunoGen for $10.1 billion on February 12, 2024 (AbbVie 2024). ImmunoGen (www.immunogen.com), now a part of AbbVie, is headquartered in Waltham, Massachusetts, with its manufacturing facility in Norwood, Massachusetts. Founded in 1981, ImmunoGen's ADC technology comprises highly potent maytansinoids that can be attached to the surface lysine groups of antibodies via novel thioether linkers. The stable linkers keep the payload attached to the antibody in the bloodstream and then the drug is released inside a cancer cell after internalization (Lambert and Chari 2014).

Genentech/Roche licensed ImmunoGen's ADC technology to develop Kadcyla (ado-trastuzumab emtansine, T-DM1). The ADC consists of ImmunoGen's proprietary maytansinoid agent DM1 attached to trastuzumab, the human epidermal growth factor receptor 2 (HER2)-binding antibody. It was approved by the US FDA for patients with HER2-positive early breast cancer who have residual invasive disease after neoadjuvant treatment with taxane and trastuzumab-based therapy, or patients with HER2-positive metastatic breast cancer who have previously received trastuzumab and a taxane, separately or in combination. These patients should have either received prior therapy for metastatic disease or developed disease recurrence during or within 6 months of completing adjuvant therapy (FDA 2013, 2019b).

With the experience gained through the development of Kadcyla, ImmunoGen was developing at least three ADC clinical candidates in-house. The company was evaluating mirvetuximab soravtansine (IMGN853) for folic acid alpha receptor positive solid tumors such as epithelial ovarian cancer and non-small-cell lung cancer, IMGN529 for NHL, and IMGN289 for epidermal growth factor receptor (EGFR)-positive lung, head and neck, and other EGFR-positive solid tumors. Elahere (mirvetuximab soravtansine) received accelerated approval by the FDA for FRα-positive, platinum-resistant epithelial ovarian, fallopian tube, or peritoneal cancer in November 2022 (FDA 2022).

Currently, ImmunoGen/AbbVie is conducting clinical trials to expand the indications for Elahere. There are at least three additional ADC clinical candidates in its pipelines, including anti-CD123 ADC pivekimab sunirine in Phase 2 trials (Table 9.1). In addition to its proprietary pipeline of ADC candidates, it has

Table 9.1 ADC pipeline at ImmunoGen/AbbVie

ADC program	Therapeutic area	Development atage
Elahere™	Monotherapy in FRα-high platinum-resistant ovarian cancer	Phase 3
	Doublet with Mirvetuximab + Bevacizumab maintenance in FRα-high platinum-sensitive ovarian cancer	Phase 3
	Monotherapy in FRα-high platinum-sensitive ovarian cancer	Phase 2
	Doublet with Mirvetuximab + Carboplatin in FRα-low, -medium, and -high platinum-sensitive ovarian cancer	Phase 2
Pivekimab sunirine	Monotherapy in blastic plasmacytoid dendritic neoplasma	Phase 2
Anti-CD123 ADC	Combination with azacitidine and/or venetoclax in acute myeloid leukemia	Phase 2
IMGN151	Ovarian cancer	Phase 1
Anti-FRα Biparatopic ADC	Endometrial cancer	Phase 1
IMGC936 Anti-ADAM9 ADC	Non-small cell lung cancer, gastric, pancreatic, triple-negative breast cancer, and other solid tumors	Phase 1

partnerships with CytomX Therapeutics, MacroGenics, Lilly, and Oxford BioTherapeutics to develop ADC therapeutics. It also licensed out its technology to Novartis, Bayer, Roche, Sanofi, and Amgen (ImmunoGen Pipeline 2024).

9.2.3 Genentech, Inc./Hoffmann-La Roche Ltd.

Genentech (www.gene.com), a member of the Switzerland-based Roche group, is located in South San Francisco. It probably has the broadest experience with ADCs in the pharmaceutical and biotech industry. The company licensed the ADC technology from ImmunoGen and successfully commercialized Kadcyla in 2013 (FDA 2013 and see Sect. 9.2.2). Currently it is conducting several Phase 3 trials for expanded indications, including first line HER2-positive metastatic breast cancer, early HER2-positive breast cancer in combination with pertuzumab, early and metastatic HER2-positive breast cancer in combination with atezolizumab, third-line HER2-positive metastatic breast cancer, 2 L + HER-2+ PD-L1+ mBC (KATE3), and HER2-positive early breast cancer high-risk patients (Genentech 2024). Meanwhile, Genentech licensed Seattle Gentics' ADC technology to develop several other ADCs, leading to the accelerated FDA approval in 2019 of Polivy (polatuzumab vedotin), a CD79b-directed ADC indicated in combination with bendamustine and a rituximab product for adult patients with relapsed or refractory diffuse large B-cell lymphoma (DLBCL), not otherwise specified, after at least two prior therapies (FDA 2019c).

Genentech also developed its own proprietary ADC techniques for site-specific conjugation (Junutula et al. 2008; Panowksi et al. 2014). However, there are no late-stage clinical trials on ADC candidates using this technology.

9.2.4 *Daiichi Sankyo Co. Ltd.*

Daiichi Sankyo Company, Limited (www.daiichisankyo.com), the second-largest pharmaceutical company in Japan, is a global pharmaceutical company established by the merging of Sankyo Co. Ltd. and Daiichi Pharmaceutical Co. Ltd. in 2005. Its US subsidiary, Daiichi Sankyo, Inc., headquartered in Basking Ridge, New Jersey, began its operation in 2006. Several years after the merge, it focused on its oncology drug development on ADC, with its technology based on the antibody research and protein engineering capability of the former Sankyo and the camptothecin drug payload and linker expertise from the former Daiichi Pharmaceutical.

Daiichi Sankyo collaborated with AztraZeneca in developing its first ADC clinical candidate, the HER2-targeting trastozumab dereuxtecan (DS-8201), with a $6.9 billion deal announced in March 2019 (Daiichi Sankyo 2019). On December 20, 2019, Enhertu (fam-trastuzumab deruxtecan-nxki, DS-8201, T-DXd) received accelerated approval by the FDA for the treatment of adult patients with unresectable or metastatic HER2-positive breast cancer who have received two or more prior anti-HER2-based regimens in the metastatic setting (FDA 2019d). Since then, more indications have been added for this product.

There are at least three additional ADCs in Daiichi Sankyo's pipelines. It was announced in April 2024 that the Biologics License Application (BLA) for datopotamab deruxtecan has been accepted in the United States for the treatment of adult patients with unresectable or metastatic hormone receptor (HR)-positive, HER2-negative (IHC 0, IHC 1+ or IHC 2+/ISH-) breast cancer who have received prior systemic therapy for unresectable or metastatic disease (Daiichi Sankyo 2024). An additional BLA for datopotamab deruxtecan is under review for the treatment of adult patients with locally advanced or metastatic nonsquamous non-small-cell lung cancer (NSCLC) who have received prior systemic therapy (Daiichi Sankyo 2024). In addition, ifinatamab deruxtecan is being investigated in a Phase 3 trial in subjects with relapsed small-cell lung cancer and raludotatug deruxtecan is being studied in a Phase 2/3 trial in subjects with platinum-resistant, high-grade ovarian, primary peritoneal, or fallopian tube cancer.

9.2.5 *AstraZeneca*

AstraZeneca plc (www.astrazeneca.com) with headquarters at the Cambridge Biomedical Campus in Cambridge, England, is a British-Swedish multinational pharmaceutical and biotechnology company. The company was established in 1999 by merging the Swedish Astra AB and the British Zeneca Group. It entered the cancer immunotherapy field by the acquisition of Cambridge Antibody Technology in 2006, MedImmune in 2007, and Spirogen in 2013. It also collaborates with Daiichi Sankyo and ADT Therapeutics in developing ADC therapeutics including Enhertu (trastozuman dereuxtecan) and datopotamab deruxtecan (See Sect. 9.2.4).

9.2.6 Gilead Sciences, Inc./ImmunoMedics

Gilead Sciences, Inc. (www.gilead.com) is an American biopharmaceutical company headquartered in Foster City, California. The company was founded under the name Oligogen by Michael L. Riordan in 1987. The original name was a reference to oligonucleotides used to target genetic sequences. Gradually it shifted its focuses on researching and developing antiviral drugs. On October 23, 2020, Gilead completed the acquisition of Immunomedics for $21 billion (Gilead 2020), owning the cancer treatment Trodelvy (Sacituzumab govitecan), a first-in-class Trop-2-targeted ADC approved by the FDA on February 3, 2023 for patients with unresectable locally advanced or metastatic hormone receptor-positive, human epidermal growth factor receptor 2-negative breast cancer who have received endocrine-based therapy and at least two additional systemic therapies in the metastatic setting (FDA 2020). It has since received additional approvals for other indications, including metastatic urothelial cancer and hormone receptor-positive, HER2-negative breast cancer. It is currently conducting clinical trials to expand its indications.

9.2.7 ADC Therapeutics

ADC Therapeutics (www.adctherapeutics.com) with its headquarter in Epalinges, Switzerland, was founded in 2011 as a spin-off from Spirogen Ltd. It specializes in using pyrrolobenzodiazepine as the ADC payload. Phase 1 clinical trial of its lead investigational ADC loncastuximab tesirine targeting CD19 in positive B-cell non-Hodgkin lymphoma started in 2016 and in 5 short years, Zynlonta (loncastuximab tesirine) received FDA approval in 2021 for adult patients with relapsed or refractory large B-cell lymphoma after two or more lines of systemic therapy, including diffuse large B-cell lymphoma (DLBCL) not otherwise specified, DLBCL arising from low-grade lymphoma, and high-grade B-cell lymphoma (FDA 2021b). Two additional ADCs, ADCT-602 targeting CD22 and ADCT-601 (mipasetamab uzoptirine), are in its development pipelines.

9.2.8 *Astellas Pharma Inc./Agensys, Inc.*

Astellas Pharma Inc. (www.astellas.com) is a Japanese multinational pharmaceutical company, formed on April 1, 2005 from the merger of Yamanouchi Pharmaceutical Co., Ltd. and Fujisawa Pharmaceutical Co., Ltd. It entered the ADC space by the acquisition of Agensys, Inc., located in Santa Monica, California, in 2007. Agensys specialized in developing fully human monoclonal antibodies as therapeutics to treat cancer. In addition, Agensys was co-developing with Seattle Genetics two ADCs, ASG-15ME for bladder cancer and ASG-22ME for solid tumors. It also in-licensed Seattle Genetics' technology to develop three additional ADCs in-house for cancer (AGS-16M8F, AGS-16C3F, and AGS-67E). Astellas decided to close operations at Agensys in 2017 but continue to develop ADCs. ASG-22ME eventual became Padcev (enfortumab vedotin), which was approved by the FDA in 2019 for adult patients with locally advanced or metastatic urothelial cancer who have previously received a programmed death receptor-1 (PD-1) or programmed death-ligand 1 (PD-L1) inhibitor, and a platinum-containing chemotherapy in the neoadjuvant/adjuvant, locally advanced or metastatic setting (FDA 2019a). Currently Astellas is conducting clinical trials in an effort to expand the Padcev's indications for muscle-invasive bladder cancer, non-muscle-invasive bladder cancer, and other solid tumors (Astellas R&D Pipeline 2014).

9.2.9 *ADC CDMOs*

The construct of an ADC is simple in concept but complex in practice, involving careful selection and optimization of the payload, the antibody, the linker, and the conjugation method. Making an ADC is a multistep endeavor and requires expertise in many scientific areas. This has provided business opportunities for many large and small companies. Table 9.2 summarizes some of the ADC contract development and manufacturing organizations (CDMOs) around the globe (Thayer 2014; Roots Analysis 2024).

9.3 Major ADC Clinical Trials

The combined pipelines of ADC clinical candidates from various companies are rich. There are over 150 active ADC programs at various stages of clinical investigation (Hotha 2023), for a large spectrum of both blood and solid tumors. Table 9.3 lists ongoing and recently completed late-stage clinical trials in the United States registered at www.clinicaltrials.gov. Besides those aiming to expand the indications of the approved ADCs, there are quite a few other ADCs including belantamab mafodotin, raludotatug deruxtecan, trastuzumab deruxtecan, trastuzumab

Table 9.2 Major ADC contract development and manufacturing organizations (CDMOs)

Company	Website	Headquarter location
Abzena	www.abzena.com	San Diego, CA, USA
Axplora/Novasep	www.axplora.com	Raubling, Germany
Carbogen Amcis AG	www.carbogen-amcis.com	Bubendorf, Switzerland
Catalent Pharma Solutions	www.catalent.com	Somerset, NJ, USA
CellMosaic	www.cellmosaic.com	Woburn, MA, USA
Fujifilm Diosynth	www.fujifilmdiosynth.com	Morrisville, NC, USA
GBI Biomanufacturing	www.gbibio.com	Plantation, FL, USA
Lonza	www.lonza.com/	Basel, Switzerland
MabPlex	www.mabplexinc.com/	Yantai, Shandong, China
Piramal Heathcare	www.piramalpharmasolutions.com	Mumbai, Maharashtra, India
Syngene	www.syngeneintl.com	Bangalore, Karnataka, India
Veranova	www.veranova.com	Wayne, PA, USA

duocarmazine, sacituzumab tirumotecan, and dato-Dxd in late-stage clinical trials. It is expected that some of these trials will be successful, and several ADCs will be approved by the FDA in the next few years to treat cancer patients.

9.4 Recent ADC Patents and Patent Applications

A search in the US patent database (https://ppubs.uspto.gov/pubwebapp/static/pages/ppubsbasic.html) using the search term ADC reveals that, not surprisingly, Pfizer/Seagen/Seattle Genetics, and AbbVie/Immunogen have the largest numbers of issued US patents and published patent applications on ADCs. This group is followed by Genentech and Daiichi Sankyo. Tables 9.4 and 9.5 list US patents issued during the 3 months of between June 25 and September 25, 2024, and published US patent applications in the 1 month period from August 20 to September 19, 2024. These numbers present a significant increase from those in 2014. A detailed analysis of these patents would be beyond the scope of this chapter. However, it is obvious to note that many more companies and research organizations are building intellectual properties in the ADC space. The future of ADC supply for targeted therapies looks extremely promising.

9.5 Cost and Outcomes Considerations

The price of cancer treatment for a single month with newly approved drugs was approximately US$196,000 in 2023 (Miljkovic et al. 2023; Prasad et al. 2017). In total, patients pay an estimated annual total of US$5.6 billion in out-of-pocket costs for cancer treatments (American Cancer Society 2020). ADCs are no exception to

Table 9.3 Ongoing and recently completed late-stage clinical trials in the US registered at www.clinicaltrials.gov (Accessed September 23, 2024)

ADC	ClinicalTrials.gov identifier	Trial name	Clinical phase	Sponsor	Status	Last update date
Belantamab mafodotin	NCT04246047	Evaluation of efficacy and safety of belantamab mafodotin, bortezomib and dexamethasone versus daratumumab, bortezomib and dexamethasone in participants with relapsed/refractory multiple myeloma (DREAMM 7)	3	GlaxoSmithKline	Active, not recruiting	2023-12-13
Mirvetuximab soravtansine	NCT04296890	A study of mirvetuximab soravtansine in platinum-resistant, advanced high-grade epithelial ovarian, primary peritoneal, or fallopian tube cancers with high folate receptor-alpha expression (SORAYA)	3	ImmunoGen	Completed	2024-08-07
Raludotatug deruxtecan	NCT06161025	A study of raludotatug deruxtecan (R-DXd) in subjects with platinum-resistant, high-grade ovarian, primary peritoneal, or fallopian tube cancer	2/3	Daiichi Sankyo	Recruiting	2024-09-20
Enfortumab vedotin	NCT03474107	A study to evaluate enfortumab vedotin versus chemotherapy in subjects with previously treated locally advanced or metastatic urothelial cancer (EV-301)	3	Astellas Pharma	Active, not recruiting	2024-09-19
Belantamab mafodotin	NCT04484623	Belantamab mafodotin plus pomalidomide and dexamethasone (Pd) versus bortezomib plus Pd in relapsed/refractory multiple myeloma (DREAMM 8)	3	GlaxoSmithKline	Recruiting	2024-05-28
Trastuzumab deruxtecan	NCT03734029	Trastuzumab deruxtecan (DS-8201a) versus investigator's choice (IC) for HER2-low breast cancer that has spread or cannot be surgically removed [DESTINY-Breast04]	3	Daiichi Sankyo	Active, not recruiting	2024-04-11
Trastuzumab duocarmazine	NCT03262935	SYD985 vs. physician's choice in participants with HER2-positive locally advanced or metastatic breast cancer (TULIP)	3	Byondis	Completed	2023-10-19
Trastuzumab deruxtecan	NCT05950945	Trastuzumab deruxtecan (T-DXd) in patients who have hormone receptor-negative and hormone receptor-positive HER2-low or HER2 IHC 0 metastatic breast cancer	3	Daiichi Sankyo	Recruiting	2024-09-19

Sacituzumab tirumotecan	NCT06132958	Sacituzumab tirumotecan (MK-2870) in post platinum and post immunotherapy endometrial cancer (MK-2870-005)	3	Merck Sharp & Dohme	Recruiting	2024-09-19
Luveltamab tazevibulin	NCT05870748	REFRaME-O1: A study to investigate the efficacy and safety of luveltamab tazevibulin versus investigator's choice (IC) chemotherapy in women with ovarian cancer (including fallopian tube or primary peritoneal cancers) expressing FOLR1	2/3	Sutro Biopharma	Recruiting	2024-08-29
Ifinatamab deruxtecan	NCT06203210	A study of ifinatamab deruxtecan versus treatment of physician's choice in subjects with relapsed small-cell lung cancer (IDeate-Lung02)	3	Daiichi Sankyo	Recruiting	2024-09-19
Brentuximab vedotin	NCT01712490	A frontline therapy trial in participants with advanced classical Hodgkin lymphoma	3	Takeda	Active, not recruiting	2024-02-20
Datopotamab deruxtecan	NCT05629585	A study of Dato-DXd with or without durvalumab versus investigator's choice of therapy in patients with stage I-III triple-negative breast cancer without pathological complete response following neoadjuvant therapy (TROPION-Breast03)	3	AstraZeneca	Recruiting	2024-08-21
Trastuzumab deruxtecan	NCT03523585	DS-8201a in pretreated HER2 breast cancer that cannot be surgically removed or has spread [DESTINY-Breast02]	3	Daiichi Sankyo	Active, not recruiting	2024-04-02
Trastuzumab deruxtecan	NCT03529110	DS-8201a versus T-DM1 for human epidermal growth factor receptor 2 (HER2)-positive, unresectable and/or metastatic breast cancer previously treated with trastuzumab and taxane [DESTINY-Breast03]	3	Daiichi Sankyo	Active, not recruiting	2024-06-12
Trastuzumab deruxtecan	NCT04494425	Study of trastuzumab deruxtecan (T-DXd) vs investigator's choice chemotherapy in HER2-low, hormone receptor positive, metastatic breast cancer (DB-06)	3	AstraZeneca	Active, not recruiting	2024-07-12
Datopotamab deruxtecan	NCT05374512	A study of Dato-DXd versus investigator's choice chemotherapy in patients with locally recurrent inoperable or metastatic triple-negative breast cancer, who are not candidates for PD-1/PD-L1 inhibitor therapy (TROPION-Breast02)	3	AstraZeneca	Active, not recruiting	2024-07-05

(continued)

Table 9.3 (continued)

ADC	ClinicalTrials.gov identifier	Trial name	Clinical phase	Sponsor	Status	Last update date
Mirvetuximab soravtansine	NCT04209855	A study of mirvetuximab soravtansine vs. investigator's choice (IC) of chemotherapy in platinum-resistant, advanced high-grade epithelial ovarian, primary peritoneal, or fallopian tube cancers with high folate receptor-alpha (FRα) expression (MIRASOL)	3	ImmunoGen	Active, not recruiting	2024-08-01
Sacituzumab tirumotecan	NCT06074588	Sacituzumab tirumotecan (MK-2870) versus chemotherapy in previously treated advanced or metastatic nonsquamous non-small-cell lung cancer (NSCLC) with EGFR mutations or other genomic alterations (MK-2870-004)	3	Merck Sharp & Dohme	Recruiting	2024-09-19
Datopotamab deruxtecan	NCT06112379	A Phase 3 randomized study to evaluate Dato-DXd and durvalumab for neoadjuvant/adjuvant treatment of triple-negative or hormone receptor-low/HER2-negative breast cancer	3	AstraZeneca	Recruiting	2024-09-19
Datopotamab deruxtecan	NCT06103864	A Phase 3 study of Dato-DXd with or without durvalumab compared with investigator's choice of chemotherapy in combination with pembrolizumab in patients with PD-L1 positive locally recurrent inoperable or metastatic triple-negative breast cancer	3	AstraZeneca	Recruiting	2024-09-19
Datopotamab deruxtecan	NCT05104866	A Phase 3, open-label, randomized study of Dato-DXd versus investigator's choice of chemotherapy (ICC) in participants with inoperable or metastatic HR-positive, HER2-negative breast cancer who have been treated with one or two prior lines of systemic chemotherapy (TROPION-Breast01)	3	AstraZeneca	Active, not recruiting	2024-08-20
Mirvetuximab soravtansine	NCT05445778	Mirvetuximab soravtansine with bevacizumab versus bevacizumab as maintenance in platinum-sensitive ovarian, fallopian tube, or peritoneal cancer (GLORIOSA)	3	ImmunoGen	Recruiting	2024-09-23

Datopotamab deruxtecan	NCT05687266	Phase 3, open-label, first-line study of Dato-DXd in combination with durvalumab and carboplatin for advanced NSCLC without actionable genomic alterations (AVANZAR)	3	AstraZeneca	Recruiting	2024-08-12
Depatuxizumab mafodotin	NCT02573324	A study of ABT-414 in participants with newly diagnosed glioblastoma (GBM) with epidermal growth factor receptor (EGFR) amplification (Intellance1)	3	AbbVie	Completed	2023-05-11
Sacituzumab govitecan	NCT05609968	Study of pembrolizumab (MK-3475) monotherapy versus sacituzumab govitecan in combination with pembrolizumab for participants with metastatic non-small-cell lung cancer (NSCLC) with programmed cell death ligand 1 (PD-L1) tumor proportion score (TPS) ≥50% (MK-3475-D46)	3	Merck Sharp & Dohme	Recruiting	2024-06-27
Brentuximab vedotin	NCT05675410	A study to compare standard therapy to treat Hodgkin lymphoma to the use of two drugs, brentuximab vedotin and nivolumab	3	National Cancer Institute	Recruiting	2024-09-20
Brentuximab vedotin	NCT03907488	Immunotherapy (nivolumab or brentuximab vedotin) plus combination chemotherapy in treating patients with newly diagnosed Stage III-IV classic Hodgkin lymphoma	3	National Cancer Institute	Active, not recruiting	2024-08-13

Table 9.4 US patents on ADCs issued from June 25 to September 25, 2024 (https://ppubs.uspto.gov/pubwebapp/static/pages/ppubsbasic.html accessed September 25, 2024)

Patent number	Title	Inventor	Assignee	Issue date
12,098,214B2	Combination therapies for treating cancer	Wan; Hong et al.	ALX Oncology Inc. (South San Francisco, CA)	2024-09-24
12,097,219B2	Single-domain antibodies against CLL1 and constructs thereof	Zhang; Wang et al.	Genentech, Inc. (South San Francisco, CA)	2024-09-24
12,098,198B2	Antibody binding specifically to ECL-2 of claudin 3, fragment thereof, and use thereof	Shin; Young Kee et al.	Abion Inc. (Seoul, KR)	2024-09-24
12,090,211B2	Methods for preparing antibody drug conjugates	Pillow; Thomas et al.	Genentech, Inc. (South San Francisco, CA)	2024-09-17
12,090,212B2	Conjugated biological molecules, pharmaceutical compositions, and methods	Chang; Michael Nientse et al.	OBI Pharma, Inc. (Taipei, TW)	2024-09-17
12,084,403B2	Linkers for antibody drug conjugates	Tsuchikama; Kyoji et al.	The Board of Regents of the University of Texas System (Austin, TX)	2024-09-10
12,084,415B2	Biomolecule-polymer-pharmaceutical agent conjugates for delivering the pharmaceutical agent	Johnson; Jeremiah A. et al.	Massachusetts Institute of Technology (Cambridge, MA)	2024-09-10
12,076,411B2	Anti-HER2 antibody-maytansine conjugates and methods of use thereof	Rabuka; David et al.	Redwood Bioscience, Inc. (Emeryville, CA)	2024-09-03
12,076,412B2	Antibody-drug conjugates through specific linker oligopeptides	Spycher; Philipp et al.	Araris Biotech AG (Au, CH)	2024-09-03
12,071,487B2	Site-specific crosslinking of antibodies	Agnew; Brian et al.	Life Technologies Corporation (Carlsbad, CA)	2024-08-27
12,071,477B2	Anti-folate receptor alpha antibody conjugated with hemiasterlins	Stafford; Ryan et al.	Sutro Biopharma, Inc. (South San Francisco, CA)	2024-08-27
12,065,504B2	Anti-glyco-MUC1 antibodies and their uses	White; Thayer	GO Therapeutics, Inc. (Cambridge, MA)	2024-08-20
12,065,489B2	Claudin-6 antibodies and drug conjugates	Conklin; Dylan et al.	The Regents of the University of California (Oakland, CA)	2024-08-20
12,065,494B2	Combination therapy comprising an fibroblast growth factor receptor inhibitor and a Nectin-4 targeting agent	Koblish; Holly K. et al.	Incyte Corporation (Wilmington, DE)	2024-08-20

(continued)

Table 9.4 (continued)

Patent number	Title	Inventor	Assignee	Issue date
12,065,496B2	Antibody molecules and conjugates	Comer; Frank Irvine et al.	AstraZeneca AB (Sodertalje, SE)	2024-08-20
12,059,473B2	Humanized anti-prostate-specific membrane antigen (PSMA) antibody drug conjugates	Bhowmik; Shivarupam et al.	Ambrx, Inc. (La Jolla, CA)	2024-08-13
12,060,427B2	Drug conjugates of cMET monoclonal binding agents, and uses thereof	Coronella; Julia et al.	Mitsubishi Tanabe Pharma Corporate (Osaka, JP)	2024-08-13
12,048,750B2	Tissue-homing peptide conjugates and methods of use thereof	Hansen; Staccy J. ct al.	Blaze Bioscience, Inc. (Scattle, WA)	2024-07-30
12,043,670B2	Anti-BCMA antibody-drug conjugates and methods of use	Henry; Ryan et al.	Eisai R&D Management Co., Ltd. (Tokyo, JP)	2024-07-23
12,037,405B2	Anti-CD70 antibody drug conjugates	Barnett; Richard S et al.	Ambrx, Inc. (La Jolla, CA)	2024-07-16
12,037,411B2	Antibodies, and bispecific antigen-binding molecules that bind HER2 and/or APLP2, conjugates, and uses thereof	Perez Bay; Andres et al.	Regeneron Pharmaceuticals, Inc. (Tarrytown, NY)	2024-07-16
12,036,286B2	Selective drug release from internalized conjugates of biologically active compounds	Bindman; Noah et al.	Seattle Genetics Inc.	2024-07-16
12,030,954B2	Antibody binding to AXL protein and application thereof	Lv; Ming et al.	Hangzhou Sumgen Biotech Co., Ltd. (Zhejiang, CN); Sumgen Mab (Beijing) Biotech Co., Ltd. (Beijing, CN)	2024-07-09
12,029,736B2	Camptothecin derivatives and conjugates thereof	Li; Wei	MediBoston Limited (Causeway Bay, HK)	2024-07-09
12,024,565B2	Targeted CD73 antibody and antibody-drug conjugate, and preparation method therefor and uses thereof	Yu; Ke et al.	Fudan University (Shanghai, CN); Bliss Biopharmaceutical (Hangzhou) Co., Ltd. (Zhejiang, CN)	2024-07-02
12,016,842B2	Methods of using anti-CD79b immunoconjugates	Polson; Andrew et al.	Genentech, Inc. (South San Francisco, CA)	2024-06-25
12,018,093B2	Hemiasterlin derivatives and antibody-drug conjugates including same	Ban; Hitoshi et al.	Sumitomo Pharma Co., Ltd. (Osaka, JP)	2024-06-25

Table 9.5 US patent applications on ADCs published from August 20 to September 19, 2024 (https://ppubs.uspto.gov/pubwebapp/static/pages/ppubsbasic.html accessed September 24, 2024)

Patent application number	Title	Inventor	Applicant/assignee	Publication date
20240309088A1	Anti-CD79B antibodies and immunoconjugates and methods of use	Chen; Yvonne et al.	Genentech, Inc. (South San Francisco, CA)	2024-09-19
20240307551A1	Antibody-drug conjugate and application thereof	Xu; Ting et al.	Jiangsu Alphamab Biopharmaceuticals Co., Ltd. (Jiangsu, CN)	2024-09-19
20240307549A1	Subcutaneously administered antibody-drug conjugates for use in cancer treatment	Kulke; Michael et al.	Heidelberg Pharma Research GmbH (Ladenburg, DE)	2024-09-19
20240299571A1	Anti-FOLR1 immunoconjugates and anti-PD-1 antibody combinations	Ruiz Soto; Rodrigo R.	ImmunoGen, Inc. (Waltham, MA); Merck Sharp & Dohme LLC (Rahway, NJ)	2024-09-12
20240299573A1	Antibody-pyrrolobenzodiazepine derivative conjugate	Toda; Narihiro et al.	Daiichi Sankyo Co., Ltd. (Tokyo, JP)	2024-09-12
20240299574A1	Antibodies to PMEL17 and conjugates thereof	Burger; Matthew et al.	Novartis AG (Basel, CH)	2024-09-12
20240302348A1	A scoring method for an anti-B7H4 antibody-drug conjugate therapy	Schmidt; Guenter et al.	Medimmune Limited (Cambridge, GB)	2024-09-12
20240299568A1	Bone-specific delivery of polypeptides	Xiao; Han et al.	William Marsh Rice University (Houston, TX); Baylor College of Medicine (Houston, TX)	2024-09-12
20240299569A1	2-Amino-4-carboxamide-benzazepine immunoconjugates, and uses thereof	Brandt; Gary et al.	Bolt Biotherapeuticals, Inc. (Redwood City, CA)	2024-09-12
20240299570A1	Combination therapies using an anti-BCMA antibody drug conjugate (ADC) in combination with a gamma secretase inhibitor (GSI)	Wu; Kaida	Celgene Corporation (Summit, NJ)	2024-09-12
20240299576A1	Targeting conjugates comprising effector molecules and uses thereof	Xiang; Jun et al.	Allygen Group (Grand Cayman, KY)	2024-09-12

(continued)

Table 9.5 (continued)

Patent application number	Title	Inventor	Applicant/assignee	Publication date
20240294661A1	Human antibody drug conjugates against tissue factor	Satijn; David et al.	Genmab A/S (Copenhagen V, DK)	2024-09-05
20240293549A1	Conjugates and conjugating reagents	Godwin; Antony et al.	Polytherics Ltd. (Cambridge, GB)	2024-09-05
20240293554A1	Quaternized nicotinamide adenine dinucleotide salvage pathway inhibitor conjugates	Neumann; Christopher Scott et al.	Seagen Inc. (Bothell, WA)	2024-09-05
20240294666A1	BCMA monoclonal antibody-drug conjugate	Kinneer; Krista et al.	MedImmune, LLC (Gaithersburg, MD)	2024-09-05
20240294644A1	Anti-VTCN1 antibodies and antibody drug conjugates	Guo; Ailan et al.	Bluefin BioMedicine, Inc. (Beverly, MA)	2024-09-05
20240293567A1	Novel cyclic dinucleotide derivative and antibody-drug conjugate thereof	Tsuda; Toshifumi et al.	Daiichi Sankyo Company, Limited (Tokyo, JP)	2024-09-05
20240293566A1	Anti-EGFRvIII antibody drug conjugates and uses thereof	Delfino; Frank et al.	Regeneron Pharmaceuticals, Inc. (Tarrytown, NY)	2024-09-05
20240293565A1	Anti-HER3 antibody, antibody drug conjugate containing the same, and use thereof	Zhong; Xiaoyan et al.	Beijing Sinotau Biopharmaceuticals Technology Co., Ltd. (Beijing, CN)	2024-09-05
20240293570A1	Highly stable targeted linker-drug conjugate	Huang; Jinkun et al.	Chengdu Scimount Pharmatech Co., Ltd. (Chengdu, CN)	2024-09-05
20240293569A1	Linker for antibody drug conjugate	Jung; Doo Young et al.	Pinotbio, Inc. (Suwon-si, KR)	2024-09-05
20240287203A1	Anti-CUB domain-containing protein 1 (CDCP1) antibodies, antibody drug conjugate, and method f use thereof	Ren; Hong et al.	Bluefin BioMedicine, Inc. (Beverly, MA)	2024-08-29
20240285788-A1	Conjugate of antibody and functional substrate or salt thereof, and antibody derivative and compound used in production of the same or salts thereof	Matsuda; Yutaka et al.	Ajinomoto Co., Inc. (Tokyo, JP)	2024-08-29
20240277672A1	Site-specific antibody-drug glycoconjugates and methods	Boons; Geert-Jan	University of Georgia Research Foundation, Inc. (Athens, GA)	2024-08-22

(continued)

Table 9.5 (continued)

Patent application number	Title	Inventor	Applicant/assignee	Publication date
20240277838A1	Combination therapy with an anti-HER2 antibody drug conjugate and a BCL-2 inhibitor	Phillips; Gail Lewis et al.	Genentech, Inc. (South San Francisco, CA)	2024-08-22
20240277860A1	Camptothecin peptide conjugates	Jeffrey; Scott C. et al.	Seagen Inc. (Bothell, WA)	2024-08-22
20240277859A1	Flow through cation exchange chromatography purification processes for antibody drug conjugates	Hendricks; Rachel et al.	Genentech, Inc. (South San Francisco, CA)	2024-08-22

the high price point of cancer treatment. Indeed, the addition of "seek-and-destroy" capability comes at a premium in price beyond already-established biologic oncology therapeutics. The complexity of these molecules and the challenges of expression, production, and purification of protein-derived therapies require, in many cases, singular manufacturing infrastructure and expertise. These demands figure prominently in the pricing scheme of these agents. Moreover, these new targeted agents display the characteristics of other specialty biologic pharmaceuticals necessitating additional effort to distribute, handle, educate, and mitigate risks (Sullivan 2008). Additional justifications of the pricing include inordinately high costs of bringing drugs to market in the current environment and funding necessary to support future research and development of new molecules. It is important to recall that most investigational drugs do not reach the market and yet consume a large amount of research and development spending. The funds expended in the research program of these abandoned molecules by firms have been linked to pricing structure of agents that do see market life (DiMasi et al. 2003).

The argument for the increasing price point of newer cancer drugs is predicated on bolstered survival benefits of these new agents, and in the case of ADCs, improved safety over traditional chemotherapeutic agents (Chen et al. 2019; Zhang et al. 2022). The most persuasive justification in price may lie in the purported safety profile of these agents. ADCs are designed to locate and deliver anticancer payload solely to cancerous cells, deftly avoiding health tissue and the adverse effects concomitant with damaging noncancerous cells (Goldberg 2013). Traditional oncology agents are virtually synonymous with quality of life depending on adverse events, including hair loss, muscle pain, weakness, and blood cell disorders. Specifically, the adverse side effects of diarrhea, fatigue, nausea, neuropathy, and vomiting have been statistically tied to reductions in quality-of-life metrics in prior cancer trials (Boyd et al. 2011). The notion of potential willingness to pay for quality-of-life enhancement in oncology treatment has been previously described and factored into pharmacoeconomics (Tonia et al. 2012). The simplified premise being that 1 year of feeling perfectly healthy is equal to 2 years of feeling half of perfectly healthy. If drug A and drug B equally extend a person's life by 1 year, but drug A results in better quality of life for that additional year, the value of drug A is considered greater than drug B. For years, validated means of measuring quality of life have been broadly used in health care via rating scale surveys that evaluate patient preferences for different states of health (Drummond et al. 1997). The American Society of Clinical Oncology incorporates patient quality of life into its conceptual framework in assessment of the value of cancer treatment options along with clinical efficacy, patient convenience, and cost (Schnipper et al. 2015). These additional patient-specific considerations contribute to the observed premium in costs among ADCs versus (vs.) traditional small-molecule oncology agents, leading to heterogeneity among cancer patients when prioritizing treatments of choice. The unique addition of patient quality of life in considerations of value thus increase the importance of determining perspective for value assessments (e.g. patient perspective, health system, third party payer, etc.). These justifications have fallen under greater scrutiny as a significant portion of the often-cited estimate of US$1 billion to bring a drug to market includes non-research spending as well as evidence that a portion of the preliminary research is taxpayer- or government-funded (Goozner 2004).

Quality of life, life expectancy, and outcomes analysis in the oncology space are complicated by the application of several different endpoints that are also commonly used. The gold standard measurement is the straightforward outcome of overall survival (OS) and is defined as the time from randomization in a clinical trial to death. The true virtue of OS is the absence of clinician interpretation in rating the outcome. Application of the other commonly used endpoints in cancer studies can be frustrated by additional challenges in subjective interpretation of the endpoint. Tumor size reduction equal to or greater than a prespecified amount can also serve as an endpoint in cancer studies. The percentage of patients who experience a tumor size reduction of a minimum magnitude for a prespecified minimum amount of time is called the objective response rate (ORR). The time from first clinical response until documented tumor progression is deemed the response duration. The endpoint of progression-free survival (PFS) is the time from randomization to objective tumor progression or death, whichever occurs first. PFS is a subcomponent of event-free survival (EFS), denoted as the time from randomization to an event which may include disease progression, discontinuation of the treatment for any reason, or death (Gyawali et al. 2020). Finally, complete response (CR) is defined by lack of detectable evidence of a tumor, typically measured using imaging studies or histopathology (Delgado and Guddati 2021). The challenge presented to policymakers and payers, therefore, is to understand which of these clinical endpoints hold the most value to cancer patients. However, the cancer population is highly heterogeneous with respect to cancer severity, and one endpoint may be of value to one patient, but not another. A patient with severe end-stage pancreatic cancer may place higher value on additional life years (OS), than relief of symptoms compared to patients with less severe cancers (Schnipper et al. 2015). This challenging task that payers confront is further complicated by the wide range of FDA-approved indications for the ADCs (Table 9.6).

Table 9.6 Summary of FDA-approved ADC therapeutics

	Antibody drug conjugate	Indications	Pivotal Phase 3 clinical trials	Main clinical efficacy outcomes measured	Comparators	Main findings	Drug cost (VA FSS pricing in 2024 USD)
Hematologic cancers	Gemtuzumab ozogamicin (Mylotarg)	Newly diagnosed, de novo AML (combination or single agent), relapse or refractory AML (single agent)	ALFA-0701 (Castaigne et al. 2012)	EFS (primary), OS (secondary)	Daunorubicin + Cytarabine	Improved EFS by 34%, non-significantly improved OS	$9603.32 per 5 mg vial
	Brentuximab vedotin (Adcertis)	Hodgkin lymphoma (stage III or IV)	ECHELON-1 (Ansell et al. 2022)	PFS (primary), OS	Doxorubicin + Bleomycin + Vinblastine + Dacarbazine	Improved PFS, non-significantly improved OS	$7576.77 per 5 mg/10.5 mL vial
	Inotuzumab ozogamicin (Besponsa)	Relapsed or refractory ALL	INO-VATE ALL (Kantarjian et al. 2019)	OS (primary), PFS, ORR	Bendamustine, Rituximab + Gemcitabine,	No significant difference in OS, PFS, or OS, but greater rate of stem cell transplants	$16,277.63 per 0.9 mg vial
	Polatuzumab vedotin (Polivy)	Relapsed or refractory diffuse large B-cell lymphoma	POLARIX (Tilly et al. 2021)	PFS (primary), EFS, OS	Rituximab + Cyclophosphamide + Doxorubicin + Vincristine + Prednisone	Improved PFS, no significant difference in OS	$3492.97 per 30 mg vial
	Loncastuximab tesirine (Zynlonta)	Relapsed or refractory large B-cell lymphoma	LOTIS-5 (Kwiatek et al. 2024)	PFS (primary), OS, ORR, CRR	Rituximab + Gemcitabine + Oxaliplatin	Ongoing	$18,485.56/10 mg vial

Solid tumors	Trastuzumab emtansine (Kadcyla)	Metastatic breast cancer	KATHERINE (von Minckwitz et al. 2019), TH3RESA (Krop et al. 2017)	DFS (primary), OS, recurrence-free survival	Trastuzumab	Improved DFS, recurrence-free survival, no significant difference in OS. Significantly improved OS in open-label trial (TH3RESA)	$5897.04 per 160 mg/20 mL vial
	Enfortumab vedotin (Padcev)	Locally advanced or metastatic urothelial cancer	EV-301 (Powles et al. 2021), EV-302 (with pembrolizumab, Powles et al. 2024)	EV-301: OS (primary)	EV-301: Docetaxel, Paclitaxel, or Vinflunine	EV-301: Improved OS, PFS	$1892.70 per 20 mg vial
				EV-302: PFS (primary), OS (primary)	EV-302: Gemcitabine + Cisplatin or Carboplatin	EV-302: Improved PFS, OS	
	Trastuzumab deruxtecan (Enhertu)	Unresectable or metastatic breast cancer, non-small-cell lung cancer, gastric or gastroesophageal junction adenocarcinoma	DESTINY-Breast03 (Cortés et al. 2022)	PFS (primary), OS	Trastuzumab emtansine	Improved PFS and OS	$2435.71 per 100 mg vial
	Sacituzumab govitecan (Trodelvy)	Locally advanced or metastatic breast cancer, urothelial cancer	ASCENT (breast cancer, Bardia et al. 2021)	ASCENT: PFS (primary), OS	ASCENT: Eribulin, vinorelbine, capecitabine, or gemcitabine	ASCENT: Improved PFS, OS	$2129.51 per 180 mg vial
			TROPiCS-04 (urothelial cancer, Gilead 2024)	TROPiCS-04: OS (primary), PFS, ORR	TROPiCS-04: Paclitaxel, Docetaxel, Vinflunine	TROPiCS-04: No difference in OS, non-significant improvement in PFS, ORR	

(continued)

Table 9.6 (continued)

	Antibody drug conjugate	Indications	Pivotal Phase 3 clinical trials	Main clinical efficacy outcomes measured	Comparators	Main findings	Drug cost (VA FSS pricing in 2024 USD)
	Tisotumab vedotin (Tivdak)	Recurrent or metastatic cervical cancer	innovaTV 301 (Vergote et al. 2024)	OS (primary), PFS, ORR	Topotecan, Vinorelbine, Gemcitabine, Irinotecan, or Permetrexed	Improved OS, PFS, ORR	$4530.96 per 40 mg vial
	Mirvetuximab soravtansine (Elahere)	Ovarian, fallopian tube, or primary peritoneal cancer	FORWARD I (ovarian cancer, Moore et al. 2021)	FORWARD I: PFS (primary), ORR	FORWARD I, MIRASOL: Paclitaxel, pegylated liposomal doxorubicin, topotecan	FORWARD I: No significant difference in PFS, significantly improved ORR	$4732.45 per 100 mg injection
			MIRASOL (high grade ovarian cancer, Moore et al. 2023)	MIRASOL: PFS (primary), OS		MIRASOL: Improved PFS, OS	

Note: *EFS* Event-Free Survival, *OS* Overall Survival, *PFS* Progression-Free Survival, *CRR* Complete Remission, *DFS* Disease-Free Survival, *ORR* Objective Response Rate

Currently, there are 11 FDA-approved ADCs, indicated for either hematologic cancers (Mylotarg, Adcertis, Besponsa, Polivy, Zynlonta) or for solid tumor cancers (Kadcyla, Padcev, Enhertu, Trodelvy, Tivdak, Elahere). Almost all have been previously tested in landmark Phase 3 clinical trials, typically against traditional chemotherapeutic agents (See Table 9.6). The exception to this is Zylonta, which was granted accelerated FDA approval based on the results of LOTIS-2, a single-arm trial (ADC Therapeutics 2023). The head-to-head study design of the controlled Phase 3 trials facilitates comparative effectiveness analyses that enable value assessments for ADCs to proceed. It is important to recognize that most of these trials were powered to detect significant differences in PFS as a primary outcome. While a commonly used endpoint, PFS has not always been found to be predictive of survival. This has made use of PFS as an endpoint a subject of scrutiny (Gyawali et al. 2020). With this acknowledged, many ADCs have found a significant improvement in PFS. Of those measuring OS, only the ADCs used for solid tumors found significant improvements. However, while some ADCs may not necessarily improve OS, they may improve the rate of alternative, non-pharmaceutical treatments. For example, Besponsa did not improve OS or PFS compared to traditional chemotherapy in the ECHELON-1 trial but did improve rate of stem cell transplants. This is important, as stem cell transplants have greatly extended overall survival for patients not achieving an adequate response with chemotherapy (Maranzano and Mead 2022). Therefore, while Besponsa may not contribute to improved life expectancy, it may confer other benefits to patients that are worth consideration in value assessments. Modern decision-making approaches will often include factoring in additional clinical consequences, outcomes, and adverse events avoided when evaluating the value proposition of ADCs specifically.

Several considerations should be made when considering the cost of ADCs to payers and patients. For value assessments, the Second Panel in Cost Effectiveness in Health and Medicine recommended the use of the publicly available Federal Supply Schedule (FSS) pricing from the Department of Veteran Affairs (VA), (Sanders et al. 2016). Per statute, the VA can purchase prescription drugs at discounted price (McCaughan 2017). Hence, these prices denote the best possible acquisition prices for prescription drugs for practical purposes (McCaughan 2017). However, these prices do provide a more standardized representation of drug costs than other drug pricing metrics (e.g. average wholesale price, etc., (McCaughan 2017).

While the United States currently lacks government organizations that oversee the specific negotiation of ADC pricing based on the value assessment of these drugs, other countries do have mechanisms to complete these evaluations and apply them in price negotiations. The assessments published by these organizations may provide some insight into the value of ADCs. The National Institute for Health and Care Excellence (NICE) from the United Kingdom, for example, uses both clinical trial data and economic data, usually in the form of a cost-effectiveness model, to make formulary decisions for new drugs. Moreover, NICE also uses feedback from patients and patient representatives when to incorporate into decision-making. Of the 12 FDA-approved ADCs, seven are currently recommended as cost-effective for the United Kingdom National Health Service (NHS), with the exception of Enhertu and Padcev. Tivdak, and Elahere are currently undergoing a value assessment. Padcev was retracted from consideration, as Astellas Pharmaceuticals did not provide enough evidence for a determination (NICE 2022). Moreover, Astellas confirmed there was additional evidence needed to prove cost-effectiveness in the United Kingdom. Enhertu was rejected from consideration due to a lack of comparative effectiveness data to Trodelvy for metastatic breast cancer patients as well as an unacceptable expected cost-effectiveness return (NICE 2022). This decision highlights that while most ADCs have been evaluated compared to traditional chemotherapy, few have been tested against other ADCs that may be used for the same indication. This constitutes a gap in value assessment for ADCs, which future trials may address. Conversely, the Canadian Agency for Drugs and Technologies in Health (CADTH) concluded that Enhertu was cost-effective for metastatic breast cancer patients, citing that patients prefer more treatment options for unresectable or metastatic HER2-positive breast cancer that extend survival and have a manageable toxicity profile (CADTH 2022). Ultimately, while these health value assessments provide insight relevant to determining if ADCs are cost-effective, these assessments described were made in the context of health systems outside of the United States, which are subject to a different drug pricing and reimbursement structure. To ascertain value in the context of the current US drug pricing of these agents, ADCs will need to undergo evaluations on the parameters of quality of life, life expectancy, and cost offsets from the patient, payer, and societal perspective in the United States. Moreover, the rapidly increasing prices for these agents and specialty medications in general are demanding multi-stakeholder discussions (payer, manufacturer, patient, health system, etc.) to ensure sustainable financing and patient access are possible. This will continue to expand the roles of value-based reimbursement and outcomes-based contracting models that put a premium on current and relevant data. Moreover, the importance of study designs and statistical approaches that allow appropriate capture of data that informs value has become paramount. The improved need for validation of biomarkers and surrogate endpoints and approaches to ensure studies are appropriately powered to assess outcomes that are relevant to patients, payers, and regulators grow each day. The rapid development of ADC therapeutics offers the potential for great promise for improving the lives of patients, but complex challenges continue to mount.

9.6 Additional ADC Resources

Published materials on ADC discovery and development stemmed from the 1960s. Conjugation chemistry and preclinical research results are usually published in biomedical and pharmaceutical journals such as *Journal of Medicinal Chemistry* and *Bioconjugate Chemistry*. Clinical findings usually appear in clinical journal such as *New England Journal of Medicine*. Timely findings are presented in annual meetings of pharmaceutical and medical associations (American Association of Pharmaceutical Association, American Association for Cancer Research, American Society of Clinical Oncology, etc.). Several ADC workshops have been organized by the American Association of Pharmaceutical Association. A specialized conference, World ADC Summit (worldadc-usa.com), is held annually with its 2024 meeting site in San Diego, California. Root Analysis (www.rootsanalysis) is a UK-based business research and consulting firm, providing ADC marketing analysis and projection.

9.7 Conclusion

With 11 FDA-approved ADCs, the concept of ADC has been widely accepted by the pharmaceutical and medical fields. These twenty-first-century "magic bullets" are redefining the practice of cancer therapy and greatly benefiting many cancer patients. With the increasing numbers of pharmaceutical and biotech companies joining the trend, clinical candidates in the pipeline and clinical trials, issued patents and published patent applications, it is a certainty that in the next decade, many more ADCs will be added to approved list and cancer therapy outcomes will be vastly improved to benefit cancer patients.

References

AbbVie (2022) AbbVie announces U.S. FDA granted breakthrough therapy designation (BTD) to telisotuzumab vedotin (Teliso-V) for previously treated non-small cell lung cancer. https://news.abbvie.com/2022-01-04-AbbVie-Announces-U-S-FDA-Granted-Breakthrough-Therapy-Designation-BTD-to-Telisotuzumab-Vedotin-Teliso-V-for-Previously-Treated-Non-Small-Cell-Lung-Cancer. Accessed September 21, 2024

AbbVie (2024) AbbVie completes acquisition of ImmunoGen. https://news.abbvie.com/2024-02-12-AbbVie-Completes-Acquisition-of-ImmunoGen. Accessed September 21, 2024

ADC Therapeutics (2023). https://ir.adctherapeutics.com/press-releases/press-release-details/2023/ADC-Therapeutics-Announces-Updated-LOTIS-2-Results-Demonstrating-Durable-Long-Term-Responses-of-ZYNLONTA-in-RelapsedRefractory-DLBCL/default.aspx. Accessed Aug 27, 2024

American Cancer Society (2020). https://www.fightcancer.org/sites/default/files/National%20Documents/Costs-of-Cancer-2020-10222020.pdf. Accessed August 27, 2024

Ansell SM, Radford J, Connors JM, Długosz-Danecka M, Kim WS, Gallamini A, Ramchandren R, Friedberg JW, Advani R, Hutchings M, Evens AM, Smolewski P, Savage KJ, Bartlett NL, Eom HS, Abramson JS, Dong C, Campana F, Fenton K, Puhlmann M, Straus DJ (2022) Overall survival with brentuximab vedotin in Stage III or IV Hodgkin's lymphoma. N Engl J Med 387:310–320

Arnon R, Sela M (1982) In vitro and in vivo efficacy of conjugates of daunomycin with anti-tumor antibodies. Immunol Rev 62:5–27

Astellas R&D Pipeline (2014). https://www.astellas.com/en/system/files/382e51ce75/1q2024_pip_en.pdf. Accessed September 28, 2024

Bardia A, Hurvitz SA, Tolaney SM, Loirat D, Punie K, Oliveira M, Brufsky A, Sardesai SD, Kalinsky K, Zelnak AB, Weaver R, Traina T, Dalenc F, Aftimos P, Lynce F, Diab S, Cortés J, O'Shaughnessy J, Diéras V, Ferrario C, Schmid P, Carey LA, Gianni L, Piccart MJ, Loibl

S, Goldenberg DM, Hong Q, Olivo MS, Itri LM, Rugo HS (2021) Sacituzumab govitecan in metastatic triple-negative breast cancer. N Engl J Med 384:1529–1541

Borthakur G, Cortes JE, Estey EE, Jabbour E, Faderl S, O'Brien S, Garcia-Manero G, Kadia TM, Wang X, Patel K, Luthra R, Koller C, Brandt M, Ravandi F, Kantarjian H (2014) Gemtuzumab ozogamicin with fludarabine, cytarabine, and granulocyte colony stimulating factor (FLAGGO) as front-line regimen in patients with core binding factor acute myelogenous leukemia. Am J Hematol 89:964–968

Boyd KA, Briggs AH, Paul J, Iveson T, Midgely R, Harkin A, Bates G, Alexander L, Cassidy J (2011) Analysis of adverse events and quality of life data for an economic evaluation of adjuvant chemotherapy in colorectal cancer: when can we stop collecting? Trials 12:A41

CADTH Reimbursement Review (2022). https://www.cda-amc.ca/sites/default/files/DRR/2022/PC0285%20Enhertu%20-%20Draft%20CADTH%20Recommendation%20September%201,%202022_For%20posting.pdf. Accessed August 27, 2024

Castaigne S, Pautas C, Terré C, Raffoux E, Bordessoule D, Bastie JN, Legrand O, Thomas X, Turlure P, Reman O, de Revel T, Gastaud L, de Gunzburg N, Contentin N, Henry E, Marolleau JP, Aljijakli A, Rousselot P, Fenaux P, Preudhomme C, Chevret S, Dombret H (2012) Effect of gemtuzumab ozogamicin on survival of adult patients with novo acute myeloid leukemia (ALFA-0701): a randomized, open-label phase 3 study. Lancet 379:1508–1516

Chen AJ, Hu X, Conti RM, Jena AB, Goldman DP (2019) Trends in the price per median and mean life-year gained among nwly approved cancer therapies 1995 to 2017. Value Health 22:1387–1395

Cortés J, Kim SB, Chung WP, Im SA, Park YH, Hegg R, Kim MH, Tseng LM, Petry V, Chung CF, Iwata H, Hamilton E, Curigliano G, Xu B, Huang CS, Kim JH, Chiu JWY, Pedrini JL, Lee C, Liu Y, Cathcart J, Bako E, Verma S, Hurvitz SA (2022) Trastuzumab deruxtecan versus trastuzumab emtansine for breast cancer. N Engl J Med 386:1143–1154

Daiichi Sankyo (2019) Daiichi Sankyo and AstraZeneca announce global development and commercialization collaboration for Daiichi Sankyo's HER2 targeting antibody drug conjugate [Fam-] trastuzumab deruxtecan (DS-8201). https://www.daiichisankyo.com/media/press_release/detail/index_3199.html. Accessed September 30, 2024

Daiichi Sankyo (2024) Datopotamab deruxtecan Biologics License Application accepted in the U.S. for patients with previously treated metastatic HR positive, HER2 negative breast cancer. https://www.daiichisankyo.com/files/news/pressrelease/pdf/202404/20240402_E.pdf. Accessed September 30, 2024

Delgado A, Guddati AK (2021) Clinical endpoints in oncology—a primer. Am J Cancer Res 11:1121–1131

DiMasi JA, Hansen RW, Grabowski HG (2003) The price of innovation: new estimates of drug development costs. J Health Econ 22:151–185

Drummond MF, O'Brien B, Stoddart GL, Torrance GW (1997) Methods for the economic evaluation of health care programmes. Oxford Medical Publications

FDA (2011) FDA approval package: Adcetris (brentuximab vedotin). https://www.accessdata.fda.gov/drugsatfda_docs/nda/2011/125399_adcetris_toc.cfm. Accessed September 30, 2024

FDA (2013) Drug approval package: Kadcyla (ado-trastuzumab emtansine) injection. https://www.accessdata.fda.gov/drugsatfda_docs/nda/2013/125427Orig1s000TOC.cfm. Accessed September 21, 2024

FDA (2017a) FDA approves Mylotarg for treatment of acuter myeloid leukemia. https://www.fda.gov/news-events/press-announcements/fda-approves-mylotarg-treatment-acute-myeloid-leukemia. Accessed September 21, 2024

FDA (2017b) FDA approves inotuzumab ozogamicin for relapsed or refractory B-cell precursor ALL. https://www.fda.gov/drugs/resources-information-approved-drugs/fda-approves-inotuzumab-ozogamicin-relapsed-or-refractory-b-cell-precursor-all. Accessed September 21, 2024

FDA (2019a) FDA grants accelerated approval to enfortumab vedotin-ejfv for metastatic urothelial cancer. https://www.fda.gov/drugs/resources-information-approved-drugs/fda-grants-accelerated-approval-enfortumab-vedotin-ejfv-metastatic-urothelial-cancer. Accessed September 21, 2024

FDA (2019b) FDA approves ado-trastuzumab emtansine for early breast cancer. https://www.fda.gov/drugs/resources-information-approved-drugs/fda-approves-ado-trastuzumab-emtansine-early-breast-cancer. Accessed September 21, 2024

FDA (2019c) FDA D.I.S.C.O. Burst Edition: FDA approval of Polivy (polatuzumab vedotin-piiq) for previously untreated diffuse large B-cell lymphoma, not otherwise specified, and high-grade B-cell lymphoma. https://www.fda.gov/drugs/resources-information-approved-drugs/fda-disco-burst-edition-fda-approval-polivy-polatuzumab-vedotin-piiq-previously-untreated-diffuse. Accessed September 21, 2024

FDA (2019d) FDA approves fam-trastuzumab deruxtecan-nxki for unresectable or metastatic HER2-positive breast cancer. https://www.fda.gov/drugs/resources-information-approved-drugs/fda-approves-fam-trastuzumab-deruxtecan-nxki-unresectable-or-metastatic-her2-positive-breast-cancer. Accessed September 21, 2024

FDA (2020) FDA D.I.S.C.O. Burst Edition: FDA approval of Trodelvy (sacituzumab govitecan-hziy) for HR-positive breast cancer. https://www.fda.gov/drugs/resources-information-approved-drugs/fda-disco-burst-edition-fda-approval-trodelvy-sacituzumab-govitecan-hziy-hr-positive-breast-cancer#:~:text=On%20February%203%2C%202023%2C%20the,at%20least%20two%20additional%20systemic. Accessed September 28, 2024

FDA (2021a) FDA D.I.S.C.O. Burst Edition: FDA approvals of Tivdak (tisotumab vedotin-tftv) for recurrent or metastatic cervical cancer with disease progression on or after chemotherapy, and Jakafi (ruxolitinib) for chronic graft-versus-host disease after failure of one or two lines of systemic therapy. https://www.fda.gov/drugs/resources-information-approved-drugs/fda-disco-burst-edition-fda-approvals-tivdak-tisotumab-vedotin-tftv-recurrent-or-metastatic-cervical. Accessed September 21, 2024

FDA (2021b) FDA grants accelerated approval to loncastuximab tesirine-lpyl for large B-cell lymphoma. https://www.fda.gov/drugs/resources-information-approved-drugs/fda-grants-accelerated-approval-loncastuximab-tesirine-lpyl-large-b-cell-lymphoma. Accessed September28, 2024

FDA (2022) FDA grants accelerated approval to mirvetuximab soravtansine-gynx for FRα positive, platinum-resistant epithelial ovarian, fallopian tube, or peritoneal cancer. https://www.fda.gov/drugs/resources-information-approved-drugs/fda-grants-accelerated-approval-mirvetuximab-soravtansine-gynx-fra-positive-platinum-resistant. Accessed September 21, 2024

Federal Register (2000) Wyeth Pharmaceuticals, Inc.; Withdrawal of Approval of a New Drug Application for MYLOTARG. https://www.federalregister.gov/documents/2011/11/28/2011-30473/wyeth-pharmaceuticals-inc-withdrawal-of-approval-of-a-new-drug-application-for-mylotarg. Accessed September 21, 2024

Gamis AS, Alonzo TA, Meshinchi S, Sung L, Gerbing RB, Raimondi SC, Hirsch BA, Kahwash SB, Heerema-McKenney A, Winter L, Glick K, Davies SM, Byron P, Smith FO, Aplenc R (2014) Gemtuzumab ozogamicin in children and adolescents with de novo acute myeloid leukemia improves event-free survival by reducing relapse risk: results from the randomized Phase III children's oncology group trial AAML0531. J Clin Oncol 32:3021–3032

Genentech (2024) Pipeline. https://www.gene.com/medical-professionals/pipeline. Accessed September 21, 2024

Gilead (2020) Gilead Sciences completes acquisition of Immunomedics, Inc. https://www.gilead.com/news/news-details/2020/gilead-sciences-completes-acquisition-of-immunomedics-inc. Accessed September 30, 2024

Gilead (2024). https://www.gilead.com/news-and-press/press-room/press-releases/2024/5/gilead-provides-update-on-phase-3-tropics-04-study. Accessed August 27, 2024

Goldberg MM (2013) Pharmaceutical approval update. PT 38:246–258

Goozner M (2004) The $800 million pill: the truth behind the cost of new drugs. University of California Press

Gyawali B, Hey SP, Kesselheim AS (2020) Evaluating the evidence behind the surrogate measures included in the FDA's table of surrogate endpoints as supporting approval of cancer drugs. EClinicalMedicine 21:100332

Hills RK, Castaigne S, Appelbaum FR, Delaunay J, Petersdorf S, Othus M, Estey EH, Dombret H, Chevret S, Ifrah N, Cahn JY, Recher C, Chilton L, Moorman AV, Burnett AK (2014) Addition of gemtuzumab ozogamicin to induction chemotherapy in adult patients with acute myeloid leukaemia: a meta-analysis of individual patient data from randomised controlled trials. Lancet Oncol 15:986–996

Hotha KK (2023) The ABC of ADCs (antibody-drug conjugates): a comprehensive review of technical, regulatory, and clinical challenges. Adv Chem Eng Sci 13:363–381

Hurwitz E, Levy R, Maron R, Wilchek M, Arnon R, Sela M (1975) The covalent binding of daunomycin and adriamycin to antibodies, with retention of both drug and antibody activities. Cancer Res 35:1175–1181

Hurwitz E, Arnon R, Sahar E, Danon Y (1983) A conjugate of adriamycin and monoclonal antibodies to Thy-1 antigen inhibits human neuroblastoma cells in vitro. Ann N Y Acad Sci 417:125–136

Kantarjian HM, DeAngelo DJ, Stelljes M, Liedtke M, Stock W, Gökbuget N, O'Brien SM, Jabbour E, Wang T, Liang White J, Sleight B, Vandendries E, Advani AS (2019) Inotuzumab ozogamicin versus standard of care in relapsed or refractory acute lymphoblastic leukemia: final report and long-term survival follow-up from the randomized, phase 3 INO-VATE study. Cancer 125:2474–2487

ImmunoGen Pipeline (2024). https://www.immunogen.com/what-we-do/our-pipeline/. Accessed September 22, 2024

Junutula JR, Raab H, Clark S, Bhakta S, Leipold DD, Weir S, Chen Y, Simpson M, Tsai SP, Dennis MS, Lu Y, Meng YG, Ng C, Yang J, Lee CC, Duenas E, Gorrell J, Katta V, Kim A, McDorman K, Flagella K, Venook R, Ross S, Spencer SD, Lee Wong W, Lowman HB, Vandlen R, Sliwkowski MX, Scheller RH, Polakis P, Mallet W (2008) Site-specific conjugation of a cytotoxic drug to an antibody improves the therapeutic index. Nat Biotechnol 26:925–932

Kharfan-Dabaja MA (2014) A new dawn for gemtuzumab ozogamicin? Lancet Oncol 15:913–914

Krop IE, Kim SB, Martin AG, LoRusso PM, Ferrero JM, Badovinac-Crnjevic T, Hoersch S, Smitt M, Wildiers H (2017) Trastuzumab emtansine versus treatment of physician's choice in patients previously treated HER2-positive metastatic breast cancer (TH3RESA): final overall survival results from a randomized open-label phase 3 trial. Lancet Oncol 18:743–754

Kwiatek M, Carlo-Stella C, Urban A, Niewiarowski A, Wang L, Hamadani M (2024) LOTIS-5: an ongoing, phase 3, randomized study of loncastuximab tesirine with rituximab (Lonca-R) versus immunochemotherapy in patients with R/R DLBCL. J Clin Oncol 42:Suppl TPS7097

Lambert JM, Chari RV (2014) Ado-trastuzumab emtansine (T-DM1): an antibody-drug conjugate (ADC) for HER2-positive breast cancer. J Med Chem 57:6949–6964

Maranzano M, Mead M (2022) The role of transplantation in Hodgkin lymphoma. Front Oncol 12:1054314

McCaughan M (2017) Veterans Health Administration. Health Affairs Prescription Briefing

Miljkovic MD, Tuia J, Olivier T, Haslam A, Prasad V (2023) Cancer drug price and novelty in mechanism of action. JAMA Netw Open 6:e2347006

Moore KN, Oza AM, Colombo N, Oaknin A, Scambia G, Lorusso D, Konecny GE, Banerjee S, Murphy CG, Tanyi JL, Hirte H, Konner JA, Lim PC, Prasad-Hayes M, Monk BJ, Pautier P, Wang J, Berkenblit A, Vergote I, Birrer MJ (2021) Phase III, randomized trial of mirvetuximab soravtansine versus chemotherapy in patients with platinum-resistant ovarian cancer: primary analysis of FORWARD I. Ann Oncol 32:757–765

Moore KN, Angelergues A, Konecny GE, García Y, Banerjee S, Lorusso D, Lee JY, Moroney JW, Colombo N, Roszak A, Tromp J, Myers T, Lee JW, Beiner M, Cosgrove CM, Cibula D, Martin LP, Sabatier R, Buscema J, Estévez-García P, Coffman L, Nicum S, Duska LR, Pignata S, Gálvez F, Wang Y, Method M, Berkenblit A, Bello Roufai D, Van Gorp T (2023) Mirvetuximab soravtansine in FRa-positive, platinum-resistant ovarian cancer. N Engl J Med 389:2162–2174

NICE (2022). https://www.nice.org.uk/guidance/ta797. Accessed August 27, 2024

O'Hear C, Rubnitz JE (2014) Recent research and future prospects for gemtuzumab ozogamicin: could it make a comeback? Expert Rev Hematol 7:427–429

Panowksi S, Bhakta S, Raab H, Polakis P, Junutula JR (2014) Site-specific antibody drug conjugates for cancer therapy. MAbs 6:34–45

Pfizer (2009) Pfizer to acquire Wyeth, creating the world's premier biopharmaceutical company. https://www.pfizer.com/news/press-release/press-release-detail/pfizer_to_acquire_wyeth_creating_the_world_s_premier_biopharmaceutical_company. Accessed September 21, 2024

Pfizer (2023) Pfizer completes acquisition of Seagen. https://www.pfizer.com/news/press-release/press-release-detail/pfizer-completes-acquisition-seagen. Accessed September 21, 2024

Powles T, Rosenberg JE, Sonpavde GP, Loriot Y, Durán I, Lee JL, Matsubara N, Vulsteke C, Castellano D, Wu C, Campbell M, Matsangou M, Petrylak DP (2021) Enfortumab vedotin in previously treated advanced urothelial carcinoma. N Engl J Med 384(12):1125–1135

Powles T, Valderrama BP, Gupta S, Bedke J, Kikuchi E, Hoffman-Censits J, Iyer G, Vulsteke C, Park SH, Shin SJ, Castellano D, Fornarini G, Li JR, Gümüş M, Mar N, Loriot Y, Fléchon A, Duran I, Drakaki A, Narayanan S, Yu X, Gorla S, Homet Moreno B, van der Heijden MS (2024) Enfortumab vedotin and pembrolizumab in untreated advanced urothelial cancer. N Engl J Med 390:875–888

Prasad V, De Jesus K, Mailankody S (2017) The high price of anticancer drugs: origins, implications, barriers, and solutions. Nat Rev Clin Oncol 14(6):381–390

Roots Analysis (2024). https://www.rootsanalysis.com/key-insights/top-5-adc-manufacturers-offering-contract-services.html. Accessed September 28, 2024

Ross HJ, Hart LL, Swanson PM, Rarick MU, Figlin RA, Jacobs AD, McCune DE, Rosenberg AH, Baron AD, Grove LE, Thorn MD, Miller DM, Drachman JG, Rudin CM (2006) A randomized, multicenter study to determine the safety and efficacy of the immunoconjugate SGN-15 plus docetaxel for the treatment of non-small cell lung carcinoma. Lung Cancer 54:69–77

Rowland GF, O'Neill GJ, Davies DA (1975) Suppression of tumor growth in mice by a drug antibody conjugate using a novel approach to linkage. Nature 255:487–488

Sanders GD, Neumann PJ, Basu A, Brock DW, Feeny D, Krahn M, Kuntz KM, Meltzer DO, Owens DK, Prosser LA, Salomon JA, Sculpher MJ, Trikalinos TA, Russell LB, Siegel JE, Ganiats TG (2016) Recommendations for conduct, methodological practices, and reporting of cost-effectiveness analyses: second panel on cost-effectiveness in health and medicine. JAMA 316:1093–1103

Schnipper LE, Davidson NE, Wollins DS, Tyne C, Blayney DW, Blum D, Dicker AP, Ganz PA, Hoverman JR, Langdon R, Lyman GH, Meropol NJ, Mulvey T, Newcomer L, Peppercorn J, Polite B, Raghavan D, Rossi G, Saltz L, Schrag D, Smith TJ, Yu PP, Hudis CA, Schilsky RL (2015) American Society of Clinical Oncology statement: a conceptual framework to assess the value of cancer treatment options. J Clin Oncol 33:2563–2577

Senter PD, Sievers EL (2012) The discovery and development of brentuximab vedotin for use in relapsed Hodgkin lymphoma and systemic anaplastic large cell lymphoma. Nat Biotechnol 30:631–637

Smith S (2001) Technology evaluation: SGN-15, Seattle Genetics Inc. Curr Opin Mol Ther 3:295–302

Sullivan SD (2008) The promise of specialty pharmaceuticals: are they worth the price? J Manag Care Pharm 14:S3–S6

Thayer AM (2014) Building antibody-drug conjugates. Chem Eng News 92:13–21

Tilly H, Morschhauser F, Sehn LH, Friedberg JW, Trněný M, Sharman JP, Herbaux C, Burke JM, Matasar M, Rai S, Izutsu K, Mehta-Shah N, Oberic L, Chauchet A, Jurczak W, Song Y, Greil R, Mykhalska L, Bergua-Burgués JM, Cheung MC, Pinto A, Shin HJ, Hapgood G, Munhoz E, Abrisqueta P, Gau JP, Hirata J, Jiang Y, Yan M, Lee C, Flowers CR, Salles G (2021) Polatuzumab vedotin in previously untreated diffuse large B-cell lymphoma. N Engl J Med 386:351–363

Tonia T, Mettler A, Robert N, Schwarzer G, Seidenfeld J, Weingart O, Hyde C, Engert A, Bohlius J (2012) Erythropoietin or darbepoetin for patients with cancer. Cochrane Database Syst Rev 12:CD003407

Uprety D, Remon J, Adjei AA (2021) All that glitters is not gold: the story of rovalpituzumab tesirine in SCLC. J Thorac Oncol 16:1429–1433

Vergote I, González-Martín A, Fujiwara K, Kalbacher E, Bagaméri A, Ghamande S, Lee JY, Banerjee S, Maluf FC, Lorusso D, Yonemori K, Van Nieuwenhuysen E, Manso L, Woelber L, Westermann A, Covens A, Hasegawa K, Kim BG, Raimondo M, Bjurberg M, Cruz FM, Angelergues A, Cibula D, Barraclough L, Oaknin A, Gennigens C, Nicacio L, Teng MSL, Whalley E, Soumaoro I, Slomovitz BM (2024) Tisotumab vedotin as second- or third-line therapy for recurrent cervical cancer. N Engl J Med 391:44–55

von Minckwitz G, Huang CS, Mano MS, Loibl S, Mamounas EP, Untch M, Wolmark N, Rastogi P, Schneeweiss A, Redondo A, Fischer HH, Jacot W, Conlin AK, Arce-Salinas C, Wapnir IL, Jackisch C, DiGiovanna MP, Fasching PA, Crown JP, Wülfing P, Shao Z, Rota Caremoli E, Wu H, Lam LH, Tesarowski D, Smitt M, Douthwaite H, Singel SM, Geyer CE Jr (2019) Trastuzumab emtansine for residual invastive HER2-positive breast cancer. N Engl J Med 380:617–628

Zhang L, Shen D, Yu L, Yan Y, Wasan HS, Yu J, Zhang S, Sun L (2022) Is antibody-drug conjugate a rising star for clinical treatment of solid tumors? A systematic review and meta-analysis. Crit Rev Oncol Hematol 177:103758

Part V
Case Studies

Chapter 10
Mylotarg®: Revisiting Its Clinical Potential Post-withdrawal

Samantha Marcotte and Jennica L. Zaro

Abstract Mylotarg® (gemtuzumab ozogamicin) was the first antibody-drug conjugate approved for treatment of acute myeloid leukemia (AML) by the Food and Drug Administration via the accelerated approval process in 2000. Mylotarg® consists of an antibody directed toward the CD33 antigen conjugated to the antitumor antibiotic, calicheamicin. Mainly due to concerns about the safety profile and lack of improvement of clinical benefit in post-approval clinical trials, the drug was voluntarily withdrawn from the market by Wyeth in October 2010. Following its withdrawal from the market, results from several clinical trials warranted revisiting the clinical use of the drug when used at a low-dose range in newly diagnosed AML with favorable cytogenetics. The significant clinical evidence led to the eventual reapproval of Mylotarg® in 2017, following a new dosing regimen and lower recommended dose.

Keywords Mylotarg® · Gemtuzumab ozogamicin · Acute myeloid leukemia · CD33 antigen · Calicheamicin

10.1 Introduction

Under the accelerated approval process by the US Food and Drug Administration (FDA), gemtuzumab ozogamicin (GO) (Mylotarg®) was approved in May 2000 for the treatment of acute myeloid leukemia (AML) (Fig. 10.1). However, due to concerns about the product's safety, and its lack of improving clinical benefit in post-approval clinical trials, GO was voluntarily withdrawn from the market by Wyeth in

S. Marcotte · J. L. Zaro (✉)
Department of Pharmacology and Pharmaceutical Sciences, Alfred E. Mann School of Pharmacology and Pharmaceutical Sciences, University of Southern California, Los Angeles, CA, USA
e-mail: zaro@usc.edu

J. L. Zaro, J. Wang (eds.), *Antibody-Drug Conjugates*, AAPS Advances in the Pharmaceutical Sciences Series 17, https://doi.org/10.1007/978-3-031-83005-1_10

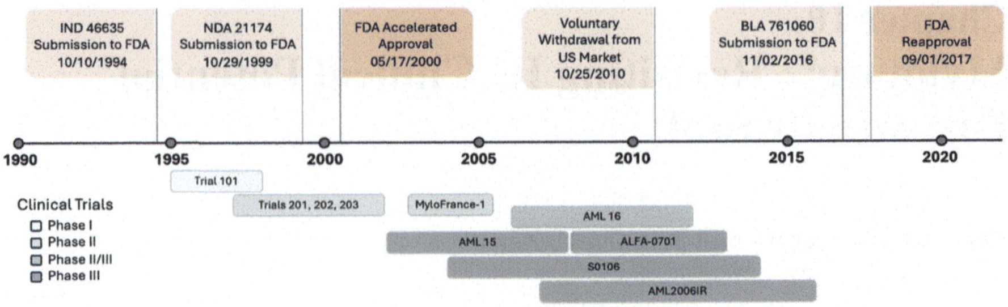

Fig. 10.1 Timeline of FDA events and clinical trials involved in the approval of Mylotarg®

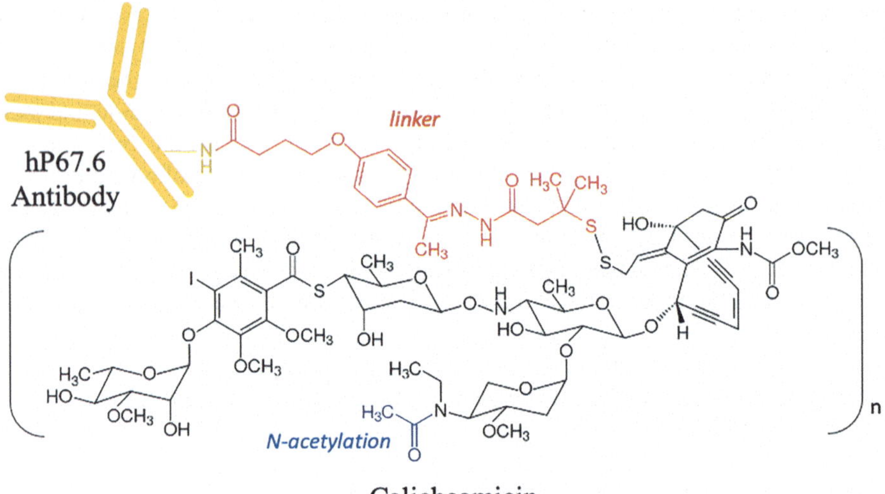

Fig. 10.2 Structure of GO. GO consists of the drug, calicheamicin, linked to the anti-CD33 hP67.6 antibody with an average drug-antibody ratio (DAR) of 3:4. The drug is linked to the lysine residues of the antibody via a pH-sensitive hydrazone linker. *GO* gemtuzumab ozogamicin

October 2010. The antibody-drug conjugate (ADC) GO consists of an antibody targeting the CD33 antigen expressed on most myeloid and monocytic leukemia cells linked to an antitumor antibiotic, calicheamicin, via a pH-sensitive hydrazone linker (Fig. 10.2). Following binding to CD33 antigens, GO is internalized and routed to the acidic lysosome, where the hydrazone linker is cleaved, freeing the conjugated drug. The drug is then able to be internalized into the nucleus, where it induces breaks in DNA. The activity of GO depends on several factors, including the expression and saturation levels of CD33, binding of GO at the cell surface, release of calicheamicin and induction of DNA breaks in the nucleus, and the efflux of calicheamicin via multidrug resistance (MDR)-related proteins (Fig. 10.3). Over

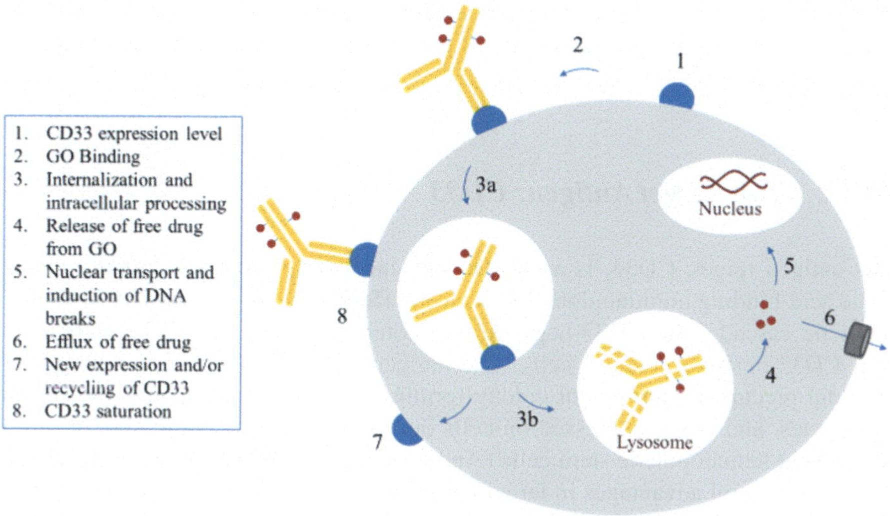

Fig. 10.3 Important criteria for the activity of GO. The activity of GO at the cellular level depends on several factors including the expression level of the antigen at the cell surface of target cells (1), the binding (2), internalization (3a), and intracellular processing (3b) of GO to the acidic lysosomal environment, where the hydrazone linkage is cleaved. Once freed from the antibody (4), the drug can enter the nucleus (5), where it then induces DNA breaks. The drug is also susceptible to efflux via MDR proteins (6). The CD33 antigens are continuously renewed at the cell surface (7). Following infusion with GO, the CD33 antigens are rapidly saturated (8), which also affects activity. *GO* gemtuzumab ozogamicin, *MDR* multidrug resistance

the past decade, numerous review articles have been published discussing the development and approval of Mylotarg® for AML treatment (Cowan et al. 2013; Sievers 2003; Tanimoto et al. 2013). Therefore, the details of the ADC, its approval process, and clinical trials will only be briefly introduced. The main focus of this case study will be on the revival of Mylotarg® in the clinics through clinical trials leading to its eventual second approval.

10.1.1 The Antibody: Gemtuzumab

Mylotarg® employs a monoclonal antibody, gemtuzumab (hP67.6), which is targeted against the CD33 antigen. Gemtuzumab is a humanized antibody derived from the complementarity-determining regions of the murine antibody p67.6. (Hamann et al. 2002). The IgG4 isotype influences the effector function of the antibody, where the effector functions correlate with the Fc-receptor-binding affinity (IgG1 > IgG3 > IgG4 > IgG2) (Charles et al. 2001). Therefore, consistent with its isotype, Mylotarg® does not exhibit any effector functions. Although engagement of secondary immune functions is considered beneficial for antitumor activity, the lack of binding to effector cells could conceivably lead to higher tumor localization. Another advantage of IgG4 is that it has the longest circulating half-life compared to the other isotypes.

10.1.2 The Target Antigen: CD33

The antigen target, CD33, is a 67-kDa transmembrane protein belonging to the sialic acid-binding immunoglobulin-like lectin (Siglec) family. As part of the hematopoietic cascade, the CD33-negative pluripotent stem cells (PSCs) differentiate into CD33-positive myeloid cells. The antigen is found on normal multipotent myeloid precursors, uni-potent colony-forming cells, maturing granulocytes, and monocytes, and is not expressed on cells outside the hematopoietic system or on pluripotent hematopoietic stem cells (Andrews et al. 1989; Robertson et al. 1992). There are several advantages in targeting the population of cells expressing CD33. AML has been shown to derive predominantly from malignant CD33-positive myeloid cells, and therefore a large majority of patients (>80%) express CD33 antigen on leukemic cells (Dinndorf et al. 1986; Griffin et al. 1984; Peiper et al. 1988). Another advantage of the CD33 target is that any normal CD33-positive cells that are destroyed by treatment can be regenerated by differentiation of the CD33-negative PSCs. CD33 antigens are also continuously renewed on the cell surface of myeloid cells (van Der Velden et al. 2001), thereby replenishing the population of available binding sites for GO. Finally, studies have suggested that CD33 stimulation is inhibitory. Binding and phosphorylation of the transmembrane protein leads to suppression of signals generated by receptor systems containing the immunoreceptor tyrosine-based activation motif (Paul et al. 2000). Therefore, gemtuzumab binding may also exert anti-leukemia effects on its own; however, this has not been confirmed.

Disadvantages of the CD33 antigen target include its relatively low expression level and slow internalization following antibody binding. Quantitative flow cytometry studies have estimated that leukemic cells express a relatively low number of CD33 molecules (approximately 10^4 CD33 molecules/cell) (Jilani et al. 2002; Scheinberg et al. 1989). Consequently, the CD33 antigenic sites are saturated after treatment with GO. As shown by van Der Velden et al., the CD33 sites were saturated within 3 h after the start of a 2-h infusion of 9 mg/m^2 of Mylotarg® (van Der Velden et al. 2001). Following binding to bivalent antibodies, the CD33–antibody complex internalizes, but the rate is relatively slow compared to other antigens, including the transferrin receptor (van der Jagt et al. 1992; Audran et al. 1995). Following internalization in myeloid blast cells and monocytes, results have shown that the complex is routed to the lysosomes where it is proteolytically degraded (van der Jagt et al. 1992; Press et al. 1996; Caron et al. 1992; van Der Velden et al. 2001). Taken together, the low receptor number, combined with the slow internalization, limits the amount of drug that can be delivered by the ADC. Therefore, it is crucial that a highly potent drug is utilized.

10.1.3 The Drug: Calicheamicin

Calicheamicin is an antitumor antibiotic that binds to the minor groove of DNA and causes cleavage in a sequence-dependent manner. The DNA damage causes cell cycle arrest followed by cell death via apoptosis predominantly via the mitochondrial pathway involving release of cytochrome c and involvement of Bcl-2 proteins and caspase activation (Walter et al. 2012; Linenberger 2005). Studies have shown that the treatment of CD33-expressing cells with GO results in induction of apoptosis (Naito et al. 2000; van Der Velden et al. 2001) within 72–96 h of culturing cells. Due to its high potency and delayed toxicity in animal models, this drug has not been utilized therapeutically on its own as a single-agent therapy.

10.1.4 The Conjugation Chemistry and Hydrazone Linker

In GO, a semisynthetic derivative of calicheamicin, N-acetyl-γ-calicheamicin 1,2-dimethyl hydrazine) is utilized. For conjugation, the lysine residues of the antibody are first activated to the amine-reactive N-hydroxysuccinimide (NHS)-ester. The cytotoxic drug is then conjugated to the NHS-antibody by covalent linkage of a bifunctional linker, 4-(4-acetylphenoxy) butanoic acid (AcBut linker), resulting in an acid-labile hydrazone linkage. The average drug–antibody ratio (DAR) of Mylotarg® is ~4 mol/mol, and approximately 50% of the antibody is unconjugated. Following binding to CD33 and receptor-mediated endocytosis, the ADC is presumably exposed to an acid lysosomal compartment, where the calicheamicin derivative is hydrolytically released intracellularly (Ricart 2011) (Fig. 10.3).

10.2 Approval Process, Post-marketing Analysis, and Withdrawal

As part of the FDA's accelerated approval process, Mylotarg® was the first ADC to be approved in the United States in 2000. The approval of Mylotarg® was based on three open-label, multicenter single-arm Phase 2 clinical trials (Trials 201, 202 and 203, Table 10.1) that enrolled a combined total of 142 patients at >60 years old, where the clinical data showed an overall response rate (ORR) of 26% (complete remission [CR], of 13% and CR with incomplete platelet count recovery [CRp] of 13%) (Sievers et al. 2001; Bross et al. 2001). The patients typically received two 9 mg/m^2 doses of GO over 14 days apart, which was shown in earlier phase 1 studies (Trial 101, Table 10.1) to saturate CD33 binding sites without dose-limiting non-hematologic toxicity (Sievers et al. 1999). It is important to point out that half of the CRs were CRps (13%), supporting the activity of the drug, but questioning the effect on survival rate. A subsequent confirmatory Phase 3 trial (S0106) by the

Table 10.1 Summary of clinical trial results considered in the 2017 FDA approval of GO (US FDA 2017)

Study ID#, Phase Study years	Treatment	Primary endpoint(s)	CD33+ AML patient population	CR% GO vs. No GO OR [95%CI]	EFS/RFS% GO vs. No GO HR [95%CI]	OS (%) GO vs. No GO HR [95% CI]	Comments
Trial 101[a] 0903A1–101 Phase 1 1995–1998	GO 0.25–9 mg/m^2 × 3, ≥14 d apart	Safety	Age 16–70 with R/R, $n = 41$	n/a	n/a	n/a	Dose escalation trial included in the 1999 NDA submission; 9 mg/m^2 identified as the MTD
Trials 201–203[a] 0903B1–201, 0903B1–202, 0903B1–203 Phase 2 1997–2002	GO 9 mg/m^2 × 2–3, 14–28 d apart	CR rate and safety	Adults in first relapse after CR ≥6 months, age 18+, $n = 84$ (201), $n = 95$ (202) or CR ≥3 months, adults ≥60, $n = 98$ (203)	17% [11%, 26%] (OR n/a)	n/a	n/a	Safety and efficacy trial with interim report results from 104 patients included in the 1999 NDA submission and supported GO's accelerated approval
AML16[b] NCT00454480 Phase 1/2 2002–2012	GO 3 mg/m^2 D1 + DA or DClo	OS	Adults ≥60y with untreated de novo or 2° AML or high risk MDS, $n = 1115$	62% vs. 58% 0.84 [0.66, 1.06] $p = 0.04$	3y RFS 21% vs. 16% 0.84 [0.71, 0.99] $p = 0.04$	3y OS 25% vs. 20% 0.87 [0.76, 1.00] $p = 0.5$	Study focused on treatment of elderly patients and supported the lower dose of GO with chemotherapy
AML15[c] ISRCTN17161961 Phase 3 2002–2008	GO 3 mg/m^2 D1 + DA or ADE or FLAG-Ida	OS, CR rate	Age 0–73 with untreated de novo or 2° AML, $n = 1113$	82% vs. 83% 1.04 [0.76,1.42] $p = 0.8$	5y RFS 39% vs. 35% 0.87 [0.73, 1.02] $p = 0.09$	5y OS 43% vs. 41% 0.92 [0.79, 1.08] $p = 0.3$	Supported the lower dose of GO with standard- and high-dose cytarabine and anthracyclines
Mylofrance-1[d] ID# n/a Phase 2 2003–2005	GO 3 mg/m^2 D1, 4, 7 + AraC	CR + CRp rate	Adults 22-80y after CR ≥ 3 ≤ 18 mo, $n = 57$	26% [16%, 40%] (OR n/a)	n/a	n/a	Pivotal trial assessing the activity of GO at the lower fractionated dose

SWOG S0106[e] NCT00085709 2004–2014	GO 6 mg/m^2 D4 + DA (3 + 7) + GF	DFS, CR rate	Adults <60 with untreated de novo AML, $n = 595$	69% vs. 70% 1.06 [0.75, 1.50] $p = 0.75$	5y RFS 43% vs. 42% 0.97 [0.75, 1.26] $p = 0.40$	5y OS 46% vs. 50% 1.13 [0.90, 1.42] $p = 0.59$	Confirmatory study terminated early after interim analysis showed lack of improvement in CR and increased deaths; Recognized as a major reason for the 2010 withdrawal
ALFA0701[f] NCT00927498 Phase 3 2007–2013	GO 3 mg/m^2 D1, 4, 7 + DA (3 + 7)	EFS	Adults 50-70y with untreated de novo AML, $n = 271$	70% vs. 70% 0.98 [0.58, 1.64] $p = 0.93$	3y EFS 40% vs. 14% 0.56 [0.42, 0.76] $p = <0.001$	3y OS 46% vs. 37% 0.81 [0.60, 1.09]	Trial was ongoing at the time of the S0106 trial closure and supported a lower GO dose
AML2006 IR[g] NCT00860639 2007–2016	GO 6 mg/m^2 D4 + DA (3 + 7)	EFS	Adults <60y with untreated de novo AML, $n = 254$	92% vs. 87% 0.60 [0.26, 1.34] $p = 0.22$	3y EFS 51% vs. 33% (HR n/a)	3y OS 53% vs. 46% (HR n/a)	Study failed to improve OS, but a better EFS was observed

ADE daunorubicin, cytarabine and etoposide, *ALFA* Acute Leukemia French Association, *AML* Acute myeloid leukemia, *AraC* cytarabine, *CI, 95%* 95% confidence interval, *CR* complete remission, *CRp* CR with incomplete platelet recovery, *D* day, *DA* daunorubicin+cytarabine, *DClo* daunorubicin+clofarabine, *DFS* disease-free survival, *EFS* event-free survival, *FLAG-Ida* fludarabine, cytarabine, granulocyte colony-stimulating factor, idarubicin, *GF* growth factor, *GO* gemtuzumab ozogamicin, *HR* hazard ratio, *n/a* not available, *OS* overall survival, *PCD* primary completion date, *RFS* relapse-free survival, *R/R* relapsed or refractory, *y* years, *6-MP* 6-mercaptopurine.

[a]US FDA (2000)
[b]Burnett et al. (2012)
[c]Burnett et al. (2011)
[d]Taksin et al. (2007)
[e]Petersdorf et al. (2013)
[f]Kirby (2012)
[g]Delaunay et al. (2011)

Southwest Oncology Group (SWOG) failed to demonstrate a clinical benefit compared to standard chemotherapy and also showed an increased mortality in the group of patients who received the ADC (Petersdorf et al. 2013).

This SWOG0106 study has received several critiques regarding the results obtained and the design of the experimental groups. First, the results showed a significantly worse mortality in GO (6 mg/m^2) than control (5.8% vs. 0.8%). However, the induction death rate obtained with GO treatment is within the normal range (5–7%) usually found in the younger patients age group represented in the trial. Also, the SWOG trial compared a daily dose of daunorubicin at 60–45 mg/m^2 daunorubicin + GO. Studies have shown that, in younger patients, a high-dose daunorubicin (90 mg/m^2) improves overall survival (OS) and CR compared to a low dose (45 mg/m^2). Therefore, the data could be interpreted as a compensation by GO for the reduction of the daunorubicin dose in the GO treatment group. Nevertheless, based on this study, the FDA requested that Wyeth voluntarily withdraw Mylotarg® from the US market in May 2000. Subsequently, Mylotarg® was voluntarily withdrawn on October 25, 2010. Following the SWOG study, there were several randomized trial reports that suggested Mylotarg® improved clinical outcomes at a lower fractionated dose, leading to the revisiting of its clinical use.

10.3 Clinical Studies Considered in the FDA Reapproval of Mylotarg®

Following withdrawal of Mylotarg® in 2010, the results from several ongoing clinical studies had been reported (Table 10.1). The main difference in the newer studies was the lower dose of GO administered to the patients. As previously mentioned, new CD33-binding sites are continuously renewed at the cell surface, with CD33 levels returning to pretreatment levels within 72 h after GO administration (van Der Velden et al. 2001; Caron et al. 1994). Therefore, these data indicated that the intracellular drug accumulation was enhanced by administration of lower doses of GO every 3 days. The more recent clinical trials, summarized below, indicated that a lower dose with more frequent administration improved the efficacy and toxicity profile of GO. Based on these studies, a Biologics License Application (BLA) was submitted to the FDA in 2016 (US FDA 2017).

10.3.1 MyloFrance-1 Trial

The MyloFrance-1 study, completed in 2005, was used to evaluate the safety and efficacy of the fractionated dosing regimen proposed in the BLA submission (US FDA 2017). In the MyloFrance-1 Phase 2 study, adult AML patients (median age 64 years, $n = 57$) in first relapse, with intermediate or poor risk cytogenetics, were given a single induction course consisting of three fractionated GO doses (3 mg/m^2) on days 1, 4, and 7. Patients receiving the GO monotherapy had a median overall survival of 8.4 months, a 33% OR rate, 26% CR (median 11.6 months duration), and 7% CRp. Compared to pivotal Phase 2 studies (Sievers et al. 2001), where patients were given higher GO doses (9 mg/m^2) on days 1 and 14, similar efficacy was observed while demonstrating an improved safety profile with no grade 3 or 4 liver function abnormalities observed. Additionally, better toleration of the fractionated low GO doses allowed 18 of 19 patients to receive consolidation chemotherapy (cytarabine) after achieving CR/CRp (Taksin et al. 2007).

10.3.2 AML15 Trial

The Phase 3 AML15 trial evaluated the benefit of adding GO to induction and/or consolidation chemotherapy. The trial included 1113 patients who were predominantly <60 years old and who received two induction courses with and without a single dose of GO (3 mg/m^2) in combination with one of three induction schedules (daunorubicin and cytarabine; cytarabine, daunorubicin and etoposide; fludarabine, cytarabine, granulocyte colony-stimulating factor, and idarubicin). The results showed that in a predefined subgroup of patients with favorable cytogenetics, there was an improved OS at 5 years (79% vs. 70%), but no benefit for patients with unfavorable cytogenetics (Burnett et al. 2011).

10.3.3 AML16 Trial

The AML16 Phase 2/3 trial from the UK Medical Research Council also assessed the benefit of adding GO to standard induction chemotherapy. AML patients (51–84 years, n = 1115) were randomized to receive different chemotherapy regimens (daunorubicin or daunorubicin with clofarabine), with and without a 3 mg/m^2 dose of GO on day one. The lower dose was established in a pilot study in younger patients (Burnett et al. 2006) as well as the AML15 trial described above (Burnett et al. 2011). Although patients receiving GO induction showed no overall difference in the rate of relapse, OS, or relapse-free survival (RFS), the 3-year cumulative incidence of relapse was significantly lower with GO (68% vs. 76%, $p = 0.007$), and the 3-year survival was significantly better (25% vs. 20%, $p = 0.05$). Additionally, patients with favorable cytogenetics whose leukemia was CD33+ showed an overall survival benefit (OS 79% vs. 51%). There was no difference in 30- or 6- day mortalities and no major increase in toxicity with GO treatment (Burnett et al. 2012).

10.3.4 ALFA0701 Trial

In the Acute Leukemia French Association trial, ALFA0701, fractionated doses of GO (3 mg/m^2 on days 1, 4, and 7) were administered to older AML patients (age 50–70 years, $n = 280$). In this study, both EFS (17.1% vs. 40.8%, $p = 0.0003$) and OS (53.2% vs. 41.9%, $p = 0.0368$) were improved at 2 years in the GO group. Subgroup analysis showed that the improvements were predominantly in those patients with cytogenetically favorable or intermediate-risk disease (Renneville et al. 2014). Hematological toxicity was found to be significantly higher in the GO treated versus the control group (16% vs. 3%, $p < 0.0001$); however, there was no increase in the risk of death from toxicity (Kirby 2012).

10.3.5 AML2006 Trial

In the study by the Groupe Ouest Est d'Etude des Leucémies et Autres Maladies du Sange, AML2006 IR, 254 patients with an intermediate risk karyotype (age 18–60 years) were randomized to receive 6 mg/m^2 with both induction and consolidation chemotherapy. The patients in the GO showed an improved 3-year event-free survival (EFS) (53.7% vs. 27%, $p = 0.0308$), but there was no improvement in OS (Delaunay et al. 2011).

10.4 Additional Studies Supporting the Clinical Use of Mylotarg

10.4.1 AML17 Trial

To investigate the feasibility of a sequential induction strategy, a Phase 2 trial was first carried out by the European Organization for Research and Treatment of Cancer (EORTC) and Gruppo Italiano Malittie Ematologiche dell'Adulto (GIMEMA) consortium. In this trial, 9 mg/m^2 of GO (days 1 and 15) was administered in older patients with previously untreated AML, followed by standard chemotherapy. The trial resulted in approximately 70% of patients being able to complete the induction sequence and an overall response rate of 54%. However, the induction mortality was high (14% of patients dying as a result of toxicity), and therefore a lower dose was considered for the subsequent trial.

In the follow-up Phase 3 trial, patients received a course of standard chemotherapy with or without a preceding course of a lower 6 mg/m^2 dose of GO (days 1 and 15). The results of the trial ($n = 472$ patients, ages 61–75) showed that the ORR was comparable between the two arms (with GO, 45%; without GO, 49%). However, the induction and 60-day mortality rates, as well as hematologic and liver toxicities, were higher in the GO-treated patients. Therefore, the trial failed to demonstrate an advantage of GO before standard chemotherapy and showed a higher risk of early mortality in the older age subgroups >70 years (Amadori et al. 2013).

10.4.2 AML19 Trial

In another Phase 3 EORTC-GIMEMA trial, AML19, newly diagnosed, older AML patients (ages >60, $n = 237$) ineligible for intensive chemotherapy were randomized to receive GO as a monotherapy with best supportive care (BSC) (blood transfusions, antimicrobials, other symptomatic therapies) ($n = 118$) or BSC including hydroxyurea or low-dose cytarabine ($n = 119$). In the treatment arm, GO was given as a fractionated, single induction dose (6 mg/m^2 on day 1 and 3 mg/m^2 on day 8) with a continuation course of 2 mg/m^2 every 4 weeks for up to 8 months if no evidence of significant toxicities or disease progression occurred (CR, PR, SD). Patients in the GO group had higher median OS (4.9 months vs. 3.6 months) and a significant one-year survival rate percentage (24.3% vs. 9.7%) compared to the BSC group (Amadori et al. 2016).

10.4.3 AAML0531 Trial

The AAML0531 trial evaluated the addition of GO to standard chemotherapy dose escalation to improve EFS and OS in pediatric patients with newly diagnosed CD33-positive AML ($n = 1063$, ages 0–29). Patients randomly received a standard five-cycle chemotherapy treatment ($n = 531$) or the same chemotherapy plus two doses of GO (3 mg/m^2) in two courses ($n = 532$). OS did not significantly improve (3 years: 69.4% vs. 65.4%). However, the addition of GO significantly increased EFS (3 years: 53.1% vs. 46.9%) compared to chemotherapy alone and significantly reduced relapse risk (3 years: 32.8% vs. 41.3%) (Aplenc et al. 2013; Gamis et al. 2014).

10.5 Post-withdrawal

Overall, the response to the voluntary withdrawal of GO had been mixed. When comparing the ongoing randomized trials at the time, they all found that newly diagnosed AML patients with a favorable cytogenetic risk profile were likely to benefit from GO treatment. Furthermore, those with an intermediate-risk AML had the same benefit in most of the trials. Cytogenetic analysis has allowed for the identification of several gene mutations that have a prognostic significance in comparison to cytogenetically normal (CN) AML. Identification of these mutations has allowed a prognostic classification of CN-AML and improved the risk stratification in this subset of patients. Furthermore, results have shown a great benefit of GO in treatment of acute promyelocytic leukemia (APL), likely due to the higher surface expression of CD33. GO has been shown to result in durable molecular remissions in patients with relapsed APL (Lo-Coco et al. 2004) and can effectively replace anthracyclines in the management of APL (Breccia et al. 2007; Ravandi et al. 2009). Therefore, in line with moving toward more personalized medicine, it is clear that there are certain well-defined molecular subgroups that could benefit from GO treatment.

The optimal timing and dose of GO administration was another area that needed further consideration. The AML17 and ALFA0701 trials showed that coadministration of low doses of GO with induction therapy improves the outcome and is a safer alternative to sequential administration of higher doses. Therefore, it is better to use lower, concomitant doses rather than the higher sequential doses used in several of the trials. Taken together, the data support the efficacy of GO when used at a dose range of 3–6 mg/m^2 in APL, and in newly diagnosed AML with favorable cytogenetics. Unfortunately, these data were not available when GO was withdrawn from the market. Since AML is a highly heterogeneous disease, basing the potential clinical benefit on average patient data may not fully reflect the true therapeutic potential of the drug. Therefore, GO may be an active anti-AML with an acceptable toxicity profile when using a more optimal dose in a sub-patient population with favorable to intermediate cytogenetic risk profiles.

10.6 Reapproval and Summary

Following the withdrawal of GO in 2010, the FDA reapproved GO, with orphan drug designation, on September 1, 2017, for treating newly diagnosed adult CD33+ AML patients and relapsed or refractory CD33+ AML patients (ages 2+). The approval of Mylotarg was based on the ALFA-0710, S0106, AML15, AML16, and AML2006IR trials, which were included as supporting evidence in the BLA submission. The submission included a meta-analysis using individual patient data ($n = 3331$ patients) from these five trials. The results from the meta-analysis indicated that although the CR was not improved with the addition of GO, the risk of relapse (RR) was reduced (OR 0.81 [95% CI 0.73–0.9, $p = 0.0001$]) and the 5-year OS increased (OR 0.9 [95% CI 0.82–0.98, $p = 0.01$]) (US FDA 2017). In addition, the results from the MyloFrance-1 study supported the lower fractionated dosing schedule. Generally, patients receive an induction dose of 3 mg/m^2 or 6 mg/m^2 followed by a consolidation course of 3 mg/m^2 or 2 mg/m^2. Regardless of the lower dose, a black box warning for hepatotoxicity, including severe or fatal veno-occlusive disease (VOD), remains on GO's label, and patients with a history of stem cell transplant, who are candidates for stem cell transplant, or have adverse risk cytogenetics should use GO with caution. Overall, however, fractionated dosing for optimal GO delivery in patients allows for better GO tolerability while maintaining efficacy. As of February 2018, GO is a recommended frontline treatment for younger CD33+ AML patients by the NCCN Clinical Practice Guidelines for AML (NCCN

2025). GO is also recommended for acute promyelocytic leukemia (APL) intervention when all-trans retinoic acid (ATRA), arsenic trioxide (ATO), and chemotherapy present a high risk to the patient (NCCN 2025). Additionally, supported by clinical evidence from AAML0531, GO indications further expanded to newly diagnosed AML pediatric patients 1 month and older on June 16, 2020. GO's history warrants investigation of alternative dosing schedules and doses to improve tolerability and efficacy in various patient populations to fully reveal a drug's potential clinical benefit.

References

Amadori S, Suciu S, Stasi R, Salih HR, Selleslag D, Muus P, De Fabritiis P, Venditti A, Ho AD, Lubbert M, Thomas X, Latagliata R, Halkes CJ, Falzetti F, Magro D, Guimaraes JE, Berneman Z, Specchia G, Karrasch M, Fazi P, Vignetti M, Willemze R, de Witte T, Marie JP (2013) Sequential combination of gemtuzumab ozogamicin and standard chemotherapy in older patients with newly diagnosed acute myeloid leukemia: results of a randomized phase III trial by the EORTC and GIMEMA consortium (AML-17). J Clin Oncol Off J Am Soc Clin Oncol 31(35):4424–4430. https://doi.org/10.1200/JCO.2013.49.0771

Andrews RG, Singer JW, Bernstein ID (1989) Precursors of colony-forming cells in humans can be distinguished from colony-forming cells by expression of the CD33 and CD34 antigens and light scatter properties. J Exp Med 169(5):1721–1731

Aplenc R, Alonzo TA, Sung L, Meshinchi S, Gerbing RB, Raimondi SC, Hirsch B, Kahwash S, Heerema-McKenney A, Winter L, Glick K, Davies SM, Byron P, Smith FO (2013) Gemtuzumab Ozogamicin (GO) In Children With De Novo Acute Myeloid Leukemia (AML) Improves Event-Free Survival (EFS) By Reducing Relapse Risk—Results From The Randomized Phase III Children's Oncology Group (COG) Trial, AAML0531. Blood 122(21):355

Audran R, Drenou B, Wittke F, Gaudin A, Lesimple T, Toujas L (1995) Internalization of human macrophage surface antigens induced by monoclonal antibodies. J Immunol Methods 188(1):147–154

Breccia M, Cimino G, Diverio D, Gentilini F, Mandelli F, Lo Coco F (2007) Sustained molecular remission after low dose gemtuzumab-ozogamicin in elderly patients with advanced acute promyelocytic leukemia. Haematologica 92(9):1273–1274

Bross PF, Beitz J, Chen G, Chen XH, Duffy E, Kieffer L, Roy S, Sridhara R, Rahman A, Williams G, Pazdur R (2001) Approval summary: gemtuzumab ozogamicin in relapsed acute myeloid leukemia. Clin Cancer Res 7(6):1490–1496

Burnett AK, Hills RK, Milligan D, Kjeldsen L, Kell J, Russell NH, Yin JA, Hunter A, Goldstone AH, Wheatley K (2011) Identification of patients with acute myeloblastic leukemia who benefit from the addition of gemtuzumab ozogamicin: results of the MRC AML15 trial. J Clin Oncol Off J Am Soc Clin Oncol 29(4):369–377. https://doi.org/10.1200/JCO.2010.31.4310

Burnett AK, Kell WJ, Hills RK, Russell NH, Yin J, Hunter A, Culligan D (2006) The feasibility of combining daunorubicin, clofarabine and gemtuzumab ozogamicin is feasible and effective. A pilot study. Blood 108:1950

Burnett AK, Russell NH, Hills RK, Kell J, Freeman S, Kjeldsen L, Hunter AE, Yin J, Craddock CF, Dufva IH, Wheatley K, Milligan D (2012) Addition of gemtuzumab ozogamicin to induction chemotherapy improves survival in older patients with acute myeloid leukemia. J Clin Oncol Off J Am Soc Clin Oncol 30(32):3924–3931. https://doi.org/10.1200/JCO.2012.42.2964

Caron PC, Co MS, Bull MK, Avdalovic NM, Queen C, Scheinberg DA (1992) Biological and immunological features of humanized M195 (anti-CD33) monoclonal antibodies. Cancer Res 52(24):6761–6767

Caron PC, Jurcic JG, Scott AM, Finn RD, Divgi CR, Graham MC, Jureidini IM, Sgouros G, Tyson D, Old LJ et al (1994) A phase 1B trial of humanized monoclonal antibody M195 (anti-CD33) in myeloid leukemia: specific targeting without immunogenicity. Blood 83(7):1760–1768

Charles A, Janeway J, Travers P, Walport M, Shlomchik MJ (2001) Immunobiology. In: The immune system in health and disease. Garland Science, New York

Cowan AJ, Laszlo GS, Estey EH, Walter RB (2013) Antibody-based therapy of acute myeloid leukemia with gemtuzumab ozogamicin. Front Biosci 18:1311–1334

Delaunay J, Recher C, Pigneux A, Witz F, Vey N, Blanchet O, Lefebvre P, Luquet I, Guillerme I, Volteau C, Gyan E, Lioure B, Jourdan E, Bouscary D, Guieze R, Randriamalala E, Ojeda Uribe ME, Dreyfus F, Lacombe C, Béné M-c, Cahn J-Y, Harousseau J-L, Ifrah N (2011) Addition of gemtuzumab ozogamycin to chemotherapy improves event-free survival but not overall survival of AML patients with intermediate cytogenetics not eligible for allogeneic transplantation. Results of the GOELAMS AML 2006 IR study. Blood 118(21):79

Dinndorf PA, Andrews RG, Benjamin D, Ridgway D, Wolff L, Bernstein ID (1986) Expression of normal myeloid-associated antigens by acute leukemia cells. Blood 67(4):1048–1053

Gamis A, Alonzo T, Meshinchi S, Sung L, Gerbing R, Raimondi SC, Hirsch BA, Kahwash SB, Heerema-McKenny A, Winter L, Glick K, Davies SM, Byron P, Smith FO, Aplec R (2014) Gemtuzumab Ozogamicin in Children and Adolescents With De Novo Acute Myeloid Leukemia Improves Event-Free Survival by Reducing Relapse Risk: Results From the Randomized Phase III Children's Oncology Group Trial AAML0531 Journal of Clinical Oncology 32(27):3021–3032. https://doi.org/10.1200/JCO.2014.55.3628

Griffin JD, Linch D, Sabbath K, Larcom P, Schlossman SF (1984) A monoclonal antibody reactive with normal and leukemic human myeloid progenitor cells. Leuk Res 8(4):521–534

Hamann PR, Hinman LM, Beyer CF, Lindh D, Upeslacis J, Flowers DA, Bernstein I (2002) An anti-CD33 antibody-calicheamicin conjugate for treatment of acute myeloid leukemia. Choice of linker. Bioconjug Chem 13(1):40–46

Jilani I, Estey E, Huh Y, Joe Y, Manshouri T, Yared M, Giles F, Kantarjian H, Cortes J, Thomas D, Keating M, Freireich E, Albitar M (2002) Differences in CD33 intensity between various myeloid neoplasms. Am J Clin Pathol 118(4):560–566. https://doi.org/10.1309/1WMW-CMXX-4WN4-T55U

Kirby T (2012) Louisa Degenhardt: hooked on addiction research. Lancet 379(9810):21. https://doi.org/10.1016/S0140-6736(12)60010-5

Linenberger ML (2005) CD33-directed therapy with gemtuzumab ozogamicin in acute myeloid leukemia: progress in understanding cytotoxicity and potential mechanisms of drug resistance. Leukemia 19(2):176–182. https://doi.org/10.1038/sj.leu.2403598

Lo-Coco F, Cimino G, Breccia M, Noguera NI, Diverio D, Finolezzi E, Pogliani EM, Di Bona E, Micalizzi C, Kropp M, Venditti A, Tafuri A, Mandelli F (2004) Gemtuzumab ozogamicin (Mylotarg) as a single agent for molecularly relapsed acute promyelocytic leukemia. Blood 104(7):1995–1999. https://doi.org/10.1182/blood-2004-04-1550

National Comprehensive Cancer Network (NCCN) (2025) Acute Myeloid Leukemia Clinical Guidelines. Version 2.2025

Naito K, Takeshita A, Shigeno K, Nakamura S, Fujisawa S, Shinjo K, Yoshida H, Ohnishi K, Mori M, Terakawa S, Ohno R (2000) Calicheamicin-conjugated humanized anti-CD33 monoclonal antibody (gemtuzumab zogamicin, CMA-676) shows cytocidal effect on CD33-positive leukemia cell lines, but is inactive on P-glycoprotein-expressing sublines. Leukemia 14(8):1436–1443

Paul SP, Taylor LS, Stansbury EK, McVicar DW (2000) Myeloid specific human CD33 is an inhibitory receptor with differential ITIM function in recruiting the phosphatases SHP-1 and SHP-2. Blood 96(2):483–490

Peiper SC, Ashmun RA, Look AT (1988) Molecular cloning, expression, and chromosomal localization of a human gene encoding the CD33 myeloid differentiation antigen. Blood 72(1):314–321

Petersdorf SH, Kopecky KJ, Slovak M, Willman C, Nevill T, Brandwein J, Larson RA, Erba HP, Stiff PJ, Stuart RK, Walter RB, Tallman MS, Stenke L, Appelbaum FR (2013) A phase 3 study of gemtuzumab ozogamicin during induction and postconsolidation therapy in younger patients with acute myeloid leukemia. Blood 121(24):4854–4860. https://doi.org/10.1182/blood-2013-01-466706

Press OW, Shan D, Howell-Clark J, Eary J, Appelbaum FR, Matthews D, King DJ, Haines AM, Hamann P, Hinman L, Shochat D, Bernstein ID (1996) Comparative metabolism and retention of iodine-125, yttrium-90, and indium-111 radioimmunoconjugates by cancer cells. Cancer Res 56(9):2123–2129

Ravandi F, Estey E, Jones D, Faderl S, O'Brien S, Fiorentino J, Pierce S, Blamble D, Estrov Z, Wierda W, Ferrajoli A, Verstovsek S, Garcia-Manero G, Cortes J, Kantarjian H (2009) Effective treatment of acute promyelocytic leukemia with all-trans-retinoic acid, arsenic trioxide, and gemtuzumab ozogamicin. J Clin Oncol Off J Am Soc Clin Oncol 27(4):504–510. https://doi.org/10.1200/JCO.2008.18.6130

Renneville A, Abdelali RB, Chevret S, Nibourel O, Cheok M, Pautas C, Dulery R, Boyer T, Cayuela JM, Hayette S, Raffoux E, Farhat H, Boissel N, Terre C, Dombret H, Castaigne S, Preudhomme C (2014) Clinical impact of gene mutations and lesions detected by SNP-array karyotyping in acute myeloid leukemia patients in the context of gemtuzumab ozogamicin treatment: results of the ALFA-0701 trial. Oncotarget 5(4):916–932

Ricart AD (2011) Antibody-drug conjugates of calicheamicin derivative: gemtuzumab ozogamicin and inotuzumab ozogamicin. Clin Cancer Res 17(20):6417–6427. https://doi.org/10.1158/1078-0432.CCR-11-0486

Robertson MJ, Soiffer RJ, Freedman AS, Rabinowe SL, Anderson KC, Ervin TJ, Murray C, Dear K, Griffin JD, Nadler LM et al (1992) Human bone marrow depleted of CD33-positive cells mediates delayed but durable reconstitution of hematopoiesis: clinical trial of MY9 monoclonal antibody-purged autografts for the treatment of acute myeloid leukemia. Blood 79(9):2229–2236

Scheinberg DA, Tanimoto M, McKenzie S, Strife A, Old LJ, Clarkson BD (1989) Monoclonal antibody M195: a diagnostic marker for acute myelogenous leukemia. Leukemia 3(6):440–445

Sievers EL (2003) Antibody-targeted chemotherapy of acute myeloid leukemia using gemtuzumab ozogamicin (Mylotarg). Blood Cells Mol Dis 31(1):7–10

Sievers EL, Appelbaum FR, Spielberger RT, Forman SJ, Flowers D, Smith FO, Shannon-Dorcy K, Berger MS, Bernstein ID (1999) Selective ablation of acute myeloid leukemia using antibody-targeted chemotherapy: a phase I study of an anti-CD33 calicheamicin immunoconjugate. Blood 93(11):3678–3684

Sievers EL, Larson RA, Stadtmauer EA, Estey E, Lowenberg B, Dombret H, Karanes C, Theobald M, Bennett JM, Sherman ML, Berger MS, Eten CB, Loken MR, van Dongen JJ, Bernstein ID, Appelbaum FR, Mylotarg Study G (2001) Efficacy and safety of gemtuzumab ozogamicin in patients with CD33-positive acute myeloid leukemia in first relapse. J Clin Oncol Off J Am Soc Clin Oncol 19(13):3244–3254

Taksin AL, Legrand O, Raffoux E, de Revel T, Thomas X, Contentin N, Bouabdallah R, Pautas C, Turlure P, Reman O, Gardin C, Varet B, de Botton S, Pousset F, Farhat H, Chevret S, Dombret H, Castaigne S (2007) High efficacy and safety profile of fractionated doses of Mylotarg as induction therapy in patients with relapsed acute myeloblastic leukemia: a prospective study of the alfa group. Leukemia 21:66–71

Tanimoto T, Tsubokura M, Mori J, Pietrek M, Ono S, Kami M (2013) Differences in drug approval processes of 3 regulatory agencies: a case study of gemtuzumab ozogamicin. Investig New Drugs 31(2):473–478. https://doi.org/10.1007/s10637-012-9877-8

van der Jagt RH, Badger CC, Appelbaum FR, Press OW, Matthews DC, Eary JF, Krohn KA, Bernstein ID (1992) Localization of radiolabeled antimyeloid antibodies in a human acute leukemia xenograft tumor model. Cancer Res 52(1):89–94 van Der Velden VH, te Marvelde JG, Hoogeveen PG, Bernstein ID, Houtsmuller AB, Berger MS, van Dongen JJ (2001) Targeting of the CD33-calicheamicin immunoconjugate Mylotarg (CMA-676) in acute myeloid leukemia: in vivo and in vitro saturation and internalization by leukemic and normal myeloid cells. Blood 97(10):3197–3204

Walter RB, Appelbaum FR, Estey EH, Bernstein ID (2012) Acute myeloid leukemia stem cells and CD33-targeted immunotherapy. Blood 119(26):6198–6208. https://doi.org/10.1182/blood-2011-11-325050

U.S. Food and Drug Administration (2000) Center for Drug Evaluation and Research New Drug Application (NDA) Medical Review, Application Number NDA 21174. Drug Name Mylotarg® (Gemtuzumab Ozogamicin); Applicant: Wyeth Ayerst Research

U.S. Food and Drug Administration (2017) Oncologic Drugs Advisory Committee (ODAC) Briefing Document for Biologics License Application (BLA) 761060. Drug Name Mylotarg® (Gemtuzumab Ozogamicin); Applicant: Wyeth Pharmaceuticals Inc., a subsidiary of Pfizer Inc.

Chapter 11
ADCETRIS: A Regulatory Case Study of a New-Generation Antibody-Drug Conjugate

Bruce W. Hart Jr.

Abstract ADCETRIS® (brentuximab vedotin) is a CD30-directed antibody-drug conjugate (ADC). When it was approved by the Food and Drug Administration (FDA) in August 2011, it became only the second ADC approved by the FDA and the only one on the market at that time following the withdrawal of Mylotarg (gemtuzumab ozogamicin) in June of 2010. The development pathway incorporated nearly all of the FDA incentive development pathways available at that time including fast track and orphan drug designations, priority review, accelerated approval, and special protocol assessment and was the subject of the Oncologic Drugs Advisory Committee. This chapter will describe the US regulatory development path for ADCETRIS, pointing out the key discussion points with the agency and the outcomes.

Keywords Brentuximab vedotin · Regulatory · Food and Drug Administration · Biologics license application

11.1 ADCETRIS Description

The ADCETRIS antibody-drug conjugate (ADC) consists of the antibody cAC10 and the small molecule microtubule inhibitor monomethyl auristatin E (MMAE) attached to the antibody via a protease cleavable linker (Fig. 11.1).

The cAC10 antibody is targeted specifically to CD30, an antigen with limited expression on normal human tissues but preferential expression on select hematologic malignancies including Hodgkin lymphoma (HL), anaplastic large cell lymphoma (ALCL), and some other non-Hodgkin lymphomas. Nonclinical data suggest

B. W. Hart Jr. (✉)
Independent Consultant, Seattle, WA, USA

J. L. Zaro, J. Wang (eds.), *Antibody-Drug Conjugates*, AAPS Advances in the Pharmaceutical Sciences Series 17, https://doi.org/10.1007/978-3-031-83005-1_11

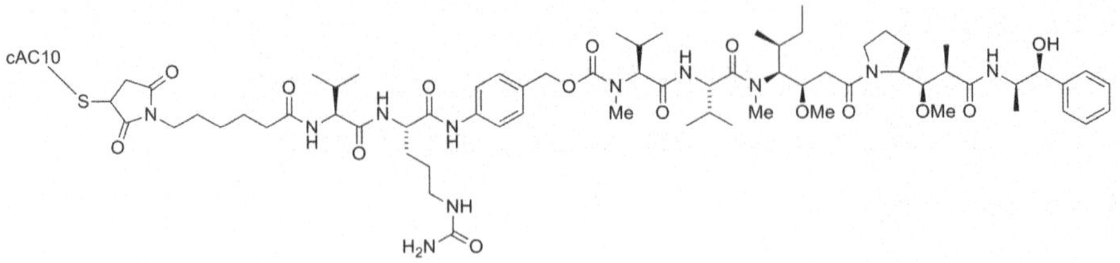

Fig. 11.1 Structure of brentuximab vedotin

that the anticancer activity of brentuximab vedotin is due to the binding of the ADC to CD30-expressing cells, followed by internalization of the ADC CD30 complex, and the release of MMAE via proteolytic cleavage. Binding of MMAE to tubulin disrupts the microtubule network within the cell, subsequently inducing cell cycle arrest and apoptotic death of the cells. Brentuximab vedotin embodies several important ADC design principles, including an internalizing antigen that is preferentially expressed on tumor cells rather than on normal tissues, a potent cytotoxic component that is rapidly released inside the tumor cell, and a stable linker that limits systemic release of the cytotoxic.

11.2 SGN-30

Prior to the clinical development of ADCETRIS, the naked antibody cAC10, also known as SGN-30, entered clinical development in 2003 and was studied in two Phase 1 trials and two Phase 2 trials. SGN-30, administered as weekly doses as high as 12 mg/kg, exhibited antitumor activity in systemic ALCL (sALCL) and CD30+ cutaneous T-cell lymphoma (CTCL). In Phase 2 trials, the overall response rate (ORR) in ALCL patients was 17% and 70% in CTCL patients (Forero-Torres et al. 2009; Duvic et al. 2009). However, the activity of SGN-30 in HL (0% ORR) did not warrant further development and therefore a corporate decision was made to discontinue the program.

11.3 ADCETRIS Regulatory Timeline

Listed in Table 11.1 are the major regulatory interactions during the development of ADCETRIS listed in chronological order. All of the regulatory information contained in this chapter is publically available on the Food and Drug Administration (FDA) website at drugs@fda.com and http://www.fda.gov/AdvisoryCommittees/default.htm.

Table 11.1 Major regulatory interactions

Investigational new drug (IND) submission	June 2006
Orphan drug designation for HL	January 2007
End of Phase 1 meeting with FDA	July 2008
Orphan drug designation for ALCL	October 2008
Special protocol assessment granted for HL pivotal trial	January 2009
Fast track designation for HL	March 2009
Nonclinical/clinical pharmacology meeting with FDA	March 2009
End of Phase 2 chemistry, manufacturing, and controls (CMC) meeting with FDA	January 2010
Clinical pre-biologics licensing application (BLA) meeting with FDA	November 2010
CMC pre-BLA meeting with FDA	December 2010
BLA submission	February 2011
Oncologic Drugs Advisory Committee meeting	July 2011
Confirmatory trial meeting with FDA	July 2011
BLA approval by FDA	August 2011

11.4 Early Days: Phase 1 and the Food and Drug Administration

SGN-35 began its US regulatory life cycle in June 2006 with the submission of an investigational new drug (IND) application to the Division of Drug Oncology Products in the Office of Oncology Drug Products. The first Phase 1 trial of SGN-35, as it was designated then, was designed to test the safety and antitumor activity of a dose of drug administered once every three weeks (q3w). Enrollment began in November 2006 and the last patient completed the study in July 2009. Forty-five (45) patients were enrolled and received study drug. The ORR in this trial was 36% and was dose-level-dependent. SGN-35 was also generally well tolerated (Younes et al. 2010).

A second Phase 1 trial was submitted to the IND in September 2007. This trial was designed to explore the safety and antitumor activity of a once weekly dose schedule of SGN-35 on a 28-day cycle (3 weeks on, 1 week off). Enrollment began in March 2008 and the last patient completed the study in February 2010. Forty-four (44) patients were enrolled and received study drug. The ORR in this trial was 53% and SGN-35 was generally well tolerated (Fanale et al. 2012).

Based on the compelling early tumor response data, Seattle Genetics requested a Type B End of Phase 1 (EOP1) meeting with the FDA. This meeting occurred in July 2008 and during the meeting, the design of two single-arm pivotal trials was discussed. For HL, the FDA agreed that patients who had progressive or relapsed HL following autologous stem cell transplant (ASCT) were an unmet need. When asked if a single-arm trial with ORR as the primary endpoint would be acceptable for accelerated approval in this patient population, the FDA responded that a controlled randomized trial is preferred but a single-arm trial may be acceptable provided there is sufficient evidence of efficacy, i.e., high response rate with prolonged duration and with acceptable safety. For sALCL, FDA agreed that patients who had progressive or relapsed sALCL following at least one prior therapy were an unmet need. When asked if a single-arm trial with ORR as the primary endpoint would be acceptable for accelerated approval in this patient population, the FDA responded that the rarity of a disease does not necessarily preclude one from performing a randomized trial but a single-arm Phase 2 study may support accelerated approval depending on the response rate, including proportion of complete response (CR)s, duration of response (DOR), and the risk-benefit ratio. A sufficiently large CR rate and duration might support full approval. Seattle Genetics also asked for FDA feedback on a proposed Phase 3 trial in patients at high risk of progression following ASCT that was to act as a confirmatory trial in the case of an accelerated approval. The FDA had several questions regarding the patient population and the endpoints and no agreements were gained at the meeting. The FDA did agree that 1.8 mg/kg of SGN-35 every 3 weeks was appropriate for future single-agent development and that a safety database of at least 175 patients would be sufficient for registrational purposes in the intended indications. The FDA reminded Seattle Genetics of pre-IND comments regarding the clinical pharmacology plan and specifically pointed out that the potential for QT/QTc interval prolongation would need to be addressed and requested the submission of an ECG evaluation plan for review.

11.5 Pivotal Trial Regulatory Agreements

During the EOP1 meeting, the FDA requested that the proposed pivotal trials be submitted for special protocol assessment (SPA). Seattle Genetics submitted a request for SPA for the single-arm Phase 2 pivotal trial for relapsed/refractory HL. In January 2009, the FDA granted an SPA for the HL protocol, agreeing to trial design aspects including the patient population (patients who had relapsed after or were refractory to an ASCT), the number of patients (100), and the primary endpoint (ORR).

Hodgkin lymphoma and ALCL are both rare diseases and orphan drug designation was granted for HL in January 2007 and for ALCL in October 2008. Based on the potential for ADCETRIS to treat a serious condition and to fill an unmet medical need, fast track status was granted for HL in March 2009.

11.6 Nonclinical and Clinical Pharmacology Development

The nonclinical and clinical pharmacology development plans were discussed with the FDA in a teleconference held in March of 2009. The FDA agreed with the proposed nonclinical plan including completed rat and monkey toxicology and safety pharmacology studies and in vitro metabolism and transport studies and an as yet to be conducted panel of genotoxicity studies of MMAE, a single chronic toxicology study of the ADC and MMAE in cynomolgus monkeys, and a single reproductive toxicology study of MMAE in rats. The FDA confirmed that no carcinogenicity study was required.

The FDA disagreed with Seattle Genetics's proposal to not conduct several human clinical pharmacology studies. The FDA did not agree to a proposal to conduct a mass balance study of SGN-35 in cynomolgus monkeys and to extrapolate the results to cancer patients. Seattle Genetics then proposed to collect MMAE and metabolites in urine and feces of healthy subjects. The FDA responded that the plan may be acceptable provided the Agency concurred with the dose and the results of genetic toxicology and safety pharmacology studies.

The FDA stated that drug-drug interaction (DDI) studies of SGN-35 with a strong CYP3A4 inhibitor, a strong CYP3A4 inducer, and a sensitive CYP3A4 substrate should be conducted during drug development and be included in the new drug application (NDA) to provide important safety information. Seattle Genetics queried whether a study in healthy subjects can be conducted. FDA responded that results of the genetic toxicology and safety pharmacology studies need to be submitted for review and FDA concurrence prior to initiating the studies.

The FDA stated that special population (renal or hepatic impairment) studies should be conducted during drug development and be included in the NDA when it is filed to provide important safety information on the use of your drug. If a population pharmacokinetic (PK) approach was used, the FDA recommended a sufficient number of patients with a wide range of hepatic and renal function be enrolled in the studies and enough PK samples collected to characterize the PK.

Seattle Genetics proposed to perform intensive ECG monitoring in a subset of patients treated with SGN-35 as a single-agent to address the potential for QT/QTc prolongation. The FDA responded that the proposed ECG sub-study was adequate to characterize large effects on the QT interval due to SGN-35 but noted there would be significant confounding due to comorbidities and prior therapies. The FDA agreed that if there were no QTc signal in this assessment, no further QT studies would be required. Seattle Genetics stated that to enroll 24 subjects from the pivotal trial will be challenging as this study would enroll only 100 patients in total and enrollment had already commenced. There were three potential alternatives proposed to FDA in order to obtain the necessary number of patients for the QT study: (1) enroll patients from other single-agent studies where SGN-35 is given at the same dose and schedule as on the pivotal study, (2) include patients from a retreatment protocol, and (3) include patients who have already received one or more doses of SGN-35. All three scenarios were considered potentially acceptable to the FDA.

For the BLA submission, Seattle Genetics conducted two dedicated clinical pharmacology studies and submitted the data in support of prescribing information (PI) language. SGN35-007 was designed to characterize the potential for ADCETRIS to prolong the QT interval. Administration of brentuximab vedotin to 46 evaluable patients did not prolong the mean QTc interval > 10 ms from baseline. SGN35-008 was designed to characterized MMAE excretion, DDI, and the effects of renal or hepatic impairment (special populations). Coadministration of ADCETRIS with a potent CYP3A4 inhibitor increased exposure to MMAE by approximately 34% and coadministration of ADCETRIS with a potent CYP3A4 inducer reduced exposure to MMAE by approximately 46%. The final US prescribing information (PI) included language advising that patients who are receiving strong CYP3A4 inhibitors concomitantly with ADCETRIS should be closely monitored for adverse reactions. Coadministration of ADCETRIS did not affect exposure to a CYP3A4 substrate and ADCETRIS is not expected to alter the exposure to drugs that are metabolized by CYP3A4 enzymes.

11.7 Phase 2 Registrational Trials and AETHERA

The Phase 2 HL trial (SGN35-003) enrolled its first patient in February 2009. All patients were required to have relapsed or refractory disease following an ASCT and the target enrollment was 100 patients. The Phase 2 sALCL trial (SGN35-004) enrolled its first patient in June 2009. All patients were required to have relapsed or refractory sALCL following at least one prior therapy and the target enrollment was 50 patients.

A request for SPA was submitted for the AETHERA Phase 3 confirmatory trial in April 2009. After review, the FDA did not agree to the SPA. Seattle Genetics requested a Type A meeting to discuss the areas of disagreement with the FDA. This meeting occurred in October of 2009. At the center of the disagreement were the two issues identified at the EOP1 meeting, namely the patient population to be enrolled and the appropriate primary endpoint. The FDA did not agree that even with proposed modifications to the protocol that PFS was an appropriate primary endpoint for the trial. They noted that the clinical study as defined appeared to be enrolling a heterogeneous population with up to 40% of patients being in partial response (PR) or unconfirmed CR (CRu) (CR unconfirmed) and 60% in CR. The FDA recommended that the trial focus on a single population. Progression-free survival (PFS) could not be recommended as an endpoint in a trial with patients who are already in CR because the study would be a maintenance trial in patients who may not need additional therapy and overall survival (OS) was considered to be the appropriate endpoint for those patients. Additionally, the FDA questioned whether achieving stable disease for those patients who are in PR/CRu would represent a clinical benefit. The Agency recommended that the sponsor carefully write the exclusion/inclusion criteria to avoid enrollment of patients who are less than a PR and if so, the proposal to use PFS as the primary endpoint and OS as a key secondary endpoint might be acceptable.

11.8 Phase 2 Clinical Data

The Phase 2 HL trial's last patient visit was in August 2010. One hundred and two (102) patients were enrolled and received study drug. The median age was 31 years with a range of 15–77. In total 47% of patients were male. Patients had received a median of 3.5 prior therapies (range 1–13). The results of the trial described significant antitumor activity and a generally well-tolerated safety profile. Brentuximab vedotin demonstrated an ORR of 73% and a CR rate of 32%. The median duration of response (DOR) was 6.7 months. The most common adverse reactions (≥20%), regardless of causality, were neutropenia, peripheral sensory neuropathy, fatigue, upper respiratory tract infection, nausea, diarrhea, anemia, pyrexia, thrombocytopenia, rash, abdominal pain, cough, and vomiting (Younes et al. 2012).

The Phase 2 sALCL trial's last patient visit was in June 2011. Fifty-eight (58) patients were enrolled and received study drug. The median age was 52 years with a range of 14–76. In total 57% of patients were male; 72% of patients had anaplastic lymphoma kinase (ALK)-negative disease. Patients had received a median of two prior therapies (range 1–6) and 26% had received a prior ASCT. The results of the trial described significant antitumor activity and a generally well-tolerated safety profile. Brentuximab vedotin demonstrated an ORR of 86% and a CR rate of 57%. The median DOR was 12.6 months. The most common adverse reactions (≥20%), regardless of causality, were neutropenia, anemia, peripheral sensory neuropathy, fatigue, nausea, pyrexia, rash, diarrhea, and pain (Pro et al. 2012).

11.9 Chemistry, Manufacturing, and Controls Plan

Seattle Genetics had two major discussions with the FDA regarding the chemistry, manufacturing, and controls (CMC) for a brentuximab vedotin marketing authorization application. The first was an End of Phase 2 meeting that occurred in January of 2010. At that time, the FDA agreed in general that the proposed control strategy for the good manufacturing practice (GMP) intermediate SGD-1006, including raw materials, in-process controls, and release testing, was acceptable for commercial manufacture. However, they did not agree with the designation of only three starting materials and proposed that there were actually six potential starting materials. These included three custom-synthesized starting materials and three commercially available starting materials. The FDA stated that these three additional compounds significantly contribute to the structure of the drug substance (DS) and might be considered as more appropriate starting materials. For the commercially available starting materials, the FDA requested at least two methods of identification, assay, impurity profile, chiral purity, diastereomer content (if applicable), acceptance criteria, test methods, and certificates of analysis from the supplier and the applicant. For the custom-synthesized starting materials, the FDA requested a brief description of synthesis, and literature reference or reference to a drug master file (DMF) with purging studies to show the removal of carryover impurities in the downstream process, plus all of the attributes mentioned for the commercially available starting materials. Seattle Genetics stated that they would consider this proposal. The FDA agreed that the proposed control strategy for cAC10, brentuximab bulk DS and drug product (DP), including raw materials, in-process controls, release and stability testing, was acceptable for commercial manufacture, and the characterization plans for cAC10 and bulk DS and DP were adequate to support filing of a new drug application (NDA) for brentuximab vedotin.

The FDA agreed with the proposed structure of Module 3 for the planned NDA submission, which included cAC10 and SGD-1006 being described in separate DS sections as intermediates in the manufacture of brentuximab vedotin bulk drug substance (BDS). They added that the section describing cAC10 should essentially be a stand-alone unit, containing all of the information that would be expected for a biologics licensing application (BLA) for a monoclonal antibody.

That meeting was followed nearly a year later by a pre-BLA meeting in December 2010. The key topics of discussion for that meeting were the establishment of release specifications and the adequacy of the justification of the specifications. The FDA stated that the proposed plan appeared reasonable and added that using tolerance intervals (TI) to justify specifications is acceptable. However, as the number of lots manufactured increases, the TI should decrease and the sponsor should continue to analyze lots and reevaluate specifications as additional lots of cACl0 and SGN-35 DS and DP are manufactured. Seattle Genetics stated that they planned to reevaluate specifications as additional lots of cAC10 and SGN-35 DS and DP were manufactured. The FDA also agreed that proposed designations and controls for the three previously identified additional starting materials appear to be acceptable to support the registration. The FDA agreed that submission of a simple stability update for DP within the planned time frame would not extend the Prescription Drug User Fee Act (PDUFA) review date of the application.

11.10 After Phase 2 and Onward to the Biologics License Application

A clinical pre-NDA meeting was held with the FDA in November 2010. A key discussion topic was whether the submission would qualify for regular or accelerated approval. The FDA responded that response rate in a single-arm trial is generally not adequate for regular approval and whether a high response rate with an acceptable duration will be considered a surrogate for clinical benefit (and acceptable for accelerated approval) will be a review issue. The FDA continued by stating that regular approval of an oncologic new molecular entity (NME) usually requires two adequate and well-controlled clinical trials establishing that the NME provides clinical benefit and has an acceptable benefit-to-risk ratio. Accelerated approval, covered under Subpart E 21 CFR 601.40–46, may be applicable to brentuximab vedotin and would take into account magnitude and durability of response and require a confirmatory trial for regular approval. The FDA stated that the confirmatory study for conversion to regular approval does not necessarily have to be in the exact indication that holds accelerated approval. FDA reiterated that PFS is not an acceptable efficacy endpoint in a single-arm trial and whether the PFS analysis is utilized for regulatory action will be a review issue. During the meeting, the FDA indicated that the application would likely go to the Oncologic Drugs Advisory Committee (ODAC), which would be scheduled after the BLA submission. Specific for the sALCL submission, the FDA agreed that a single update of response duration could be submitted early in the BLA review without triggering an action date extension.

Subsequent to the pre-NDA meeting, Seattle Genetics was informed that the marketing application would be reviewed as a BLA rather than an NDA. An important outcome of this change in designation is the term of data exclusivity which is 12 years for BLAs and 7 years for NDAs.

11.11 A Changing Landscape

On February 8, 2011, the FDA held a meeting of ODAC to discuss requirements and expectations for the accelerated approval regulatory pathway. Overall, members agreed that randomized controlled trials should be the standard and that single-arm trials should be the exception except for rare diseases and agents with a high level of activity or pronounced treatment effect. Overall, members also agreed that at least two controlled trials should be needed for accelerated approval commitments, with the caveat that in rare diseases and pediatrics this may not be feasible. Members then noted that a prospective, well-designed development plan is also needed to decrease the time to meet accelerated approval commitments. Overall, members felt that a well-designed development plan is needed prior to the application being filed. Most also preferred that the sponsor have confirmatory studies already ongoing at the time of application.

11.12 The Biologics Licensing Application

The ADCETRIS BLA was submitted on February 28, 2011. The application requested priority review based on the potential for ADCETRIS to be a significant improvement in the safety and effectiveness of the treatment of HL and sALCL. The application also requested a waiver from the Prescription Drug User Fee Act (PDUFA) fee and pediatric data requirements based on the designation of ADCETRIS as an orphan drug for both indications. The BLA was administratively split into two by the FDA to allow for separate regulatory actions if needed. Reponses to requests for information were submitted to both BLAs during the review. The two BLA applications were officially filed on April 29 and priority review was granted with an action date (PDUFA date) of August 30, 2011.

11.13 Meeting with the Committee

ADCETRIS was the subject of two sessions at ODAC on July 14. The morning session involved the HL indication and followed a standard advisory committee agenda with the sponsor presenting first and the FDA presenting after. This was followed by a question-and-answer period, public hearing, discussion, and finally the vote. The vote for accelerated approval for HL was 10 for and 0 against. The vote was followed by a discussion about the AETHERA trial as a confirmatory study to convert the accelerated approval to regular approval. There was general discomfort by the advisory committee members with the design of the AETHERA trial as a confirmatory trial. The afternoon session involved the sALCL indication and again followed the standard agenda. The vote for accelerated approval for sALCL was 10 for and 0 against. The vote was followed again by a discussion about confirmatory trials required to convert the accelerated approval to regular approval.

11.14 Confirmatory Trials

On July 21, a face-to-face meeting was held between Seattle Genetics and the FDA to discuss the design of potential confirmatory trials. In addition to the previously discussed AETHERA trial, two other designs were discussed including a Phase 3 trial of ADCETRIS in combination with Adriamycin, vinblastine, and dacarbazine (AVD) in newly diagnosed HL patients and a Phase 3 trial of ADCETRIS in combination with cyclophosphamide, doxorubicin, and prednisone (CHP) in patients with newly diagnosed mature T-cell lymphomas (MTCLs). The FDA considered both of these designs to be acceptable as confirmatory trials.

11.15 Chemistry, Manufacturing, and Controls Issues During the Review

During the BLA review, several teleconferences were held to discuss CMC issues that the FDA had identified. These included (1) deficiencies and harmonization of the drug master file for the drug-linker, (2) the introduction of new products into contract manufacturing facilities, (3) the potential for glass lamella formation, (4) specification reassessment, and (5) criteria for determination of product quality criticality assessment. All issues were resolved or commitments made prior to the BLA approval.

11.16 Biologics Licensing Application Approval

The BLA was granted accelerated approval on August 19, 2011 for (1) treatment of patients with HL after failure of ASCT or after failure of at least two prior multi-agent chemotherapy regimens in patients who are not ASCT candidates and (2) treatment of patients with sALCL after failure of at least one prior multi-agent chemotherapy regimen. While the pivotal HL trial did not include patients who were not candidates for ASCT, the BLA package included data from 21 patients who were enrolled on other trials. The final indication allows for these patients to receive ADCETRIS.

11.17 After Biologics Licensing Application Approval

As a condition of the accelerated approval, Seattle Genetics was required to meet a number of post-market requirements for additional clinical trials (confirmatory trials). The two Phase 3 trials in newly diagnosed HL and MTCL patients discussed with the FDA in July were listed as post-market requirements in the approval letter for confirmation of the accelerated approval. The success of either trial would be considered sufficient to convert the BLA to regular approval for both of the initial approval indications. Both of these trials completed enrollment on the projected timelines, were submitted to the FDA, and were approved as new indications in March 2018 for HL and November 2018 for ALCL. In addition, the approval letter listed several other post-approval commitments including the completion of the AETHERA clinical trial and submission of the clinical study report characterizing the severity, duration, and reversibility of treatment-emergent neuropathy. Orphan drug status was conferred for both indications at the time of approval, granting Seattle Genetics 7 years of market exclusivity (superseded by the BLA exclusivity) and tax credits up to 50% for qualified clinical research costs.

Subsequent to the approval of the initial BLA, the US PI for ADCETRIS has been revised multiple times, including five times to add new indications (Table 11.2).

11.18 Other ADC Approvals

Since the approval of ADCETRIS in 2011, ten new ADCs have been approved (Table 11.3). Mylotarg, which had been removed from the market in 2010, was reapproved with a new dose and schedule in September 2017.

Table 11.2 Post-approval changes to ADCETRIS PI

8/17/2015	As single agent for patients with classical HL at high risk of relapse or progression as post-autologous hematopoietic stem cell transplant (auto-HSCT) consolidation[a]
11/09/2017	As single agent for adult patients with primary cutaneous ALCL or CD30-expressing mycosis fungoides who have received prior systemic therapy
3/20/2018	In combination with chemotherapy for adult patients with previously untreated Stage III or IV classical HL[a]
11/16/2018	In combination with chemotherapy for adult patients with previously untreated sALCL or other CD30-expressing peripheral T-cell lymphoma (PTCL)
11/10/2022	In combination with chemotherapy for pediatric patients with previously untreated high-risk classical HL

[a]Fulfilled Subpart E requirement

Table 11.3 FDA ADC approvals post ADCETRIS

Trade (nonproprietary) name	Sponsor	Indication	Approval date
Kadcyla (ado-trastuzumab emtansine)	Genentech	Breast cancer	2/22/2013
Besponsa (inotuzumab ozogamicin)	Wyeth Pharmaceuticals	Leukemia	8/17/2017
Polivy (polatuzumab vedotin-piiq)	Genentech	Lymphoma	6/10/2019
Padcev (enfortumab vedotin-ejfv)	Astellas Pharma US	Urothelial cancer	12/18/2019
Enhertu (fam-trastuzumab deruxtecan-nxki)	Daiichi Sankyo	Breast cancer	12/20/2019
Trodelvy (sacituzumab govitecan-hziy)	Immunomedics	Breast cancer	4/22/2020
Blenrep (belantamab mafodotin-blmf)	GlaxoSmithKline	Myeloma	8/05/2020
Zynlonta (loncastuximab tesirine-lpyl)	ADC Therapeutics SA	Lymphoma	4/23/2021
Tivdak (tisotumab vedotin-tftv)	Seagen	Cervical cancer	9/20/2021
Elahere (mirvetuximab soravtansine-gynx)	ImmunoGen	Ovarian cancer	11/14/2022

A number of these approvals were facilitated through Breakthrough Therapy Designation, which was introduced in the 2012 PDUFA renewal. Several of the more recent approvals have also utilized the Assessment Aid, a voluntary submission from the applicant to facilitate FDA's assessment of the BLA, in their review.

11.19 Conclusion

ADCETRIS, when it was approved in August of 2011, became the first ADC approved since the Mylotarg approval in 2000. The ADCETRIS development pathway involved frequent interactions with the FDA and utilization of many regulatory incentive programs available at the time. As the first of a new generation of ADCs, the ADCETRIS case continues to be informative for others, considering the development of novel ADCs. Since ADCETRIS's approval in 2011, there have been ten new ADC approvals in the United States. Antibody-drug conjugates have continued and will continue to grow as a group of entities within the larger class of pharmaceuticals.

References

Duvic M, Reddy SA, Pinter-Brown L et al (2009) A phase II study of SGN-30 in cutaneous anaplastic large cell lymphoma and related lymphoproliferative disorders. Clin Cancer Res 15:6217–6224

Fanale MA, Forero-Torres A, Rosenblatt JD et al (2012) A phase I weekly dosing study of brentuximab vedotin in patients with relapsed/refractory CD30-positive hematologic malignancies. Clin Cancer Res 18:248–255

Forero-Torres A, Leonard JP, Younes A et al (2009) A phase II study of SGN-30 (anti-CD30 mAb) in Hodgkin lymphoma or systemic anaplastic large cell lymphoma. Br J Haematol 146:171–179

Pro B, Advani R, Brice P, et al (2012) Brentuximab vedotin (SGN-35) in patients with relapsed or refractory systemic anaplastic large-cell lymphoma: results of a phase II study. J Clin Oncol 30:2190–2196

Younes A, Bartlett NL, Leonard JP et al (2010) Brentuximab vedotin (SGN-35) for relapsed CD30-positive lymphomas. N Engl J Med 363:1812–1821

Younes A, Gopal AK, Smith SE et al (2012) Results of a pivotal phase II study of brentuximab vedotin for patients with relapsed or refractory Hodgkin's lymphoma. J Clin Oncol 30:2183–2189

Chapter 12
Ado-Trastuzumab Emtansine

Yun Luo, Jérôme J. Lacroix, and Sunil Prabhu

Abstract Ado-trastuzumab emtansine (T-DM1 or Kadcyla™) is an antibody-drug conjugate (ADC) designed to deliver the antimitotic mertansine (DM1) drug to epithelial cancer cells overexpressing the oncoprotein human epidermal growth factor receptor 2 (HER2). T-DM1 has been approved in many countries for HER2-positive metastatic breast cancer (MBC) patients and has recently entered a Phase III clinical trial for advanced HER2-positive gastric cancer patients. The success of T-DM1 lies in the optimization of drug delivery to the targeted cells by the antibody while the non-reducible thioether bonds holding DM1 molecules to the antibody increase tolerability by preventing systemic release of DM1 molecules. When internalized by the targeted cell, DM1 molecules are activated and bind to tubulins, thereby inhibiting cell proliferation. Cytotoxicity of T-DM1 also arises upon binding of trastuzumab to HER2 by promoting antibody-dependent cell-mediated cytotoxicity and inhibiting HER2-dependent signaling pathways.

Keywords Ado-trastuzumab emtansine · T-DM1 · Kadcyla · ADC · HER2-positive breast cancer

Y. Luo (✉) · S. Prabhu
Department of Biotechnology and Pharmaceutical Sciences, College of Pharmacy, Western University of Health Sciences, Pomona, CA, USA
e-mail: luoy@westernu.edu

J. J. Lacroix
Department of Chemistry and Chemical Engineering, California Institute of Technology, Pasadena, CA, USA

J. L. Zaro, J. Wang (eds.), *Antibody-Drug Conjugates*, AAPS Advances in the Pharmaceutical Sciences Series 17, https://doi.org/10.1007/978-3-031-83005-1_12

12.1 Introduction

Ado-trastuzumab emtansine (T-DM1) is one of only two antibody-drug conjugates (ADCs) currently on the market. It was specifically developed by Roche (trade name Kadcyla) as anti-human epidermal growth factor receptor 2 (HER2) antibody-drug conjugate for metastatic breast cancer patients and also shows promising application against gastric cancers (Barok et al. 2011).

HER2 is a cell surface receptor member of the epidermal growth factor receptor (EGFR) family, which contains four members (HER1–4) and is normally expressed at low levels in epithelial tissues where it plays an important role by promoting normal cell proliferation and differentiation. However, amplification of the *HER2* gene and/or overexpression of the HER2 receptor are linked to the development of many human cancers in breast, ovarian, and gastrointestinal tissues and correlates with poor prognosis. This is due to the fact that, unlike HER1–3, the presence of HER2 at the cell surface is required to enhance the strength of the EGFR signaling. Hence, an estimated 15 ~ 30% of human breast cancers are linked to HER2 amplification/overexpression (Mitri et al. 2012).

The accessibility of the extracellular portion of the receptor has made HER2 a potential target for antibody therapy. Binding of the anti-HER2 humanized monoclonal antibody trastuzumab, or Herceptin, to the HER2 receptor was demonstrated to significantly prevent growth of cancer cells by preventing dimerization of the receptor and by activating antibody-mediated immunity against the targeted cell. Hence, trastuzumab was approved by the Food and Drug Administration (FDA) in 1998 for promising use for the treatment of HER2-dependent metastatic breast cancer. However, despite increased survival rates, a significant proportion of patients treated with trastuzumab either did not respond initially or relapsed after a period of clinical response (Slamon et al. 2001; Vogel et al. 2001).

HER2 overexpression does not recess when cancer cells become refractory to HER2-directed therapies (Nahta et al. 2005; Ritter et al. 2007). Thus, to improve the naked antibody therapy, trastuzumab was later engineered as a vehicle to deliver antimitotic drugs to HER2-overexpressing cells. A drug of choice was mertansine (DM1), a potent antimitotic drug. DM1 is a derivative of an old cytotoxic agent maytansine that inhibits microtubule dynamics by binding to tubulins. This targeted strategy helps to minimize the systemic side effects exhibited by maytansine-based chemotherapy. Limited side effects are also diminished by the relatively high stability of the non-reducible thioether linkers between trastuzumab and DM1, which prevents nonspecific systemic release of the drug. In the EMILIA Phase III clinical trial of women with advanced HER2-positive breast cancer who were already resistant to trastuzumab alone, it improved progression-free survival (PFS) alone with overall survival (OS) and safety compared to the combination of the chemotherapeutic agents lapatinib (EGFR inhibitor) and capecitabine (DNA synthesis inhibitor) (Verma et al. 2012). Based on that trial, the US FDA approved marketing on February 22, 2013.

12.2 Chemistry

Similarly to other antibody-drug conjugates, T-DM1 is composed of several mertansine molecules (DM1) covalently connected to the antigen-specific humanized antibody trastuzumab via a chemical linker.

Mertansine (DM1) is a derivative of cytotoxic agent maytansine. Maytansine (shown in black in Fig. 12.1) is an ansamycin family antibiotic initially isolated form the Ethiopian shrub *Maytenus serrata* but also found in other plants from the genus *Maytenus*, which grow in a wide diversity of climates around the world. The structure represents the benzoansamacrolide family that comprises an aromatic moiety bridged by a long branched unsaturated aliphatic chain. Maytansine inhibits the microtubule assembly by binding to the same site on tubulin as rhizoxin, a bacterial antimitotic agent (Partida-Martinez and Hertweck 2005).

Maytansine exhibits cytotoxicity against many tumor cell lines at subnanomolar concentrations. However, it failed as an anticancer agent in human clinical trials because of lack tumor specificity and high systemic toxicity (Iwasaki 1993; Liu et al. 2005; Lopus et al. 2010; Sawada et al. 1993; Tassone et al. 2004; Widdison et al. 2006). Mertansine is a derivative of maytansine containing a sulfhydryl (-SH) group. DM1 has an in vitro cytotoxicity threefold to tenfold greater than maytansine (Junttila et al. 2011; Kovtun et al. 2006) but still present nonselective systemic toxicity. Interestingly, the sulfhydryl group allows the creation of DM1 conjugates with crosslinking reagent through the thioether group (shown in blue in Fig. 12.1).

The crosslinking reagent used in T-DM1 is succinimidyl *trans*-4-(maleimidylmethyl)cyclohexane-1-carboxylate, also known as SMCC (Fig. 12.2). SMCC has two functional groups, a succinimide ester and a maleimide. The succinimide ester group reacts with lysine residue in the trastuzumab molecule to form an amide bond and N-hydroxysuccinimide (NHS) leaving group. The maleimide

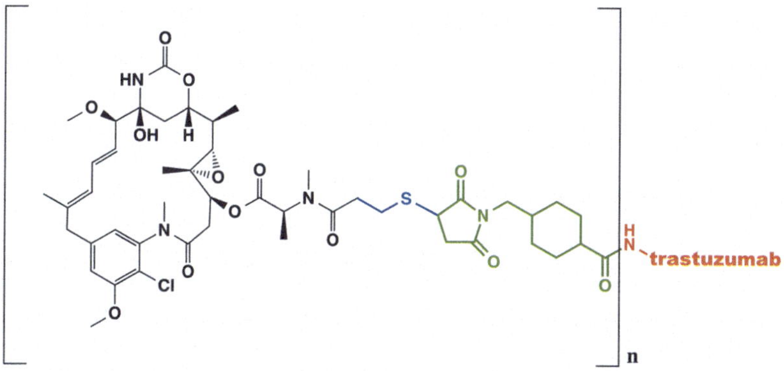

Fig. 12.1 Schematic representation of trastuzumab emtansine. The maytansine skeleton is shown on the left (black). The thioether group is shown in blue and the linker group is shown in green. The linker group is bound to the amino group of a lysine residue in the trastuzumab molecule shown in red

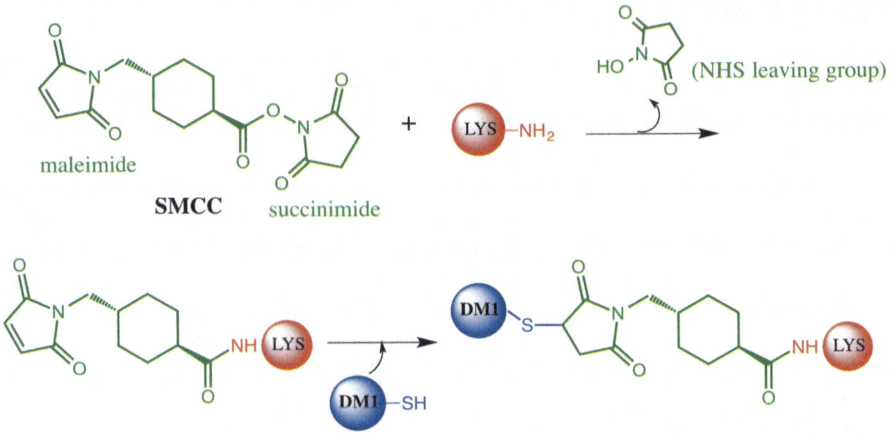

Fig. 12.2 Linking mechanism of DM1 molecules to trastuzumab. The crosslinking reagent succinimidyl trans-4-(maleimidylmethyl)cyclohexane-1-carboxylate, also known as SMCC, is shown in green. The succinimide ester group reacts with lysine residues in the trastuzumab molecule (shown in red) to form an amide bond and N-hydroxysuccinimide (NHS) leaving group. The maleimide moiety links to the free sulfhydryl group of DM1 (shown in blue), forming a thioether bond between the linker and DM1

moiety links to the free sulfhydryl group of DM1, forming a thioether bond between the linker and DM1. Each trastuzumab molecule is linked to 3.5 DM1 molecules on average with a maximum of eight payloads (Krop et al. 2010). The conjugation has been shown to preserve the activity and binding specificity of the antibody component (Junttila et al. 2011). The DM1 moiety together with the linker make emtansine, hence the generic name trastuzumab emtansine for the final antibody-drug conjugate.

Choosing a suitable crosslinking reagent is one of the most critical points in the design of an ADC. The linker should stabilize the ADC in circulation and once the compound enters the cell, it liberates the cytotoxic agent. Antibody-DM1 conjugates were originally designed with a disulfide-based linker. Indeed, after internalization of the ADC, the endosomal reduction of the disulfide bond separates the antibody from the DM1 and releases the latter into the cytoplasm. Several disulfide-linked trastuzumab ADCs were tested such as trastuzumab-SPDP-DM1, trastuzumab-SPP-DM1, trastuzumab-SSNPP-DM3, and trastuzumab-SSNPP-DM4 (Lewis Phillips et al. 2008). It was found that introducing steric hindrance near the disulfide bond would slow down the reduction rate of disulfide bond. However, even with the most hindered disulfide-linker, once the DM1 molecules are liberated from the antibody, it is cleared rapidly. Thus, with the disulfide-linked ADC, the concentration of DM1 inside the tumor cell may drop too quickly to ensure prolonged antitumor effects. The SMCC linker, in contrast, forms a thioether bond instead of a disulfide bond, and thus is non-reductible (cannot undergo enzymatic reduction of disulfide bond within the cell). Among the thioether-linked ADCs, the superior thioether stability of T-DM1 compared to the cysteine-linked ADCs has been reported

(Erickson and Lambert 2012). It is possible that for the thio-maleimide exchange reaction, the Michael donor reactivity of the sulfhydryl group of DM1 is higher than the sulfhydryl group of cysteine side chains (Baldwin and Kiick 2011). The SMCC-linked ADC is internalized upon binding to HER2 and is postulated to undergo intracellular proteolytic degradation to release active Lys-MCC-DM1 (MCC is the linker moiety) molecules into the cytoplasm of the targeted tumor cell. This postulation is supported by the blocked activity of T-DM1 in the presence of protease inhibitors in cell culture experiments (Lewis Phillips et al. 2008). The fact that Lys-MCC-DM1 molecules were detected as the major metabolites in rats and mice plasma, as well as plasma samples of patients in a Phase II clinical study (Burris et al. 2011, Krop et al. 2010, Lu et al. 2011, Shen et al. 2012) also supports this mechanism. The slow release of DM1 by proteolytic degradation inside a tumor cell guarantees the prolonged antitumor efficacy and reduces systemic toxicity. A more detailed review on the effect of T-DM1 linker can be found in Lambert et al. (2014).

12.3 HER2 and the EGFRs Family

In humans, the EGFR family consists of four transmembrane receptors: HER1, first identified and initially called "the EGF receptor"; HER2; HER3; and HER4. HER2 was identified as a homolog of the avian erythroblastosis oncogene B, or ErbB, and hence was named ErbB2 by homology. In rodents, it was independently identified in a glioblastoma cell line, a type of neuronal cancer, and was hence named Neu. Genome sequencing later confirmed that HER2, ErbB2, and Neu are in fact ortholog proteins, suggesting that they play similar functions in different vertebrate organisms (Coussens et al. 1985).

EGFR proteins share a homologous structure of a large N-terminal extracellular domain followed by a single transmembrane spanning segment and a C-terminal intracellular kinase domain (Bessman and Lemmon 2012) (Fig. 12.3). The extracellular domain of HER1 binds at least seven ligands: EGF, transforming growth factor α (TGF-α), heparin-binding EGF-like growth factor (HB-EGF), betacellulin, epiregulin, amphiregulin, and the recently identified epigen (Linggi and Carpenter 2006; Schneider and Yarden 2013). HER3 binds neuregulin 1 and neuregulin 2, while HER4 recognizes neuregulins 1 to 4, epiregulin, betacellulin, and HB-EGF.

Binding of specific ligands to HER1, HER2, or HER3 causes these receptors to dimerize, either with a receptor of the same type (homodimerization) or with another family member (heterodimerization). In contrast, HER2 does not interact with any known ligands and is thus considered an orphan receptor. HER2 cannot homodimerize and is exclusively found in heterodimers with ligand-bound HER1, HER3, or HER4 (Rubin and Yarden 2001). Hence, HER2 is thought to enhance EGFR-mediated signals by stabilizing heterodimers containing a ligand-bound EGFR. Recent molecular dynamic simulations showed that, although being asymmetric, heterodimers harboring ligand-free HER2 monomers form quite stable

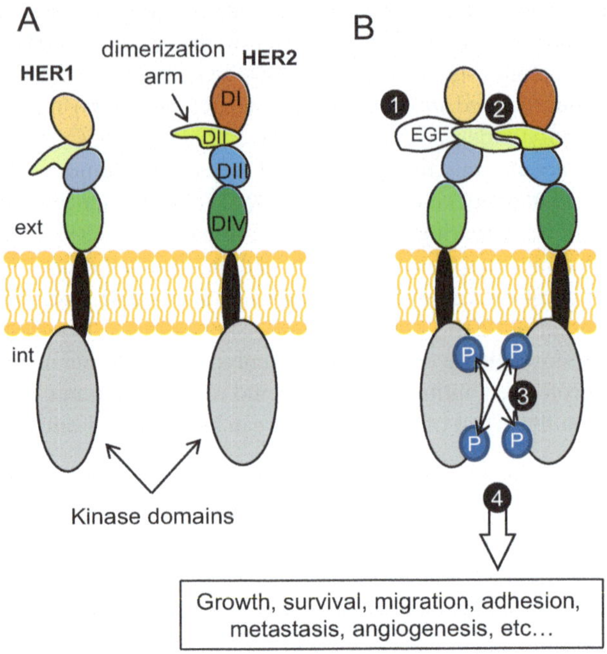

Fig. 12.3 Mechanism of cell proliferation mediated by EGFR receptors. (**a**) In absence of ligand, EGFR members are mainly present as monomers in the epithelial cell surface (here only HER1 and HER2 are shown as examples). EGFR monomers consist on a large extracellular N-terminal region composed of four structural domains named DI to DIV (indicated with different colors), a unique transmembrane segment and a cytoplasmic C-terminal catalytic domain. Ligand-dependent EGFR activation requires the interaction between two receptors via their activation arms formed by DII. In the orphan HER2 receptor, the activation arm is in an open conformation readily available for dimerization, while in HER1, HER3, and HER4 it remains in a tethered conformation in absence of ligand. (**b**) Binding of the EGF ligand to HER1 (1) opens its dimerization arm and mediates the formation of a stable heterodimer with HER2 (2). The physical proximity of the catalytic domains induce several transphosphorylations of hydroxylated side chains in the catalytic domains (3) which activate several proliferative signaling pathways (4) promoting the survival and proliferation of the cell

complex as long as one ligand molecule remains bound to the other receptor (Arkhipov et al. 2013).

The EGFR extracellular region is subdivided into four domains DI to DIV (Fig. 12.3a). Dimerization is catalyzed by the opening of a dimerization arm formed by domain II in each monomer (Ogiso et al. 2002). In HER1, HER3, and HER4, the dimerization arm swings between an inactive "tethered" conformation to an active "open" conformation upon ligand binding (Fig. 12.3b). In contrast, the dimerization arm of HER2 remains in a constitutively open conformation. Although HER2 can dimerize with any EGFR members, the HER2/HER3 pair is one of the most efficient and often observed cancer studies (Way and Lin 2005). Receptor dimerization triggers transphosphorylation of hydroxylated residues in the intracellular domain.

This event activates several proliferative intracellular signaling pathways (RAS/ERK and Pi3K/AKT). During the tumorization process, naturally selected mutations of HER2 may promote cancer growth by inducing constitutive autophosphorylation of tyrosines in the cytoplasmic kinase domain (Wang et al. 2006) or by inducing constitutive dimerization (Muller et al. 1998).

An estimated 15% to 30% of breast cancers are directly due to an increase of HER2 receptors at the cell surface (Mitri et al. 2012). Overexpression of HER2 also occurs in ovarian, stomach, and uterine cancer and causes gallbladder adenocarcinoma in mouse (Kiguchi et al. 2001). In HER2-positive cancers, the amount of HER2 proteins may be up to two orders of magnitudes higher compared to normal epithelial cells (Venter et al. 1987). More than 90% of HER2-positive breast cancers are caused by allelic amplification of the *HER2* gene (Pauletti et al. 1996). Genic amplification is also the main mechanism for HER2 overexpression in the development of ovarian and gastric tumors (Hynes and Stern 1994). Interestingly, amplification of the *HER2* gene does not seem to occur in HER2-positive cancers located in specific organs such as lungs, bladder, mesenchyme, and esophagus. In those cases, overexpression of HER2 is thought to result from alterations of transcriptional/translational and degradation/recycling processes (Slamon et al. 1989).

12.4 Trastuzumab–HER2 Interaction

To investigate the molecular determinants of the antibody–antigen interaction between trastuzumab and HER2, Cho et al. (2003) co-crystallized the extracellular domain of the human HER2 receptor with Fab fragments obtained digesting herceptin antibodies with papaine. The X-ray structure of the complex showed that trastuzumab binds the bottom part of DIV (Fig. 12.4). A closer look at the structure shows that the trastuzumab epitope in the human HER2 is formed by three adjacent loops (Fig. 12.5a) stretching from residues 557–561, 570–573, and 593–602 respectively. A sequence alignment of the three epitope loops in HER2 homologs indicates poor residue conservation between rodents and hominids (Fig. 12.5b). This explains why the trastuzumab antibody recognizes the primate HER2 receptor but not rodent Neu homologs (Poon et al. 2013).

12.5 Mechanisms of Action

T-DM1 helps fight HER2-positive cancer cells through several mechanisms. First, the binding of the antibody interferes with the transduction of HER2-mediated growth signals. Binding of trastuzumab to HER2-amplified cancer cell was shown to disrupt the ligand-independent interaction between HER2 and HER3, providing to the antibody an antiproliferative effect (Junttila et al. 2009) (Fig. 12.6a). Second, cancer cells decorated with humanized antibodies exposing Fc regions can

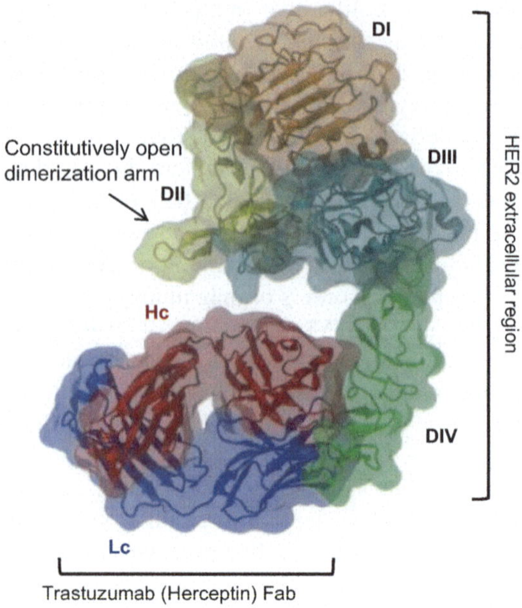

Fig. 12.4 HER2–trastuzumab interaction. Crystal structure of a trastuzumab (herceptin) antigen-binding fragment (Fab, Hc: heavy chain; Lc: light chain) complexed to the extracellular domain of HER2 showing that the antibody binds specifically to DIV (PubMed accession code: 1N8Z) (Cho et al. 2003). Note that the dimerization arm is located in DII (see Fig. 12.3) and remains in a constitutively "open conformation" ready to engage dimerization with ligand-bound EGFR members. The figure was generated using the free-access software Visual Molecular Dynamics (VMD, http://www.ks.uiuc.edu/Research/vmd). The protein backbones are shown in NewCartoon representation. The large arrows indicate the presence of beta sheets. The transparent surface delineates the external surface of the protein side chains

potentially be recognized by immune cells such as macrophages or natural killer cells and targeted for elimination through a process called antibody-dependent cell-mediated cytotoxicity (ADCC) (Fig. 12.6b). Both trastuzumab and its antibody-drug conjugate, T-DM1, trigger ADCC in clinical studies, showing the effective contribution of ADCC to fight tumor progression in vivo (Junttila et al. 2011). In principle, the Fc regions may also trigger complement-dependent cytotoxicity (CDC). Third, binding of trastuzumab favors internalization of the ADC-receptor complex leading to subsequent lysosomal proteolytic degradation of the antibody moiety. This process slowly releases Lys-MCC-DM1 molecules into the cytoplasm of the targeted HER2-amplified tumor cell (Fig. 12.6c) where they bind tubulin and inhibit microtubule polymerization.

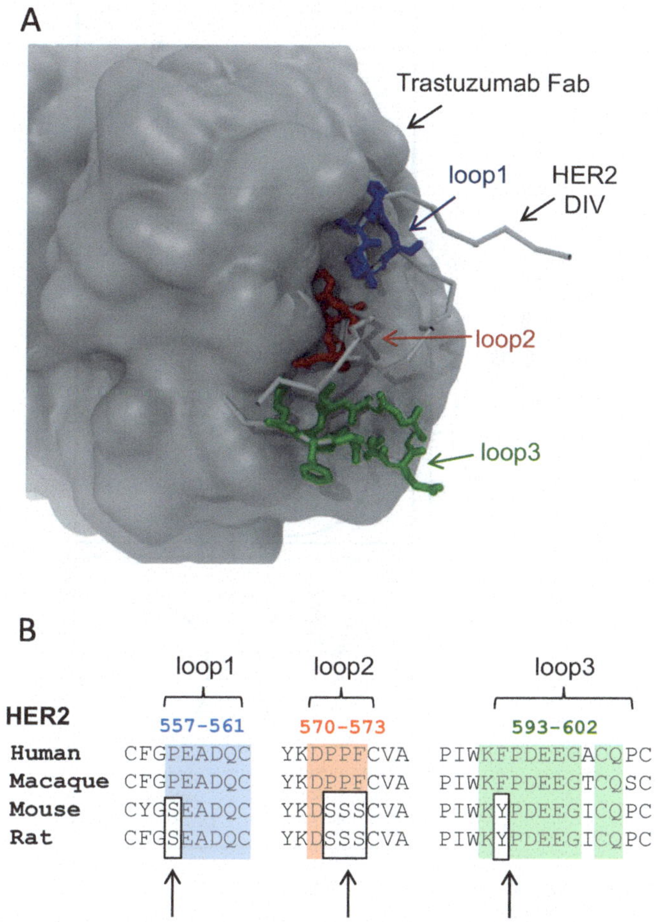

Fig. 12.5 Molecular determinants of the HER2 epitope. (**a**) Zoomed view of the structure shown in Fig. 12.4 showing the side chains from HER2 domain IV (licorice representation) making contact with the herceptin Fab (represented as a white surface). The epitope is essentially formed by three contiguous loops (loop1 in blue; loop2 in red and loop3 in green). (**b**) Sequence alignment of the three loop regions in rodents and primate HER2 homologs (human GI: 119533, macaque GI: 544499689, mouse GI: 76363513 and rat GI: 3915663). Conserved residues in the three loops are indicated by colored boxes and the residue numbers are from the human HER2 amino acid sequence. Note the presence of residue divergence for some positions in each loop between primate and rodent HER2 proteins (open boxes indicated by vertical arrows)

12.6 Metabolism and Pharmacokinetics

Several ADME studies of T-DM1 have been reported using mice and rats. The antibody component of T-DM1 do not cross-react with rodent antigens. Thus, there is no antigen-mediated contribution to the pharmacokinetics (PK). The major metabolites found in rats and mice plasma were lysine-MCC-DM1 (around 70%) (Fig. 12.7

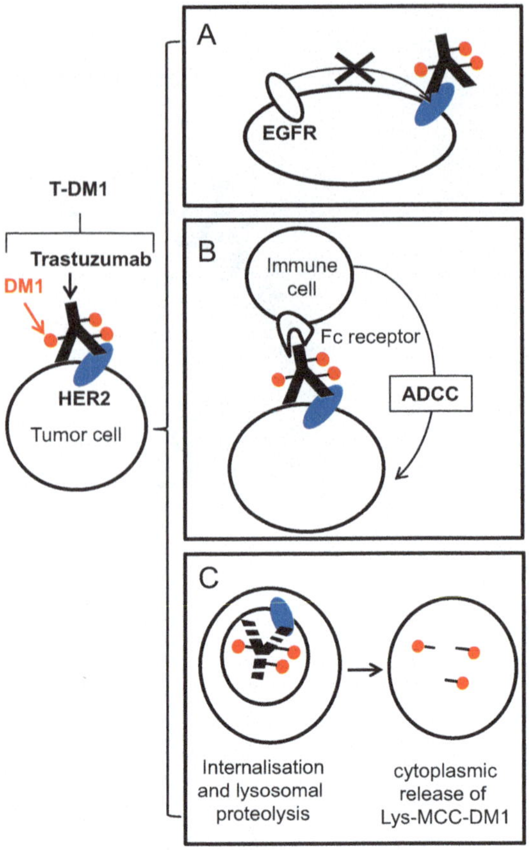

Fig. 12.6 T-DM1 mediated antimitotic mechanisms. (**a**) Binding of T-DM1 to a HER2-overexpressing tumor cell (left) prevents the formation of a stable heterodimer between HER2 and other ligand-bound members of the EGFR family and thus interferes with HER2-dependent proliferative signaling pathway. (**b**) Recognition of the Fc region of the trastuzumab antibody by Fc receptors present on macrophages or natural killer cells patrolling the surrounding tissue triggers antibody-dependent cell-mediated cytotoxicity (ADCC) directed toward the HER2-expressing tumor cell. (**c**) Binding of T-DM1 facilitates internalization of the ADC–HER2 complex. The following lysosomal proteolysis of the antibody moiety releases Lys-MCC-DM1 molecules in the cytoplasm of the tumor cell. The activated compounds bind tubulin molecules and inhibit mitotic division

top) and [*N*-maleimidylmethyl] cyclohexane-1-carboxylate-DM1 (MCC-DM1) (Fig. 12.7 bottom) (Burris et al. 2011, Krop et al. 2010, Lu et al. 2011, Shen et al. 2012). These metabolites indicate that lyosomal degradation is the main metabolism pathway for the ADC and the thioether linker is highly resistant against chemical cleavage during clearance. The major two metabolites identified in rats and mice were also detected in plasma samples from patients with HER2-positive metastatic breast cancer in a Phase II clinical study (Shen et al. 2012). DM1 was detected at low level in human plasma. The primary active metabolite, lysine-MCC-DM1, does

1

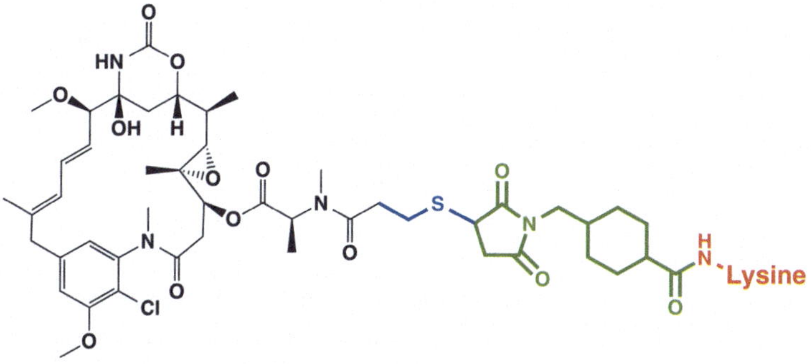

2

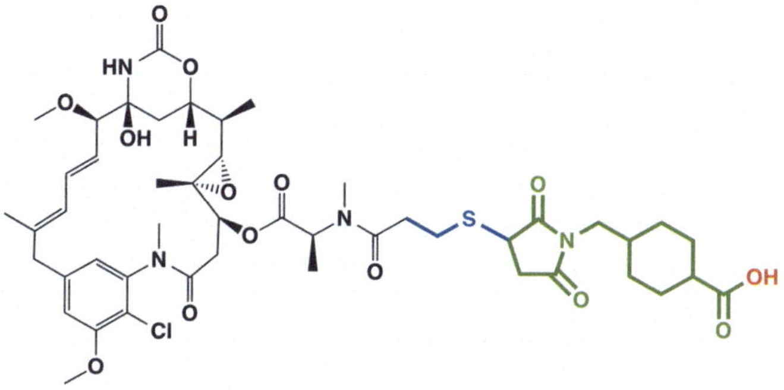

Fig. 12.7 The major metabolites found in rat plasma: (1) Lysine-MCC-DM1, (2) [N-maleimidylmethyl] cyclohexane-1-carboxylate-DM1 (MCC-DM1)

not readily cross the plasma membranes of neighboring cells and, therefore, does not induce a bystander effect (Lewis Phillips et al. 2008). In vitro studies indicate that DM1is metabolized mainly by CYP3A4 isoenzyme (Lu et al. 2011).

The clearance profiles of several DM1 conjugates are found to correlate with the in vitro stability of their chemical linkers. The clearance rate in mice was found that the regular disulfide-linked conjugate is cleared faster than the more sterically hindered disulfide-linked conjugate. The latter is cleared faster than the uncleavable thioether-linked T-DM1 (Erickson et al. 2012; Kellogg et al. 2011). In nude mice, T-DM1 showed the serum concentrations of T-DM1 measured for 1 week were not different from total serum trastuzumab concentration (Lewis Phillips et al. 2008). Study on female Sprague-Dawley rats shows that following administration of radiolabeled T-DM1, the major circulating species in plasma was T-DM1 while DM-1 concentrations were low (Shen et al. 2012).

T-DM1 shows no evidence of tissue accumulation in rats or metabolite accumulation in human plasma following multiple dosing. The average 3.5 DM1 molecules per antibody conjugating rate show no detectable impact on the pharmacokinetic

Table 12.1 Summary of T-DM1 PK parameters following single and multiple doses of T-DM1 3.6 mg/kg q3w, reproduced from (Qi and Nitin 2012)

Clinical trial	Cycle	N	C_{max} (μg/mL)	AUC (μg*d/mL)	$T_{½}$ (d)	CL (mL/d/kg)	V_d (mL/kg)
Phase I	1	15	76.2	300	3.1	12.7	58.4
Phase II	1	51–105	75.6–84.2	431–495	3.5–4	8–9.2	28.4–41.2
	3–5	39–82	68.9–80.7	456–475	4.2–4.5	6.7–8.4	33.6–45.2
Phase III	1	292	83.4	489	3.7	7.8	29.5

properties of the trastuzumab (Erickson and Lambert 2012; Mayo et al. 2005). The major pathway of DM1-containing metabolite elimination in rats was the fecal/biliary route, with the majority of T-DM1 eliminated in the feces (80%) and a small fraction in the urine. Those preclinical T-DM1 observations appear to translate to the clinic, although trastuzumab does not bind the rat or mouse HER2-like receptor.

Pharmacokinetics (PK) data of T-DM1 are available from one Phase I trial, four Phase II trials, and one Phase III trial. The proposed dose characterized by noncompartmental methods is 3.6 mg/kg as an intravenous (IV) infusion over 30 minutes once every 3 weeks (q3w). The ranges of PK parameters following single and multiple doses of T-DM1 3.6 mg/kg q3w are summarized in Table 12.1. Clinical studies show that T-DM1 has predictable pharmacokinetic properties at clinical doses. Six clinical studies reported using the accepted maximum tolerated dose of 3.6 mg/kg every 3 weeks have shown that the clearance rate in patients ranged from 6 to 13 mL/day/kg for T-DM1 and the terminal half-life is 3–4.5 days for T-DM1 and 9–11 days for total trastuzumab (Burris et al. 2011; Girish et al. 2012; Krop et al. 2010). T-DM1 accumulation was not observed following multiple dosing.

12.7 Preclinical Efficacy and Toxicity Studies

Preclinical experiments have been conducted in cells and animal models to assess the tolerability of T-DM1 and the side effects observed upon administration of the drug. Taking advantage of the fact that trastuzumab does not cross-react with the rodent HER2 homolog Neu, Poon et al. determined the antigen-dependent and antigen-independent toxicity of T-DM1 by comparing the toxicity of T-DM1 administered intravenously to cynomolgus monkeys (*Macaca fascicularis*, also known as crab-eating macaques) and rats (Poon et al. 2013). Tolerability of T-DM1 was found relatively similar for both types of animals. Toxic effects start to develop at approximately 40 mg/kg in rats (~4.4 mg DM1/body surface area in m^2) and 30 mg/kg (~6 mg DM1/m^2) in monkeys, indicating that T-DM1 toxicity mainly arises from an antigen-independent mechanism. Furthermore, the toxic effects were relatively proportional to the dose received, either acutely or chronically and corresponded to the expected effects of the emtansine molecules. In rats, toxic effects were decreased platelet, reticulocyte, and lymphocyte counts, increase in neutrophil counts, depletion or necrosis of lymphoid organs such as lymph nodes, spleen, and thymus, and various degrees of hepatocellular degenerations associated with elevated concentrations of serum alanine and glutamyl aminotransferases, alkaline phosphatase and bilirubin, the down-product of heme catabolism. Similar toxic effects on red and white blood cells counts, hepatic functions, and lymphoid system were observed in monkeys. In both animals and particularly in rats, a large number of cells in tissues beyond the liver were found in mitotic arrest, strongly suggesting the occurrence of antigen-independent T-DM1 uptake with subsequent activation of antimitotic DM1 molecules. It is hypothesized by the authors that T-DM1 may trigger endocytosis via its Fc region or that circulating DM1 molecules may be released through catabolism. In monkeys, irreversible axonal degeneration was observed after four doses of 10–30 mg/kg or eight doses of 1–10 mg/kg, although no obvious effects were noted in neurologic tests. In comparison to T-DM1, when DM1 was administered in rats, the tolerability of the dose went down to about 1.6 mg DM1/m^2, which corresponds to about half the tolerable dose of T-DM1. Hence, this study demonstrates the ability of T-DM1 to increases tolerability of the chemotherapeutic drug in animal models.

Potential benefits from T-DM1 treatments in other HER2-positive epithelial cancers have been assessed by studying the effects of T-DM1 on four HER2-positive human gastric cancer cell lines (Barok et al. 2011). In vitro assays further indicated that T-DM1 was globally more effective than trastuzumab on N-87 and OE-19 cells, while it showed moderate and limited efficacy on MKN-7 and SNU-216 cells, respectively. The authors also noted that N-87 and OE-19 cells exhibited slightly higher *HER2* gene copies and about threefold higher HER2 surface expression than MKN-7 and SNU-216 cells. Hence, N-87 and OE-19 gastric cancer cells may be more susceptible to HER2-directed ADC therapy. To address T-DM1 efficacy in vivo, the authors further xeno-transplanted N-87 and OE-19 cells into severely combined immunodeficient (SCID) mice and conducted various T-DM1 protocols. In all OE-19 and half of N-87 xenografts, T-DM1 induced a complete reversal of tumor growth leading to nearly complete remission while trastuzumab alone did not prevent tumor progression. Interestingly, these effects remained potent against large established tumors and against trastuzumab-resistant tumors. Moreover, no delay was necessary for switching between trastuzumab and T-DM1 treatments to observe the full benefits of T-DM1, suggesting a rapid turnover of HER2 proteins in these gastric cancer cells. This discovery may help to reduce delays to switch ADC treatments in HER2-positive cancer patients. Since July 2013, a Phase III clinical trial testing T-DM1 treatment for gastric cancer patients is ongoing (Fig. 12.8).

12.8 Clinical Studies

Trastuzumab emtansine (Kadcyla ™) was granted approval by the FDA in the United States in February 2013 and subsequently in Japan (September 2013) and the European Union (November 2013), thus receiving global approval of this

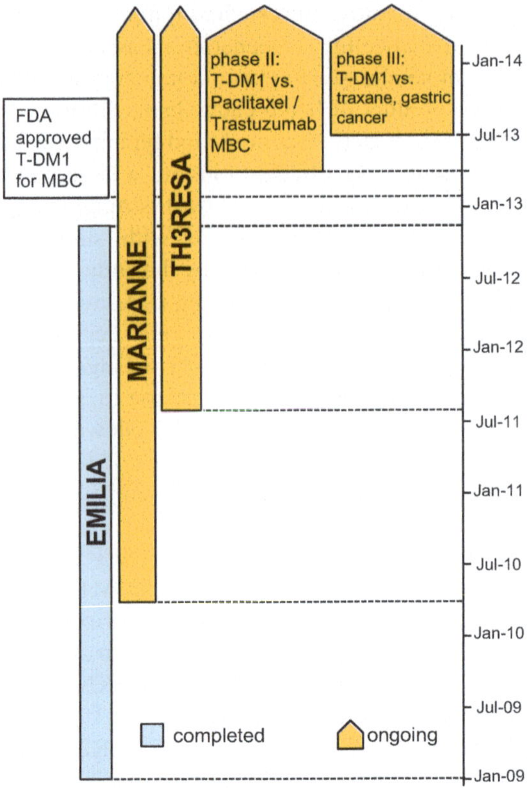

Fig. 12.8 Chronology of T-DM1 clinical trials. Four Phases II or III clinical trials are presently ongoing using various protocols to test T-DM1 efficacy against MBC and gastric cancers. Only the EMILIA clinical Phase III trials (see Sect. 12.8. Clinical Studies) have been completed in November 2012. The MARIANE trials started in April 2010 and consist of three arms: T-DM1/pertuzumab vs. T-DM1 vs. trastuzumab/taxane in HER2-positive MBC patients. The TH3RESA trials started in August 2011 and consist of two arms: T-DM1 vs. the physician's choice of therapy in HER2-positive MBC patients. T-DM1 was approved by the FDA on February 2013 for HER2-positive MBC patients based on the EMILIA study

antibody-drug conjugate. Clinical studies demonstrated increased median overall survival of MBC patients with decreased adverse effects, thus leading to an accelerated approval of this ADC (Ballantyne and Dhillon 2013).

Phase I studies evaluated the effects of trastuzumab emtansine under two different administration schedules for patients with locally advanced or metastatic HER2-positive breast cancer: (1) Once every 3 weeks ranging from 0.3 to 4.8 mg/kg (n = 24) and (2) once weekly dosing ranging from 1.2 to 2.9 mg/kg (n = 28). This was designed as an open-label, dose-escalation study. A maximum tolerated dose (MTD) was determined as 3.6 mg/kg in the first cohort dosed once every 3 weeks. The clinical benefit rate was found to be 73% (n = 15; 3.6 mg/kg;

treatment = 238 days) whereas the objective response rate (ORR) was 44% ($n = 9$) (Krop et al. 2010).

The second cohort on the once-weekly dosing schedule demonstrated an MTD lower than the first cohort (2.4 mg/kg). The partial tumor responses were reported in 46% of the patients (13 of 28) up to ~18 months of therapy. Sixteen patients received 2.4 mg/kg once weekly dosage of which 15 patients had measurable disease at baseline. Of this, 40% had an objective partial response with a median duration of response (MDR) of 5.6 months (Beeram et al. 2012).

A similar Phase I study in Japan used a single-arm, dose-escalation (Aogi et al. 2011) study, enrolling patients with HER2-positive metastatic breast cancer who had received prior therapies including trastuzumab. This study also found the MTD in Japanese patients ($n = 10$) to be 3.6 mg/kg administered once every three weeks.

Phase II studies were open-label studies in patients with HER2-positive metastatic breast cancer. Two treatment protocols followed were trastuzumab emtansine and trastuzumab with docetaxel. Patients received trastuzumab emtansine monotherapy ($n = 67$) or trastuzumab (8 mg/kg loading dose followed by 6 mg/kg every 3 weeks) plus docetaxel (75 or 100 mg/m^2, [$n = 70$]). Results from this first-line treatment with trastuzumab emtansine significantly prolonged median PFS (primary endpoint) to ~14 months compared to 9.2 months treatment with trastuzumab plus docetaxel, thus increasing PFS by 5 months (14.2 vs. 9.2 months; HR 0.59 [95% CI 0.36–0.97] (Hurvitz et al. 2013). The ORR with trastuzumab emtansine was higher than the combination dose of trastuzumab plus docetaxel (64 vs 58%). Overall survival (OS) rates showed no significant between-group differences in a preliminary analysis conducted at a median follow-up of ~23 months (HR 1.06 [95% CI 0.477–2.352]).

In an independent assessment, trastuzumab emtansine monotherapy was associated with an ORR of 25.9% (primary endpoint) and a median PFS of 4.6 months after a median follow-up of ≥12 months in a single-arm Phase II study in patients ($n = 112$) with HER2-positive breast cancer who were previously treated with trastuzumab (Burris et al. 2011).

Patients in another single-arm Phase II study in patients with HER2-positive metastatic breast cancer refractory to current therapies ($n = 110$) were required to have received prior treatment with trastuzumab, lapatinib, a taxane, an anthracycline and capecitabine in the neoadjuvant, adjuvant, locally advanced or metastatic setting (Krop et al. 2012). In addition, patients were required to have had at least two HER2-directed therapy regimens in the metastatic or locally advanced setting and progression on the most recent regimen. Independent assessments demonstrated an ORR of 34.5% (primary endpoint) and a clinical benefit rate of 48.2%, at a median follow-up of 17.4 months of trastuzumab emtansine monotherapy. The median PFS in these patients was 6.9 months and the MDR was 7.2 months.

A single-arm Phase Ib/II study in patients with HER2-positive locally advanced or metastatic breast cancer evaluated the efficacy of combination therapy with trastuzumab emtansine plus pertuzumab (Dieras et al. 2010). The Phase Ib dose-escalation phase of the study established the dosage of trastuzumab emtansine (3.6 mg/kg every 3 weeks) and pertuzumab (840 mg loading dose in cycle 1, followed by 420 mg maintenance dose every 3 weeks thereafter). Following combination treatment, 16 of 46 patients in the relapsed setting (confirmed ORR of 34.8%) and 12 of 21 patients in the first-line setting (confirmed ORR of 57.1%) (primary endpoint) showed partial or complete responses. Another 22 patients in the relapsed setting and five patients in the first-line setting had stable disease with clinical response rates of 21% and 13%, respectively. Prior treatment for recurrent locally advanced or metastatic disease (relapsed setting) or newly diagnosed or previously untreated metastatic breast cancer (first-line setting) was received by the enrolled patients.

Phase III study: The EMILIA study was a randomized, open-label, international trial, involving 991 patients from 24 countries with HER2-positive, unresectable, locally advanced or metastatic breast cancer who had previously been treated with trastuzumab and a taxane to T-DM1 or lapatinib plus capecitabine. Patients received trastuzumab emtansine ($n = 495$) or oral lapatinib (1250 mg/day on days 1–21 of a 3-weekly cycle) plus capecitabine (1000 mg/m^2 twice daily on days 1–14 of a 3-weekly cycle) ($n = 496$). The median duration of follow-up was ~13 months, with the second interim analysis of OS conducted at a median follow-up of ~19 months (Verma et al. 2012). PFS as assessed by independent review, overall survival, and safety were considered as the primary endpoints. Secondary endpoints included PFS (investigator-assessed), ORR, and the time to symptom progression. Two interim analyses of OS were conducted. The study showed that trastuzumab emtansine therapy significantly ($p < 0.001$) prolonged PFS (co-primary endpoint) by 3.2 months relative to treatment with lapatinib plus capecitabine (9.6 vs. 6.4 months, as per independent review; hazard ratio [HR] 0.65 [95 CI 0.55–0.77]) in patients with HER2-positive advanced breast cancer previously treated with trastuzumab and a taxane. An increase of median OS by 5.8 months was seen as significant ($p < 0.001$) (30.9 vs. 25.1 months at the second interim analysis; HR 0.68 [95% CI 0.55–0.85]) relative to treatment with lapatinib plus capecitabine (co-primary endpoint). The survival rate was 85.2% (estimated 1 year) in trastuzumab emtansine recipients, compared to 78.4% in lapatinib plus capecitabine recipients and the 2-year survival rates were 64.7 vs. 51.8%, respectively. In addition, trastuzumab emtansine recipients had significantly ($p \leq 0.012$) longer time to symptom progression (7.1 vs. 4.6 months) and higher objective response rates (ORRs; 43.6 vs. 30.8%) than lapatinib plus capecitabine recipients; the MDR was 12.6 and 6.5 months in the respective groups. T-DM1 significantly prolonged progression-free and overall survival with less toxicity than lapatinib plus capecitabine in patients with HER2-positive advanced breast cancer previously treated with trastuzumab and a taxane.

Safety profile: Trastuzumab emtansine was well tolerated in patients with HER2-positive advanced breast cancer previously treated with trastuzumab and a taxane. In general, most adverse events were of low grade as seen from the Phase III EMILIA study (Verma et al. 2012). As shown in Table 12.2, after a median follow-up of ~13 months of treatment, trastuzumab emtansine recipients ($n = 490$) showed lower incidence of adverse events of any grade compared with lapatinib plus capecitabine recipients ($n = 488$) (Verma et al. 2012); Both treatment groups

Table 12.2 Safety profile (Kadcyla®) Phase III EMILIA study

	Adverse events		Common adverse symptoms			
Cancer drug treatment	Overall	Grade 3 or 4	Nausea	Fatigue	Thrombocytopenia	Elevated AST
Trastuzumab emtansine (Kadcyla®) (3.6 mg/kg)	95.9%	40.8%	39.2%	35.1%	28%	4.3%
Lapatinib + Capecitabine (1250 mg/day +1000 mg/m^2)	97.7%	57%	44.7%	27.9%	2.5%	0.8%

exhibited adverse events of grade 3 or 4; however the incidence was again lower in the trastuzumab emtansine group. The most common (incidence >25%) treatment-emergent adverse events of any severity in the trastuzumab emtansine group were nausea, fatigue, and thrombocytopenia. The most common (incidence >3%) grade 3 or 4 adverse events in trastuzumab emtansine recipients were thrombocytopenia (12.9 vs. 0.2%) and elevated aspartate aminotransferase (AST; 4.3 vs. 0.8%). About 15.5% of patients in the trastuzumab emtansine group reported serious adverse events compared to 18% of patients in the lapatinib plus capecitabine group.

12.9 Perspectives

The approval of trastuzumab emtansine (Kadcyla ™) for metastatic breast cancer represents the first ADC to be approved by the FDA for HER2/neu-positive patients who have progressed on prior therapy with trastuzumab (Abraham 2013). This novel approach, whereby combining the toxic drug (emtansine) with a targeted agent (trastuzumab) via a linker molecule, provides a tremendous impetus on future drug development. Additionally, and perhaps more importantly, for patients with MBC, this ADC option improves progression-free survival and overall survival rates at safety profiles far better than its predecessors. The addition of this ADC to the arsenal of treatments have helped redefine the natural progression of metastatic HER2-positive breast cancer allowing women to live with the disease for many years and maintain an excellent quality of life.

Antibody-drug conjugates (ADCs) have generated tremendous interest for decades (Perez et al. 2013). More recently, FDA approvals of ADCETRIS® and Kadcyla® and a robust pipeline of additional ADCs have contributed to the promise of this therapeutic class of drugs. With nearly 30 additional ADCs currently in clinical development, it is anticipated that the potential of this new therapeutic class might finally be coming to fruition (Mullard 2013). While the complexities in the design of ADCs are enormous, evolving clinical data from approved products will continue to drive technological advancements in the field. Current methods for preclinical lead selection typically rely on systematic in vitro evaluation of a matrix of various mAbs, linkers, and cytotoxic payloads. However, early in vivo studies would be crucial until it is determined that in vitro models are sufficient to predict response of the ADCs. Progress in site-specific conjugation modalities, optimization of linkers with balanced stability, and identification of novel and potent cytotoxic agents should pave the way for greater insight into the contribution of these various factors to ADC efficacy, pharmacokinetics, and safety. Challenges in target tumor selection will be addressed as the roles of antigen expression, heterogeneity, and internalization rate are further elucidated. Guiding principles for the selection of an ideal anti-body Fc format are currently lacking and prompt validation of assumptions regarding antibody-dependent properties, such as specificity and immune effector functions is needed. Ongoing efforts to address these issues will continue to broaden the impact of ADCs as targeted therapeutics for the treatment of cancer and potentially other diseases (Perez et al. 2013).

References

Abraham J (2013) Trastuzumab emtansine in advanced her2-positive breast cancer. Commun Oncol 10(3):71–73

Aogi K, Masuda N, Ohno S, Oda T, Iwata H, Kashiwaba M, Fujiwara Y, Kamigaki S, Ito Y, Ueno T, Takashima S (2011) First-line bevacizumab in combination with weekly paclitaxel for metastatic breast cancer: efficacy and safety results from a large, open-label, single-arm japanese study. Breast Cancer Res Treat 129(3):829–838

Arkhipov A, Shan Y, Kim ET, Dror RO, Shaw DE (2013) Her2 activation mechanism reflects evolutionary preservation of asymmetric ectodomain dimers in the human egfr family. elife 2:e00708

Baldwin AD, Kiick KL (2011) Tunable degradation of maleimide-thiol adducts in reducing environments. Bioconjug Chem 22(10):1946–1953

Ballantyne A, Dhillon S (2013) Trastuzumab emtansine: First global approval. Drugs 73(7):755–765

Barok M, Tanner M, Köninki K, Isola J (2011) Trastuzumab-dm1 is highly effective in preclinical models of her2-positive gastric cancer. Cancer Lett 306(2):171–179

Beeram M, Krop IE, Burris HA, Girish SR, Yu W, Lu MW, Holden SN, Modi S (2012) A phase 1 study of weekly dosing of trastuzumab emtansine (t-dm1) in patients with advanced human epidermal growth factor 2-positive breast cancer. Cancer 118(23):5733–5740

Bessman NJ, Lemmon MA (2012) Finding the missing links in egfr. Nat Struct Mol Biol 19(1):1–3

Burris HA 3rd, Rugo HS, Vukelja SJ, Vogel CL, Borson RA, Limentani S, Tan-Chiu E, Krop IE, Michaelson RA, Girish S, Amler L, Zheng M, Chu YW, Klencke B, O'Shaughnessy JA (2011) Phase ii study of the antibody drug conjugate trastuzumab-dm1 for the treatment of human epidermal growth factor receptor 2 (her2)-positive breast cancer after prior her2-directed therapy. J Clin Oncol 29(4):398–405

Cho HS, Mason K, Ramyar KX, Stanley AM, Gabelli SB, Denney DW Jr, Leahy DJ (2003) Structure of the extracellular region of her2 alone and in complex with the herceptin fab. Nature 421(6924):756–760

Coussens L, Yang-Feng TL, Liao YC, Chen E, Gray A, McGrath J, Seeburg PH, Libermann TA, Schlessinger J, Francke U et al (1985) Tyrosine kinase receptor with extensive homology to egf receptor shares chromosomal location with neu oncogene. Science 230(4730):1132–1139

Dieras V, Harbeck N, Albain K, Burris HA, Awada A, Crivellari D, Andre F, Choi YJ, Huang J, Miller KD (2010) A phase ib/ii trial of trastuzumab-dm1 (t-dm1) with pertuzumab for patients with her2-positive, locally advanced or metastatic breast cancer: interim efficacy and safety results. In: Paper presented at 33rd annual San Antonio breast cancer symposium, vol 70, San Antonio, p P3-14-01

Drug Safety Data U.S. Food and Drug Administration (2012) Clinical pharmacology and biopharmaceutics review(s) bla 125427 review-t-dm1 by Liu Qi and Mehrotra Nitin

Erickson HK, Lambert JM (2012) Adme of antibody-maytansinoid conjugates. AAPS J 14(4):799–805

Erickson HK, Lewis Phillips GD, Leipold DD, Provenzano CA, Mai E, Johnson HA, Gunter B, Audette CA, Gupta M, Pinkas J, Tibbitts J (2012) The effect of different linkers on target cell catabolism and pharmacokinetics/pharmacodynamics of trastuzumab maytansinoid conjugates. Mol Cancer Ther 11(5):1133–1142

Girish S, Gupta M, Wang B, Lu D, Krop IE, Vogel CL, Burris Iii HA, LoRusso PM, Yi JH, Saad O, Tong B, Chu YW, Holden S, Joshi A (2012) Clinical pharmacology of trastuzumab emtansine (t-dm1): an antibody-drug conjugate in development for the treatment of her2-positive cancer. Cancer Chemother Pharmacol 69(5):1229–1240

Hurvitz SA, Dalenc F, Campone M, O'Regan RM, Tjan-Heijnen VC, Gligorov J, Llombart A, Jhangiani H, Mirshahidi HR, Tan-Chiu E, Miao S, El-Hashimy M, Lincy J, Taran T, Soria JC, Sahmoud T, André F (2013) A phase 2 study of everolimus combined with trastuzumab and paclitaxel in patients with her2-overexpressing advanced breast cancer that progressed during prior trastuzumab and taxane therapy. Breast Cancer Res Treat 141(3):437–446

Hynes NE, Stern DF (1994) The biology of erbb-2/neu/her-2 and its role in cancer. Biochim Biophys Acta 1198(2–3):165–184

Iwasaki S (1993) Inhibitors of tubulin assembly: specially on rhizoxin-maytansine site ligands. Tanpakushitsu Kakusan Koso 38(11):1742–1752

Junttila TT, Akita RW, Parsons K, Fields C, Lewis Phillips GD, Friedman LS, Sampath D, Sliwkowski MX (2009) Ligand-independent her2/her3/pi3k complex is disrupted by trastuzumab and is effectively inhibited by the pi3k inhibitor gdc-0941. Cancer Cell 15(5):429–440

Junttila TT, Li G, Parsons K, Phillips GL, Sliwkowski MX (2011) Trastuzumab-dm1 (t-dm1) retains all the mechanisms of action of trastuzumab and efficiently inhibits growth of lapatinib insensitive breast cancer. Breast Cancer Res Treat 128(2):347–356

Kellogg BA, Garrett L, Kovtun Y, Lai KC, Leece B, Miller M, Payne G, Steeves R, Whiteman KR, Widdison W, Xie H, Singh R, Chari RV, Lambert JM, Lutz RJ (2011) Disulfide-linked antibody-maytansinoid conjugates: optimization of in vivo activity by varying the steric hindrance at carbon atoms adjacent to the disulfide linkage. Bioconjug Chem 22(4):717–727

Kiguchi K, Carbajal S, Chan K, Beltran L, Ruffino L, Shen J, Matsumoto T, Yoshimi N, DiGiovanni J (2001) Constitutive expression of erbb-2 in gallbladder epithelium results in development of adenocarcinoma. Cancer Res 61(19):6971–6976

Kovtun YV, Audette CA, Ye Y, Xie H, Ruberti MF, Phinney SJ, Leece BA, Chittenden T, Blattler WA, Goldmacher VS (2006) Antibody-drug conjugates designed to eradicate tumors with homogeneous and heterogeneous expression of the target antigen. Cancer Res 66(6):3214–3221

Krop IE, Beeram M, Modi S, Jones SF, Holden SN, Yu W, Girish S, Tibbitts J, Yi JH, Sliwkowski MX, Jacobson F, Lutzker SG, Burris HA (2010) Phase i study of trastuzumab-dm1, an her2 antibody-drug conjugate, given every 3 weeks to patients with her2-positive metastatic breast cancer. J Clin Oncol 28(16):2698–2704

Krop IE, LoRusso P, Miller KD, Modi S, Yardley D, Rodriguez G, Guardino E, Lu M, Zheng M, Girish S, Amler L, Winer EP, Rugo HS (2012) A phase ii study of trastuzumab emtansine in patients with human epidermal growth factor receptor 2-positive metastatic breast cancer who were previously treated with trastuzumab, lapatinib, an anthracycline, a taxane, and capecitabine. J Clin Oncol 30(26):3234–3241

Lambert JM, Chari RVJ (2014) Ado-trastuzumab entansine (T-DM1): an antibody-drug conjugate (ADC) for HER2-posivite breast cancer. J Med Chem 57(16):6949–6964

Lewis Phillips GD, Li G, Dugger DL, Crocker LM, Parsons KL, Mai E, Blattler WA, Lambert JM, Chari RV, Lutz RJ, Wong WL, Jacobson FS, Koeppen H, Schwall RH, Kenkare-Mitra SR,

Spencer SD, Sliwkowski MX (2008) Targeting her2-positive breast cancer with trastuzumab-dm1, an antibody-cytotoxic drug conjugate. Cancer Res 68(22):9280–9290

Linggi B, Carpenter G (2006) Erbb receptors: new insights on mechanisms and biology. Trends Cell Biol 16(12):649–656

Liu Z, Floss HG, Cassady JM, Chan KK (2005) Metabolism studies of the anti-tumor agent maytansine and its analog ansamitocin p-3 using liquid chromatography/tandem mass spectrometry. J Mass Spectrom 40(3):389–399

Lopus M, Oroudjev E, Wilson L, Wilhelm S, Widdison W, Chari R, Jordan MA (2010) Maytansine and cellular metabolites of antibody-maytansinoid conjugates strongly suppress microtubule dynamics by binding to microtubules. Mol Cancer Ther 9(10):2689–2699

Lu D, Modi S, Elias AD, Agarwal P, Yi JH, Guardino AE, Althaus BL, Girish S (2011) P1-12-07: Pharmacokinetics (pk) of trastuzumab emtansine and paclitaxel or docetaxel in patients with her2–positive mbc previously treated with a trastuzumab-containing regimen. Paper presented at Cancer Research: December 15, 2011; Volume 71, Issue 24, Supplement 3 Dec 6–10, 2011

Mayo F, Erickson H, Wunderli P, Garrett L, Whitman K, Leece B, Lutz R (2005) Pharmacokinetics and biodistribution of huc242-dm4, an antibody-maytansinoid conjugate that targets canag-positive tumors. Paper Presented at AACR Meeting

Mitri Z, Constantine T, O'Regan R (2012) The her2 receptor in breast cancer: pathophysiology, clinical use, and new advances in therapy. Chemother Res Pract 2012:743193

Mullard A (2013) Maturing antibody-drug conjugate pipeline hits 30. Nat Rev Drug Discov 12(5):329–332

Muller WJ, Ho J, Siegel PM (1998) Oncogenic activation of neu/erbb-2 in a transgenic mouse model for breast cancer. Biochem Soc Symp 63:149–157

Nahta R, Yuan LX, Zhang B, Kobayashi R, Esteva FJ (2005) Insulin-like growth factor-i receptor/human epidermal growth factor receptor 2 heterodimerization contributes to trastuzumab resistance of breast cancer cells. Cancer Res 65(23):11118–11128

Ogiso H, Ishitani R, Nureki O, Fukai S, Yamanaka M, Kim JH, Saito K, Sakamoto A, Inoue M, Shirouzu M, Yokoyama S (2002) Crystal structure of the complex of human epidermal growth factor and receptor extracellular domains. Cell 110(6):775–787

Partida-Martinez LP, Hertweck C (2005) Pathogenic fungus harbours endosymbiotic bacteria for toxin production. Nature 437(7060):884–888

Pauletti G, Godolphin W, Press MF, Slamon DJ (1996) Detection and quantitation of her-2/neu gene amplification in human breast cancer archival material using fluorescence in situ hybridization. Oncogene 13(1):63–72

Perez HL, Cardarelli PM, Deshpande S, Gangwar S, Schroeder GM, Vite GD, Borzilleri RM (2013) Antibody–drug conjugates: current status and future directions, vol 19. Drug Discovery Today in Press, p 869

Poon KA, Flagella K, Beyer J, Tibbitts J, Kaur S, Saad O, Yi JH, Girish S, Dybdal N, Reynolds T (2013) Preclinical safety profile of trastuzumab emtansine (t-dm1): mechanism of action of its cytotoxic component retained with improved tolerability. Toxicol Appl Pharmacol 273(2):298–313

Ritter CA, Perez-Torres M, Rinehart C, Guix M, Dugger T, Engelman JA, Arteaga CL (2007) Human breast cancer cells selected for resistance to trastuzumab in vivo overexpress epidermal growth factor receptor and erbb ligands and remain dependent on the erbb receptor network. Clin Cancer Res 13(16):4909–4919

Rubin I, Yarden Y (2001) The basic biology of her2. Ann Oncol 12(Suppl 1):S3–S8

Sawada T, Kato Y, Kobayashi H, Hashimoto Y, Watanabe T, Sugiyama Y, Iwasaki S (1993) A fluorescent probe and a photoaffinity labeling reagent to study the binding site of maytansine and rhizoxin on tubulin. Bioconjug Chem 4(4):284–289

Schneider MR, Yarden Y (2013) Structure and function of epigen, the last egfr ligand. Semin Cell Dev Biol 28:57

Shen BQ, Bumbaca D, Saad O, Yue Q, Pastuskovas CV, Khojasteh SC, Tibbitts J, Kaur S, Wang B, Chu YW, LoRusso PM, Girish S (2012) Catabolic fate and pharmacokinetic characterization of trastuzumab emtansine (t-dm1): an emphasis on preclinical and clinical catabolism. Curr Drug Metab 13(7):901–910

Slamon DJ, Godolphin W, Jones LA, Holt JA, Wong SG, Keith DE, Levin WJ, Stuart SG, Udove J, Ullrich A et al (1989) Studies of the her-2/neu proto-oncogene in human breast and ovarian cancer. Science 244(4905):707–712

Slamon DJ, Leyland-Jones B, Shak S, Fuchs H, Paton V, Bajamonde A, Fleming T, Eiermann W, Wolter J, Pegram M, Baselga J, Norton L (2001) Use of chemotherapy plus a monoclonal antibody against her2 for metastatic breast cancer that overexpresses her2. N Engl J Med 344(11):783–792

Tassone P, Gozzini A, Goldmacher V, Shammas MA, Whiteman KR, Carrasco DR, Li C, Allam CK, Venuta S, Anderson KC, Munshi NC (2004) In vitro and in vivo activity of the maytansinoid immunoconjugate hun901-n2'-deacetyl-n2'-(3-mercapto-1-oxopropyl)-maytansine against cd56+ multiple myeloma cells. Cancer Res 64(13):4629–4636

Venter DJ, Tuzi NL, Kumar S, Gullick WJ (1987) Overexpression of the c-erbb-2 oncoprotein in human breast carcinomas: Immunohistological assessment correlates with gene amplification. Lancet 2(8550):69–72

Verma S, Miles D, Gianni L, Krop IE, Welslau M, Baselga J, Pegram M, Oh DY, Diéras V, Guardino E, Fang L, Lu MW, Olsen S, Blackwell K, EMILIA Study Group (2012) Trastuzumab emtansine for her2-positive advanced breast cancer. N Engl J Med 367(19):1783–1791

Vogel C, Cobleigh MA, Tripathy D, Gutheil JC, Harris LN, Fehrenbacher L, Slamon DJ, Murphy M, Novotny WF, Burchmore M, Shak S, Stewart SJ (2001) First-line, single-agent herceptin(r) (trastuzumab) in metastatic breast cancer. A preliminary report. Eur J Cancer 37(Suppl 1):25–29

Wang SE, Narasanna A, Perez-Torres M, Xiang B, Wu FY, Yang S, Carpenter G, Gazdar AF, Muthuswamy SK, Arteaga CL (2006) Her2 kinase domain mutation results in constitutive phosphorylation and activation of her2 and egfr and resistance to egfr tyrosine kinase inhibitors. Cancer Cell 10(1):25–38

Way TD, Lin JK (2005) Role of her2/her3 co-receptor in breast carcinogenesis. Future Oncol 1(6):841–849

Widdison WC, Wilhelm SD, Cavanagh EE, Whiteman KR, Leece BA, Kovtun Y, Goldmacher VS, Xie H, Steeves RM, Lutz RJ, Zhao R, Wang L, Blattler WA, Chari RV (2006) Semisynthetic maytansine analogues for the targeted treatment of cancer. J Med Chem 49(14):4392–4408

Chapter 13
The Antibody-Drug Conjugate Glembatumumab Vedotin (CDX-011) and Its Use in the Treatment of Breast Cancer

Elaine Kuhn, Nancy Chan, and Linda T. Vahdat

Abstract Breast cancer is the most common malignancy in women and the significant cause of cancer-related deaths, second only to lung cancer. Amongst different subtypes of breast cancers, triple-negative breast cancer (TNBC) is especially challenging due to fewer treatment options and more aggressive clinical course. Therefore, development of molecularly targeted therapies for TNBC is crucial. Glycoprotein nonmetastatic B (GPNMB) is overexpressed in many malignancies, including TNBC, and is associated with poor prognosis. The antibody-drug conjugate (ADC) glembatumumab vedotin (CDX-011) selectively targets GPNMB and was investigated for its efficacy in metastatic breast cancer. Despite promising initial results, development of CDX-011 was discontinued in 2018 after it failed to show benefit over capecitabine in a Phase II clinical trial. This chapter reviews the mechanisms of action, preclinical and Phase I/II results of glembatumumab vedotin, lessons learned, and insights gained in the development of ADCs for the treatment of breast cancer.

Keywords CDX011 · Breast cancer · GPNMB · ADC

13.1 Introduction

Breast cancer is the most common malignancy in women worldwide and despite advances in screening, diagnosis, and treatment, it remains one of the most significant causes of cancer-related deaths in women (Bray et al. 2013). Of the different

E. Kuhn · L. T. Vahdat (✉)
Dartmouth Geisel School of Medicine, Hanover, NH, USA
e-mail: Linda.T.Vahdat@hitchcock.org

N. Chan
NYU Langone Health, New York, NY, USA

J. L. Zaro, J. Wang (eds.), *Antibody-Drug Conjugates*, AAPS Advances in the Pharmaceutical Sciences Series 17, https://doi.org/10.1007/978-3-031-83005-1_13

subtypes of breast cancers, triple-negative breast cancer (TNBC) poses a special challenge due to its aggressive clinical course, predilection for metastases, and lack of targeted therapy (Shuen and Foulkes 2012). Molecularly targeted therapies are becoming increasingly prominent as part of the armamentarium of treatments for breast cancer (Müller et al. 2009). Glycoprotein nonmetastatic B (GPNMB) was identified as an overexpressed gene in many malignancies, including TNBC, and is associated with lower disease-free and overall survival (OS). The antibody-drug conjugate (ADC) glembatumumab vedotin (CDX-011) (GV) appears to selectively target GPNMB and was investigated in clinical trials for efficacy in patients with metastatic breast cancer and other GPNMB-expressing tumors. Despite promising initial results, development of CDX-011 was discontinued in 2018 after it failed to show benefit over capecitabine in a Phase II clinical trial for patients with metastatic, GPNMB overexpressing TNBC. This chapter reviews the mechanisms of action, preclinical and Phase I/II results of glembatumumab vedotin, lessons learned, and insights gained in the development of ADCs for the treatment of advanced breast cancer.

13.2 GPNMB as a Therapeutic Target

Glycoprotein nonmetastatic protein B (GPNMB) is a type I transmembrane protein, which is also known as osteoactivin (OA) (Saitoh et al. 1992). It was first described in 1995 by Weterman et al. as a highly expressed protein in melanoma cell line with low metastatic potential (Weterman et al. 1995). Subsequently, GPNMB was discovered to have high expression in a number of malignancies including glioblastoma multiforme (Kuan et al. 2006), cancers of the breast (Rose et al. 2007), prostate (Fiorentini et al. 2014), liver (Haralanova-Ilieva et al. 2005), and colon (Eldai et al. 2013). GPNMB has a complex role in tumor biology. Despite its originally perceived low invasive phenotype, it is associated with increased metastatic tendencies in malignancies such as breast cancer (Rose et al. 2007).

The GPNMB is located on the small arm of chromosome 7 (7q15) and belongs to the Pmel17/NMB family. Pmel17 is the major structural component of melanosomes and is essential in melanocyte pigment production (Yamaguchi and Hearing 2009). Also important is that GPNMB shares homology to the lysosomal-associated membrane protein (LAMP) glycoproteins that are implicated in cell adhesion and metastasis (Saitoh et al. 1992).

Under normal conditions, GPNMB is situated in intracellular compartments in macrophages and melanocytes (Tomihari et al. 2009; Ripoll et al. 2007). In tumor cells, however, GPNMB expression is enriched on the cell surface (Maric et al. 2013) (Fig. 13.1).

Rose et al. found that when GPNMB is overexpressed in parental 4 T1 mouse mammary carcinoma cells, they acquire a more invasive phenotype, leading to increased bone metastasis (Rose et al. 2007). High level of expression correlated

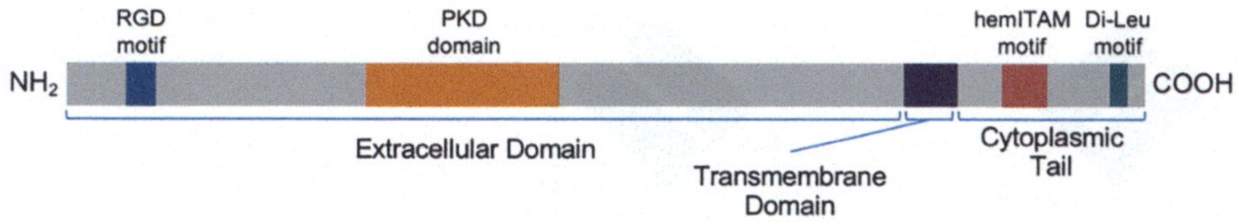

Fig. 13.1 Schematic representation of GPNMB. Abbreviations: GPNMB, glycoprotein nonmetastatic B; hemITAM, immunoreceptor tyrosine-based activation motif; Leu, Leucine; PKD, polycystic kidney disease; RGD, Arginine-Glycine-Aspartic acid

with negative estrogen receptor status and higher tumor grade. Moreover, tumors that express GPNMB have high endothelial cell density compared to those without its expression, suggesting GPNMB recruit endothelial cells to promote tumor growth and enhance the metastatic process. Quantification of apoptotic cells revealed fewer cells in GPNMB, expressing tumors were undergoing apoptosis assessed by TUNEL, which further elucidates its functional role in tumor progression (Rose et al. 2010a).

The interaction between the tumor and the microenvironment is emerging as an important component in metastasis, and GPNMB has also been implicated in this relationship. Ogawa et al. found that GPNMB can activate fibroblasts by inducing upregulation of pro-invasive metalloproteases such as MMP-3 and MMM-9 in mouse models (Ogawa et al. 2005).

13.3 Glembatumumab Vedotin

GPNMB has the ideal components required for targeted therapy. The selectively extracellular domain in tumor cells (not normal cells) enables availability to antibody targets. Glembatumumab vedotin (or CDX-011, CR011-vcMMAE) is a fully humanized monoclonal antibody-drug conjugate (ADC). The first component of the ADC is the tubulin destabilizing cytotoxin monomethyl auristatin E (MMAE). It was originally derived from peptides found in the marine shell-less mollusk *Dolabella auricularia* known as dolastatins (Bai et al. 1990). This antimitotic agent has potency of up to 200 times compared to that of vinblastine and it was found to block the growth of GPNMB-expressing melanoma cells with half-maximal inhibitory concentration (IC_{50}) of 216–300 ng/mL (Vaklavas and Forero 2014). Therefore, the toxicity profile prevents its antineoplastic usage alone, but the high potency proves to be advantageous when MMAE is linked to an antibody, at which point it is referred to as vedotin. Vedotin is also the cytotoxic agent used with brentuximab targeting CD30 in Hodgkin lymphoma and anaplastic large-cell lymphoma (Francisco et al. 2003), and more recently, enfortumab, targeting Nectin-4 in urothelial carcinoma (Powles et al. 2021). Numerous other antibodies paired with vedotin are under development, including disitamab vedotin, which targets HER2.

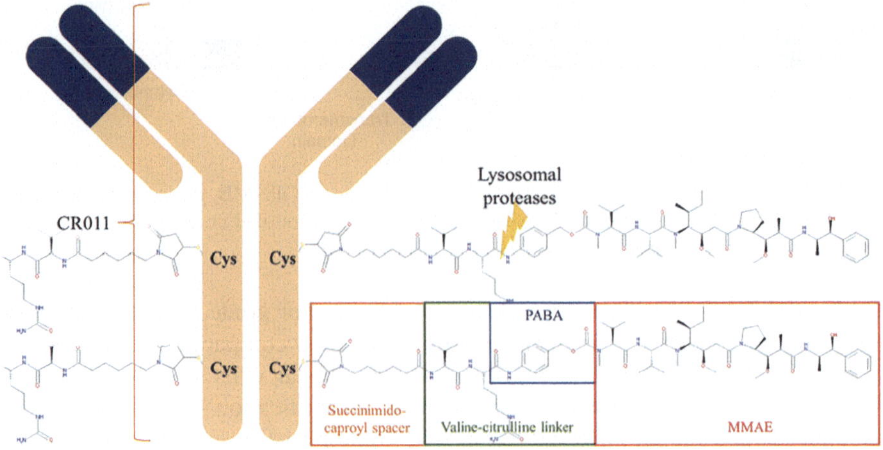

Fig. 13.2 Structure of glembatumumab vedotin. MMAE = monomethyl auristatin E, PABA = *p*-aminobenzoic acid spacer (Vaklavas and Forero 2014)

Vedotin is conjugated through a cathepsin B-sensitive valine–citrulline linker to the antibody glembatumumab, which is directed against the extracellular domain of GPNMB (Fig. 13.2). On average, four or five MMAE molecules are bound to a single antibody (Vaklavas and Forero 2014). The ADC is highly stable in serum and is released and internalized upon binding to GPNMB. After endocytosis, the synthetic dolastatin analogue MMAE is released through enzymatic cleavage into the tumor cell cytoplasm where it binds to tubulin and inhibits tubulin polymerization. This result in G2/M phase arrest and apoptosis of the tumor cells (Tse et al. 2006).

13.4 Preclinical Models

Glembatumumab vedotin was first demonstrated to be efficacious in metastatic melanoma cells in culture and in xenograft assays. Pollack et al. used the human SK-MEL-2 and SK-MEL-5 melanoma xenografts in athymic mice to assess antitumor efficacy of glembatumumab vedotin (Pollack et al. 2007). They found that it induced complete tumor regression in 100% of GPNMB-expressing xenografted melanoma cells. This was achieved at concentrations as low as 2.5 mg/kg, administered intravenously every 4 days for four treatments. Interestingly, tumors did not regrow during the nontreatment observation period of 200 days. In the breast cancer cell line MDA-MB-468, a single dose of 20 mg/kg glembatumumab vedotin was able to induce sustained tumor regression in vivo (Rose et al. 2010b).

13.5 GPNMB and Triple-Negative Breast Cancer (TNBC)

Rose et al. first investigated GPNMB as a potential therapeutic target in breast cancer by analyzing its expression in several breast cancer gene expression datasets and in primary human breast tumors. They found that GPNMB may serve as an important target, especially for TNBC patients whose clinical courses tend to be more aggressive with increased potential for metastasis. Since TNBC patients have limited therapeutic options, there is tremendous need for identification and development of precision therapy for these patients. In the Rose et al. study, the investigators first showed that GPNMB mRNA expression in 295 human breast tumors is associated with worse metastasis-free and overall survival. This was accomplished through comparing GPNMB mRNA levels with clinical outcomes in three published datasets. GPNMB protein expression in tumor epithelium was also associated with significantly inferior outcomes in human breast cancer as evaluated by immunohistochemical (IHC) staining for GPNMB using breast tissue microarray data. Intriguingly, GPNMB expression correlated with recurrence within the TNBC patients. An independent cohort of breast cancer tissue microarrays enriched for TNBC samples showed that 29% of triple-negative (TN) tumors are GPNMB-epithelial-positive, compared with only 3.6% in luminal and 11.6% in HER2 tumors. The multivariate Cox regression survival analysis revealed that GPNMB was an independent prognostic indicator of distant metastasis in TN tumors ($p = 0.0024$). Therefore, GPNMB-epithelial expression is more common in triple-negative breast cancer, and within this subtype, its expression is associated with an increased risk of recurrence (Rose et al. 2010b).

13.6 Clinical Trials of Glembatumumab Vedotin

Glembatumumab vedotin was first investigated in two multicenter Phase I/II clinical trials for patients with unresectable melanoma and advanced stage or metastatic breast cancer (Hamid et al. 2010; Saleh et al. 2010; Burris et al. 2009) (Table 13.1). In the melanoma study, tumor shrinkage was achieved in up to 56% of patients when glembatumumab vedotin was given intravenously at 1.88 mg/kg in the Phase II maximum tolerated dose (MTD) expansion at dose frequencies ranging from weekly to q3 weeks. The median progression-free survival (PFS) was highest at 3.9 months in patients treated at the q3 week dose. Importantly, the subset of patients with strong GPNMB expression, defined as 3+ by IHC with 90% staining, had higher tumor shrinkage at 86% and longer median PFS of 4.9 months. Of note, 57% of all patients developed varying degrees of skin rash, which also correlated with greater median PFS of 4.8 months versus 1.2 months in those without rash. This may be related to the presence of GPNMB in skin, acting as a predictor of drug efficacy (Hamid et al. 2010).

Table 13.1 Clinical trials of glembatumumab vedotin

Trial	Primary endpoints	Randomization	No. patients
Phase I/II in advanced melanoma	ORR = 15%–33%	No	117
Phase I/II in advanced breast cancer	12-week PFS = 33–100% in GPNMB-expressing patients	No	42
EMERGE study: Phase II in advanced breast cancer.	ORR 6% vs. 7% in overall population; 30% vs. 9% for GPNMB-overexpressing tumors	Yes[a]	124
METRIC study: Phase II in metastatic GPNMB-expressing TNBC	PFS 2.9 vs. 2.8 months in overall population	Yes[b]	327

[a]Randomization of glembatumumab vedotin vs. investigator's choice chemotherapy
[b]Randomization to receive glembatumumab vedotin or capecitabine
ORR Overall Response Rate, *PFS* Progression-free survival

Similarly, glembatumumab vedotin was used in heavily pretreated advanced breast cancer patients in Phase I standard dose escalation followed by a Phase II expansion at the MTD to assess 12-week PFS. Eligible patients for this study had at least two prior chemotherapy regimens that included taxanes, anthracycline, and capecitabine. Forty-two patients were enrolled, and 71% of the tumor specimens were positive for GPNMB, defined as >5% of malignant epithelial or stromal cells with GPNMB expression. IHC with goat polyclonal antibody to GPNMB was performed on 25% of the patient biopsies that were available for evaluation (done as a post-hoc analysis). This study demonstrated median PFS of 9.1 weeks in all patients enrolled. In GPNMB-expressing patients at the MTD/Phase II dose, PFS was doubled at 18.3 weeks. The objective response rate was also the highest among this subgroup at 29% (Burris et al. 2009). This suggestion that efficacy might be enhanced in tumors with higher GPNMB expression led to the design of future studies of glembatumumab vedotin in tumors that had expression of the target.

The Phase I/II studies demonstrated both efficacy and manageable safety profile of glembatumumab vedotin. The encouraging results led to the initiation of the EMERGE study, a Phase II, open-label, randomized study designed to evaluate the anticancer activity of glembatumumab vedotin in advanced GPNMB-expressing breast cancer (Yardley et al. 2015). Patients were stratified by GPNMB expression pattern (any tumor, low stromal or high stromal) and were randomized in 2:1 fashion to glembatumumab vedotin (1.88 mg/kg IV every 3 weeks) or investigator's choice (IC) single-agent chemotherapy. Patients were treated until progression of disease (PD) or intolerance to therapy. In this study, patients on the IC arm were allowed to crossover at PD. The primary endpoint is overall response rate (ORR) and secondary endpoints are PFS, toxicity profile, and pharmacokinetics. Final results were published in 2014, showing that glembatumumab vedotin was well tolerated in heavily pretreated patients with breast cancer. The primary endpoint was not met for the overall population of 124 patients; however, subset analysis revealed promising activity in patients with TNBC and/or GPNMB-overexpressing tumors. Specifically, an exploratory biomarker analysis suggested that patients with advanced TNBC, with at least 25% of tumor staining positive for GPNMB by IHC, were thought to be most likely to derive benefit from GV (Yardley et al. 2015).

Based on the results from the EMERGE trial, the randomized Phase II METRIC study was conducted to assess the efficacy in women with metastatic GPNMB over-expressing triple-negative breast cancer (Vahdat et al. 2021). Patients with GPNMB overexpressing TNBC (expression >25%) were randomized to receive glembatumumab vedotin or capecitabine in a 2:1 ratio. The primary endpoints were ORR and PFS. A total of 327 patients were randomized (218 to glembatumumab vedotin and 109 to capecitabine). Median PFS was 2.9 months (95% CI 2.8, 3.5) for the glembatumumab vedotin arm and 2.8 months (1.6, 3.2) for the capecitabine arm (Hazard Ratio = 0.95; 95% CI: 0.71, 1.29; $p = 0.7607$) in this heavily pretreated population. Secondary outcomes of OS, ORR, and duration of response were similar between treatment arms. The proportion of patients with any tumor shrinkage by an independent review committee was 63% in the glembatumumab vedotin arm and 39% in the capecitabine arm. A post-hoc analysis revealed a trend toward PFS benefit for patients who had been stable for more than 6 months on taxanes (3.0 vs. 1.9 months). Subgroup analyses showed no difference in response rates based on expression of GPNMB. The safety profile of glembatumumab vedotin was similar to prior studies and did not offer any particular advantage compared with capecitabine (Vahdat et al. 2021).

Overall, the study did not meet its primary endpoint. However, the PFS was similar in both arms—suggesting again that the glembatumumab vedotin has activity in this challenging population.

13.6.1 Toxicity Profile

Glembatumumab vedotin was generally well tolerated in the Phase I and II studies in advanced breast cancer patients. The most common adverse reactions (AEs) were rash (45–61%), fatigue (48–50%), and alopecia (41–50%). The most common grade 3 or 4 AEs are neutropenia (17–28%) and neuropathy (5–11%). Patients with baseline neuropathy worse than grade 1 were excluded after first two patients at 1.34 mg/kg dose experienced worsening of neuropathy (Burris et al. 2009).

13.7 Lessons Learned from Glembatumumab Vedotin

Similar to other ADCs, expression of antibody target is not predictive of response to treatment. Among the three biomarker-driven studies of GV, none demonstrated an association of the intensity of GPNMB expression with response to treatment

(Vahdat et al. 2021; Ott et al. 2019; Kopp et al. 2019). The mechanism by which GPNMB relates to worse clinical outcomes in TNBC remains elusive. One possibility is through its ability to augment the PI3K/AKT/mTOR pathway, which is enriched in 30% of TNBC. It could be that enrichment in this pathway would better predict treatment response to glembatumumab vedotin compared to the expression of the GPNMB target. Nevertheless, we are learning that the relative expression level of an antibody target does not necessarily predict response to treatment. Robust biomarkers to guide which patients will benefit most from antibody-drug conjugates are needed.

The glembatumumab vedotin linker is a protease-dependent cleavable linker that releases its payload in the presence of lysosomal enzymes. While these linkers are more stable in serum, it may be less effective in delivering drug if there is poor penetration to tumor, or inefficiency in localizing to the lysosomes. Other types of linkers such as CL2A used with sacituzumab govitecan are hydrolysable linkers that are shown to be less stable but more effective at the tumor site. These linkers allow for rapid effects on the tumor, including the release of a portion of payload outside the cell leading to bystander effects (Kopp et al. 2023), compared with protease-dependent linkers that have slower effects. It is possible that the stability of the linker did not allow for sufficient accumulation of the payload at its target. This may explain in part the short duration of response to glembatumumab vedotin as seen in the METRIC trial.

Finally, the payload, MMAE, is an antimicrotubule agent. All patients who received GV in METRIC study were previously treated with a taxane, which is also an antimicrotubule agent. There was a trend toward longer PFS in patients who had a sustained response to taxane versus those who progressed more quickly on taxane. This suggests that GV may provide greater clinical benefit in patients who have demonstrated sensitivity to taxanes.

13.8 Conclusion

GPNMB is a relevant target in breast cancer, and the antibody-drug conjugate Glembatumumab vedotin has been demonstrated in clinical trials to be efficacious in patients with advanced and metastatic disease. Triple-negative breast cancers may have higher GPNMB expression, which provides the rare opportunity of targeted therapy for this subtype. However, GPNMB expression alone does not predict response to glembatumumab vedotin. Patients who have demonstrated sensitivity to taxanes may derive greater benefit from antibody-drug conjugates that utilize a payload with similar mechanism of action such as vedotin (MMAE).

References

Bai R, Pettit GR, Hamel E (1990) Dolastatin 10, a powerful cytostatic peptide derived from a marine animal. Inhibition of tubulin polymerization mediated through the vinca alkaloid binding domain. Biochem Pharmacol 39(12):1941–1949. https://doi.org/10.1016/0006-2952(90)90613-p

Bray F, Ren J-S, Masuyer E, Ferlay J (2013) Global estimates of cancer prevalence for 27 sites in the adult population in 2008. Int J Cancer 132(5):1133–1145. https://doi.org/10.1002/ijc.27711

Burris H et al (2009) A phase (Ph) I/II study of CR011-VcMMAE, an antibody-drug conjugate, in patients (pts) with locally advanced or metastatic breast cancer (MBC). Cancer Res 69(24_Suppl):6096. https://doi.org/10.1158/0008-5472.SABCS-09-6096

Eldai H et al (2013) Novel genes associated with colorectal cancer are revealed by high resolution cytogenetic analysis in a patient specific manner. PLoS One 8(10):e76251. https://doi.org/10.1371/journal.pone.0076251

Fiorentini C et al (2014) GPNMB/OA protein increases the invasiveness of human metastatic prostate cancer cell lines DU145 and PC3 through MMP-2 and MMP-9 activity. Exp Cell Res 323(1):100–111. https://doi.org/10.1016/j.yexcr.2014.02.025

Francisco JA et al (2003) cAC10-vcMMAE, an anti-CD30-monomethyl auristatin E conjugate with potent and selective antitumor activity. Blood 102(4):1458–1465. https://doi.org/10.1182/blood-2003-01-0039

Hamid O, Sznol M, Pavlick AC, Kluger HM, Kim KB, Boasberg PD, Simantov R, Davis TA, Crowley E, Hwu P (2010) Frequent dosing and GPNMB expression with CDX-011 (CR011-vcMMAE), an antibody-drug conjugate (ADC), in patients with advanced melanoma. J Clin Oncol 28. Accessed: Oct. 22, 2024. [Online]. Available: https://ascopubs.org/doi/10.1200/jco.2010.28.15_suppl.8525

Haralanova-Ilieva B, Ramadori G, Armbrust T (2005) Expression of osteoactivin in rat and human liver and isolated rat liver cells. J Hepatol 42(4):565–572. https://doi.org/10.1016/j.jhep.2004.12.021

Kopp LM et al (2019) Phase II trial of the glycoprotein non-metastatic B-targeted antibody-drug conjugate, glembatumumab vedotin (CDX-011), in recurrent osteosarcoma AOST1521: a report from the Children's Oncology Group. Eur J Cancer 121:177–183. https://doi.org/10.1016/j.ejca.2019.08.015

Kopp A, Hofsess S, Cardillo TM, Govindan SV, Donnell J, Thurber GM (2023) Antibody–drug conjugate sacituzumab govitecan drives efficient tissue penetration and rapid intracellular drug release. Mol Cancer Ther 22(1):102–111. https://doi.org/10.1158/1535-7163.MCT-22-0375

Kuan C-T et al (2006) Glycoprotein nonmetastatic melanoma protein B, a potential molecular therapeutic target in patients with glioblastoma multiforme. Clin Cancer Res 12(7 Pt 1):1970–1982. https://doi.org/10.1158/1078-0432.CCR-05-2797

Maric G, Rose AA, Annis MG, Siegel PM (2013) Glycoprotein non-metastatic b (GPNMB): a metastatic mediator and emerging therapeutic target in cancer. Onco Targets Ther 6:839–852. https://doi.org/10.2147/OTT.S44906

Müller V, Witzel I, Stickeler E (2009) Immunological approaches in the treatment of metastasized breast cancer. Breast Care (Basel) 4(6):359–366. https://doi.org/10.1159/000262454

Ogawa T et al (2005) Osteoactivin upregulates expression of MMP-3 and MMP-9 in fibroblasts infiltrated into denervated skeletal muscle in mice. Am J Physiol Cell Physiol 289(3):C697–C707. https://doi.org/10.1152/ajpcell.00565.2004

Ott PA et al (2019) A phase 2 study of glembatumumab vedotin, an antibody-drug conjugate targeting glycoprotein NMB, in patients with advanced melanoma. Cancer 125(7):1113–1123. https://doi.org/10.1002/cncr.31892

Pollack VA et al (2007) Treatment parameters modulating regression of human melanoma xenografts by an antibody-drug conjugate (CR011-vcMMAE) targeting GPNMB. Cancer Chemother Pharmacol 60(3):423–435. https://doi.org/10.1007/s00280-007-0490-z

Powles T et al (2021) Enfortumab vedotin in previously treated advanced urothelial carcinoma. N Engl J Med 384(12):1125–1135. https://doi.org/10.1056/NEJMoa2035807

Ripoll VM, Irvine KM, Ravasi T, Sweet MJ, Hume DA (2007) Gpnmb is induced in macrophages by IFN-gamma and lipopolysaccharide and acts as a feedback regulator of proinflammatory responses. J Immunol 178(10):6557–6566. https://doi.org/10.4049/jimmunol.178.10.6557

Rose AAN, Pepin F, Russo C, Abou Khalil JE, Hallett M, Siegel PM (2007) Osteoactivin promotes breast cancer metastasis to bone. Mol Cancer Res 5(10):1001–1014. https://doi.org/10.1158/1541-7786.MCR-07-0119

Rose AAN et al (2010a) ADAM10 releases a soluble form of the GPNMB/Osteoactivin extracellular domain with angiogenic properties. PLoS One 5(8):e12093. https://doi.org/10.1371/journal.pone.0012093

Rose AAN et al (2010b) Glycoprotein nonmetastatic B is an independent prognostic indicator of recurrence and a novel therapeutic target in breast cancer. Clin Cancer Res 16(7):2147–2156. https://doi.org/10.1158/1078-0432.CCR-09-1611

Saitoh O, Wang WC, Lotan R, Fukuda M (1992) Differential glycosylation and cell surface expression of lysosomal membrane glycoproteins in sublines of a human colon cancer exhibiting distinct metastatic potentials. J Biol Chem 267(8):5700–5711

Saleh MN et al (2010) Correlation of GPNMB expression with outcome in breast cancer (BC) patients treated with the antibody–drug conjugate (ADC), CDX-011 (CR011-vcMMAE). J Clin Oncol 28(15_suppl):1095–1095. https://doi.org/10.1200/jco.2010.28.15_suppl.1095

Shuen A, Foulkes W (2012) Basal-like breast cancer—characteristics, risks, and associations. Oncol Hematol Rev 8(1):26–29

Tomihari M, Hwang S-H, Chung J-S, Cruz PD, Ariizumi K (2009) Gpnmb is a melanosome-associated glycoprotein that contributes to melanocyte/keratinocyte adhesion in a RGD-dependent fashion. Exp Dermatol 18(7):586–595. https://doi.org/10.1111/j.1600-0625.2008.00830.x

Tse KF et al (2006) CR011, a fully human monoclonal antibody-auristatin E conjugate, for the treatment of melanoma. Clin Cancer Res 12(4):1373–1382. https://doi.org/10.1158/1078-0432.CCR-05-2018

Vahdat LT et al (2021) Glembatumumab vedotin for patients with metastatic, gpNMB overexpressing, triple-negative breast cancer ('METRIC'): a randomized multicenter study. NPJ Breast Cancer 7(1):57. https://doi.org/10.1038/s41523-021-00244-6

Vaklavas C, Forero A (2014) Management of metastatic breast cancer with second-generation antibody-drug conjugates: focus on glembatumumab vedotin (CDX-011, CR011-vcMMAE). BioDrugs 28(3):253–263. https://doi.org/10.1007/s40259-014-0085-2

Weterman MA et al (1995) nmb, a novel gene, is expressed in low-metastatic human melanoma cell lines and xenografts. Int J Cancer 60(1):73–81. https://doi.org/10.1002/ijc.2910600111

Yamaguchi Y, Hearing VJ (2009) Physiological factors that regulate skin pigmentation. Biofactors 35(2):193–199. https://doi.org/10.1002/biof.29

Yardley DA et al (2015) EMERGE: a randomized phase II study of the antibody-drug conjugate glembatumumab vedotin in advanced glycoprotein NMB-expressing breast cancer. J Clin Oncol 33(14):1609–1619. https://doi.org/10.1200/JCO.2014.56.2959

Part VI
From Today to Tomorrow

Chapter 14
Summary and Future Directions of Antibody-Drug Conjugates

Jennica L. Zaro, Jeffery Wang, and Wei-Chiang Shen

Abstract Antibody-drug conjugates (ADCs) have become a powerful class of therapeutics in oncology. Since the first approval of Mylotarg® (gemtuzumab ozogamicin) by the US Food and Drug Administration (FDA) in 2000, the field has grown rapidly with 11 current FDA-approved ADCs. In 2023, the ADC market had exceeded original projections with a valuation at USD 7.2 billion (Bretz, Antibody drug conjugates market poised for unprecedented growth through 2032, 2024). By 2032, the sales of ADCs are expected to exceed USD 34 billion, with a compound annual growth rate (CAGR) of 17.5%. As of October 2023, there are over 500 active clinical trials involving approved and new ADCs (Udofa, Bioeng Transl Med 9: e10677, 2024), increasing diversity in many areas such as the disease indications, drug payloads, and linkers. In this chapter, the main components of ADCs will be reviewed with a focus on the currently approved ADCs, along with a discussion of future directions based on examination of active clinical trials.

Keywords Antibody drug conjugates · Linkers · Payloads · Molecular targets · Drug antibody ratio · Clinical trials

---

J. L. Zaro (✉) · W.-C. Shen
Department of Pharmacology and Pharmaceutical Sciences, Alfred E. Mann School of Pharmacology and Pharmaceutical Sciences, University of Southern California, Los Angeles, CA, USA
e-mail: zaro@usc.edu

J. Wang
Department of Biotechnology and Pharmaceutical Sciences, College of Pharmacy, Western University of Health Sciences, Pomona, CA, USA

J. L. Zaro, J. Wang (eds.), *Antibody-Drug Conjugates*, AAPS Advances in the Pharmaceutical Sciences Series 17, https://doi.org/10.1007/978-3-031-83005-1_14

14.1 Main Components of Antibody Drug Conjugates (ADCs)

In the design of an ADC, there are five important considerations, including (1) the molecular target; (2) the antibody; (3) the details of the chemical conjugation; (4) the linker; and (5) the cytotoxic payload (Fig. 14.1). Following cell-surface binding and internalization, the payloads are released intracellularly either by cleavage of the linkers ("cleavable linker") or by degradation of the antibody ("non-cleavable linker") resulting in cytotoxicity via various types of mechanisms (Fig. 14.2). In addition, some ADCs also exert secondary mechanisms of toxicity such as antibody-dependent cellular toxicity or diffusion of released payload into neighboring cells (i.e. bystander effect). The current portfolio of ADCs approved by the U.S. Food and Drug Administration (FDA) exhibits substantial diversity in these main components, as well as in the target indications (Table 14.1).

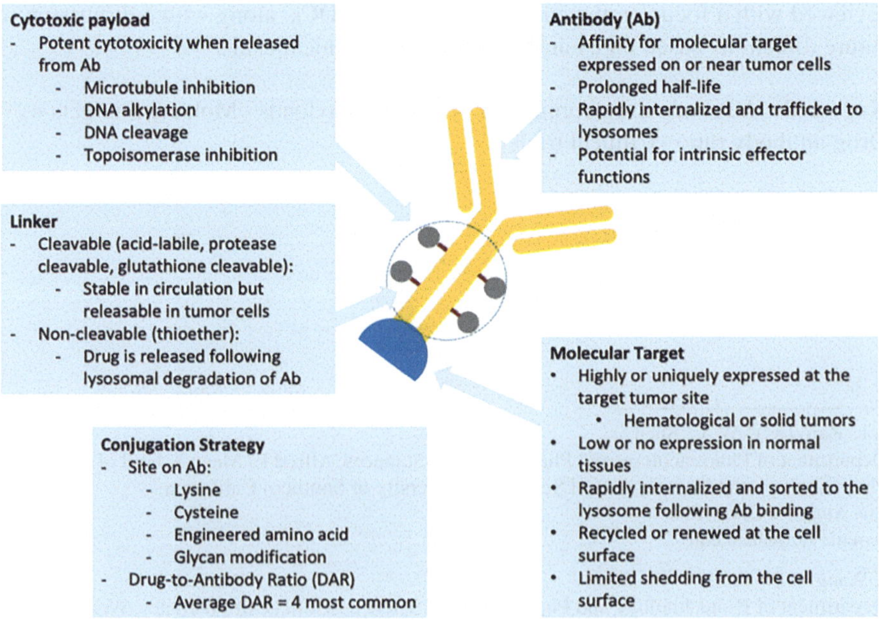

Fig. 14.1 Important components of an ADC. The main considerations in the design of ADC include the molecular target, the antibody, the details of chemical conjugation, the cytotoxic payload, and the linker between the antibody and cytotoxic payload

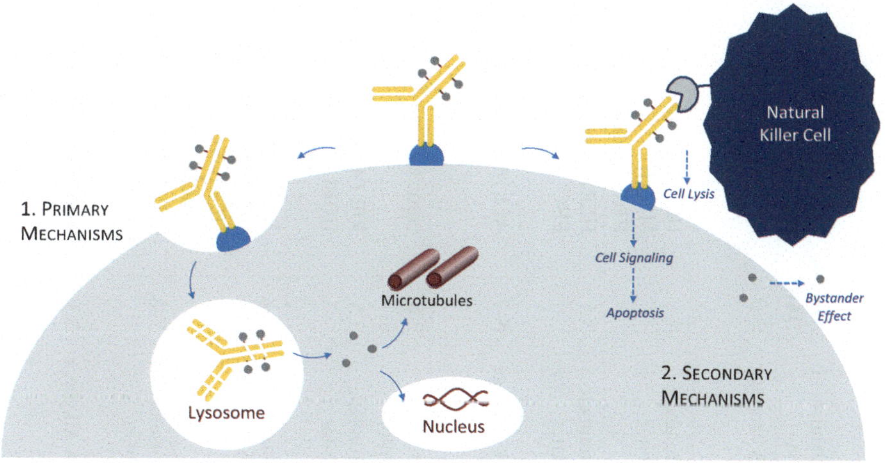

Fig. 14.2 Overview of ADC internalization and mechanisms of toxicity. Following binding to the cell surface receptor, the ADC is internalized by receptor-mediated endocytosis. The cytotoxic drug is released either by cleavage of the linker or degradation of the antibody, and then exerts its cytotoxicity (1). The main classes of drugs used in ADCs (**a**) disrupt microtubule assembly (e.g. DM1, DM4, MMAE, MMAF) or (**b**) involve DNA modifications via DNA strand cleavage (e.g. calicheamicin analogs), DNA alkylation (e.g. pyrrolobenzodiazepine [PDB] dimers), or topoisomerase inhibition (e.g. DXd, SN-38). In addition, certain ADCs also exert secondary mechanisms of toxicity (2) including antibody-related effects of certain antibodies and isotypes (e.g. IgG1 and IgG3) such as antibody-dependent cellular toxicity of complement-dependent cytotoxicity or the release of the cytotoxic payload from ADCs linked via cleavable linkers and diffusion into neighboring cells leading to cell death (bystander effect)

14.1.1 Molecular Targets

There are currently a vast range of target antigens or receptors utilized in ADCs. Of the 11 FDA approved drugs, all of the targets represent molecules that are internalized via receptor-mediated endocytosis (RME) (Table 14.2). For targets that internalize, the optimal conditions in ADC design are that they should show abundant and uniform expression on tumor cells, have no or low expression in healthy tissues and organs, exhibit rapid internalization following ADC binding and traffic to the lysosomes (Carter and Senter 2008; Teicher and Chari 2011), undergo recycling or renewal back to the cell surface following delivery of the ADC, and have low or no shedding from the cell surface. The molecular targets utilized for the currently approved ADCs do not fulfill the entire list of ideal criteria (Fig. 14.3) (Hammood 2021; Esapa 2023), and instead often rely on unique features that overcome certain disadvantages. For example, the antigen expression of CD33, the molecular target of Mylotarg®, shows low expression density, slow internalization rates, and may be shed from the cell surface (Caron 1994; Abdool 2010). However, its uniform expression on acute myeloid leukemia (AML) blast cells in the majority of patients, coupled with the lack of expression in hematopoietic stem cells, lymphoid cells, and

Table 14.1 FDA-approved ADCs

ADC	Initial year FDA approved	Applicant	Target	Antibody	Linker	Payload	Mechanism of action	Average DAR	Indication(s)
Gemtuzumab ozogamicin (Mylotarg®)	2000 (w), 2017	Wyeth Pfizer	CD33	Humanized IgG4	Acid cleavable (AcBut)	CAL	DNA cleavage	2–3	AML
Brentuximab vedotin (Adcetris®)	2011	Seattle genetics	CD30	Chimeric IgG1	Protease cleavable (vc)	MMAE	Microtubule inhibitor	4	HL, ALCL, pcALCL, MF; cHL, sALCL, PTCL; pediatric cHL
Trastuzumab emtansine (Kadcyla®)	2013	Genentech	HER2	Humanized IgG1	Non-cleavable (SMCC)	DM1	Microtubule inhibitor	3.5	HER2+ mBC; HER2+ EBC
Inotuzumab ozogamicin (Besponsa®)	2017	Pfizer	CD22	Humanized IgG4	Acid cleavable (AcBut)	CAL	DNA cleavage	5–7	ALL; pediatric ALL
Enfortumab vedotin (Padcev®)	2019	Astellas Pharma	Nectin-4	Human IgG1	Protease cleavable (vc)	MMAE	Microtubule inhibitor	3.8	la/mUC
Polatuzumab vedotin (Polivy®)	2019	Genentech	CD79b	Humanized IgG1	Protease cleavable (vc)	MMAE	Microtubule inhibitor	3.5	R/R DLBCL; DLBCL, HGBL
Trastuzumab deruxtecan (Enhertu®)	2019	Daichi Sankyo	HER2	Humanized IgG1	Protease cleavable (GGFG)	DXd	Topoisomerase inhibitor	7–8	HER2+ mBC; HER2+ GC; HER2+ low mBC, HER2 mutant NSCLC; HER2+ solid tumors
Sacituzumab govitecan (Trodelvy®)	2020	Gilead Sciences	TROP-2	Humanized IgG1	Acid cleavable (CL2A)	SN-38	Topoisomerase inhibitor	7–8	R/R TNBC; UC; HR+/HER2- mBC
Loncastuximab tesirine (Zynlonta®)	2021	ADC therapeutics	CD19	Humanized IgG1	Protease cleavable (PEG-va)	PDB dimer	DNA Alkylator	2.3	R/R LBCL, DLBCL

Tisotumab vedotin (Tivdak®)	2021	Seagen	Tissue factor	Human IgG1	Protease cleavable (vc)	MMAE	Microtubule inhibitor	4	r/m CC
Mirvetuximab soravtansine (Elahere®)	2022	ImmunoGen	FRα	Chimeric IgG1	Glutathione cleavable (sulfo-SPDB)	DM4	Microtubule inhibitor	3–5	FRα + PR OC, PR FTC, and PR PPC

Abbreviations: *AcBut* 4-(4-acetylphenoxy)butanoic acid, *ALCL* anaplastic large cell lymphoma, *ALL* acute lymphoblastic leukemia, *AML* acute myeloid leukemia, *CAL* calicheamicin, *cHL* classical Hodgkin lymphoma, *CL2A* acid-sensitive carbonate linker, *DAR* drug-to-antibody ratio, *DLBCL* diffuse large B-cell lymphoma, *DM1* mertansine, *DM4* ravtansine, *DNA* deoxyribonucleic acid, *DXd* deruxtecan, *EBC* early breast cancer, *FRα* folate receptor alpha, *FTC* fallopian tube cancer, *GC* gastric cancer, *HGBL* high-grade B-cell lymphoma, *HL* Hodgkin Lymphoma, *IgG* immunoglobulin, *HER2* human epidermal growth factor receptor 2, *HR* hormone receptor, *la/mUC* locally advanced or metastatic urothelial cancer, *LBCL* large B-cell lymphoma, *mBC* metastatic (late stage) breast cancer, *MF* Mycosis fungiodes, *MMAE* monomethyl auristatin E, *MMAF* monomethyl auristatin F, *Nectin-4* a transmembrane cell adhesion molecule, *NSCLC* non-small-cell lung cancer, *OC* ovarian cancer, *PDB* pyrrolobenzodiazepine, *pcALCL* primary cutaneous anaplastic large cell lymphoma, *PEG-va* polyethylene glycol-valine-alanine linker, *PPC* primary peritoneal cancer, *PR* platinum-resistant, *PTCL* peripheral T-cell lymphomas, *r/m CC* recurrent or metastatic cervical cancer, *R/R* relapsed or refractory, *sALCL* systemic anaplastic large cell lymphoma, *vc* valine-citrulline linker, *SN-38* 7-ethyl-10-hydroxycamptothecin, *SPDB* N-succinimidyl-4-(2-pyridyldithio)butanoate linker, *TNBC* triple-negative breast cancer, *TROP-2* trophoblast cell surface antigen 2, *UC* urothelial cancer, *(w)* withdrawn from US market

Table 14.2 Common ADC targets in FDA-approved drugs

Target	Description	Advantages	Challenges
CD33	Transmembrane glycoprotein expressed on myeloid cells	Overexpressed in AML cells Sorted to lysosomes via CME following internalization Undergoes renewed expression following internalization	Slow internalizing antigen Broad tissue expression (bone marrow, spleen)
CD30	Found on the surface of some white blood cells, including T cells and B cells	Overexpressed on various lymphoid neoplasms, including Hodgkin lymphoma and ALCL Sorted to lysosomes via CME following internalization	Slow internalizing antigen Undergoes shedding from cell surface Broad tissue expression (appendix, testes, lymph node, fallopian tube)
CD22	Restricted expression on the surface of mature B cells and regulates B-lymphocyte function	Abundant expression on malignant B cells Rapid endocytosis Undergoes recycling back to the cell surface following CME Limited shedding	Varied results in studying trafficking to lysosomes following endocytosis
CD79b	During B-cell formation, a heterodimeric complex is formed with CD79A which exhibits diverse signaling properties	Rapid internalization Sorted to lysosomes via CME following internalization Biased expression mainly in lymph nodes and spleen	Low expression levels Rapid antigen loss following treatment has been reported
HER2	Member of the EGFR family and a known driver of human malignancies and metastasis	Despite ubiquitous expression including in the kidney, skin, and small intestine, it exhibits high tumor/normal tissue ratios with high expression levels on tumor cells Rapidly recycled back to the cell surface following internalization	Internalization rates vary but are often considered slow and antibody-dependent Highly shed Dysregulation of HER2-endocytotic machinery has been linked to resistance in preclinical models
Nectin-4	Transmembrane cell adhesion molecule	Overexpressed in several tumor types but nearly absent in normal tissues Trafficked to the lysosomes following CME	Highly shed Internalization rates vary but are often considered slow Molecular mechanisms of oncogenic effects are poorly understood
TROP-2	Membrane glycoprotein that promotes calcium signal transduction	Undergoes constitutive endocytosis and trafficking to lysosomes via CME Despite ubiquitous expression, it has high tumor-to-normal ratios Undergoes recycling back to surface	Undergoes shedding Heterogeneous expression levels across different types and stages of tumors and within different regions of the same tumor

(continued)

Table 14.2 (continued)

Target	Description	Advantages	Challenges
CD19	Member of the immunoglobulin gene superfamily; expression of this cell surface protein is restricted to B-cell lymphocytes	Expressed in B-cell malignancies Not found on hematopoietic stem cells, plasma cells, or other healthy tissues Most widely and homogeneously expressed of the B-cell lymphoma antigens Limited shedding	Internalization rate is affected by various factors and is often considered inefficient
Tissue factor	Enables cells to initiate the blood coagulation cascades and functions as the high-affinity receptor for coagulation factor VI	Exhibits ubiquitous expression but is majorly localized in sequestered intracellular pools until vascular integrity is disrupted or expression induced High turnover and internalization efficiency	Need to avoid interference with the role of tissue factor in coagulation Lower antigen expression levels Limited levels of lysosomal trafficking
Folate receptor alpha	Glycoprotein receptor that binds folic acid and its reduced derivatives, and transports 5-methyltetra-hydrofolate into cells	Lower expression in normal tissues (lung, kidney) and overexpressed in tumor tissues Rapidly recycled back to surface following endocytosis	Highly shed Limited lysosomal enzyme exposure; sorted to late endosomes, which contain glutathione, before recycling back to cell surface, making a disulfide linker useful

Abbreviations: *ALCL* anaplastic large-cell lymphoma, *AML* acute myeloid leukemia, *CME* clathrin-mediated endocytosis, *EGFR* epidermal growth factor receptor, *HER2* human epidermal growth factor receptor 2, *TROP-2* trophoblast cell surface antigen 2

non-hematopoietic cells (Dinndorf et al. 1986; Peiper et al. 1988), helps it to overcome the low expression issue. In addition, CD33 undergoes renewed expression following treatment with gemtuzumab ozogamicin, enabling a sufficient fraction of the ADC to be internalized (Caron 1994). Another example is the molecular target of Kadcyla®, HER2, which has been linked to inefficient endocytosis due to dysregulation of endocytic machinery. However, the high expression levels at the cell surface of breast cancer tissue overcome this issue (Cheng 2024).

Compared to the other important considerations in ADC design, most of the diversity when comparing ADCs in clinical development is their molecular targets, especially in targeting solid tumors, which are more complicated compared to hematological tumors. Since target receptors for solid tumors are normally overexpressed, rather than uniquely expressed, at the target site, dose-limiting toxicity and off-target effects are typically more severe. In comparing the development of approved ADCs, the more recent drugs approved since 2019 target solid tumors (Fig. 14.4). These approved ADCs utilize molecular targets like the folate receptor (mirvetuximab soravtansine, Elahere®), cell adhesion molecules such as Nectin-4

Target	Optimal Criteria				
	Ideal Expression	Rapid Internalization	Lysosome Trafficking	Recycling / Renewal	Low / No Shedding
CD33	✗	✗	✓	✓	✗
CD30	✗	✗	✓	✗	✗
CD22	✓	✓	✗	✓	✓
CD79b	✗	✓	✓	✗	✓
HER2	✓	✗	✓	✓	✗
Nectin-4	✓	✗	✓	✗	✗
TROP-2	✓	✓	✓	✓	✗
CD19	✓	✗	✓	✓	✓
Tissue Factor	✗	✓	✗	✗	✓
Folate Receptor-α	✓	✓	✗	✓	✗

Fig. 14.3 Ideal characteristics of molecular targets utilized for approved ADCs. The ideal criteria for a molecular target of ADCs include ideal expression at the target cell surface (i.e., high expression levels at the tumor site and/or low expression on normal cells), rapid internalization and routing to the lysosomes following ADC binding, recycling or renewal of the target back to the cell surface following delivery of the ADC, and low or limited shedding from the cell surface. The general characteristics of molecular targets that exhibit these ideal characteristics is indicated as a check mark (✓). Details for each target are described in Table 14.2

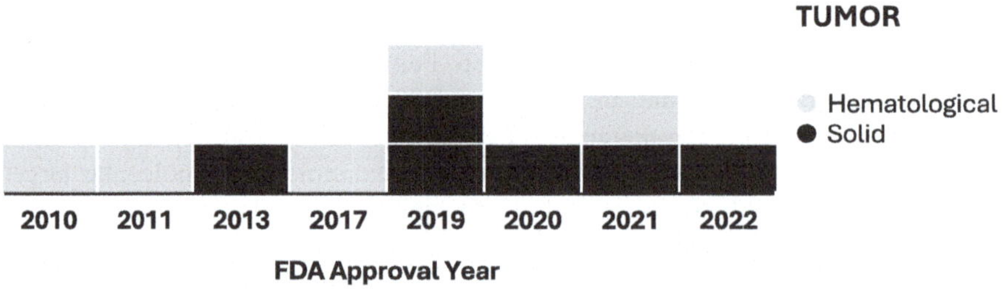

Fig. 14.4 Comparison of types of tumors targeted by approved ADCs. The currently approved ADCs are designed to target hematological tumors (Mylotarg®, Adcetris®, Besponsa®, Polivy®, and Zynlonta®) or solid tumors (Kadcyla®, Padcev®, Trodelvy®, Enhertu®, Tivdak® and Elahere®)

(enfortumab vedotin, Padcev®), and transmembrane glycoproteins such as tissue factor (tisotumab vedotin, Tivdak®). In the clinics, a recent detailed analysis of ongoing trials for ADCs revealed 46 new targets, including 14 for hematological versus 32 for solid tumors (Udofa 2024). Of these targets, the top five being tested most often in the trials target solid tumors and include HER3, epithelial growth factor receptor (EGFR), B7-H3, C-Met, and Claudin 18.2 (Table 14.3).

Table 14.3 Common ADC targets in clinical trials

Target	Description	ADC	Drug MOA	Linker	Example clinical trials	Comments
HER3	HER RTK with roles in tumorigenesis, disease progression and metastasis	U3–1402 (patritumab-GGFG-DXd)	DXd (TOPi)	Protease cleavable (GGFG)	NCT02980341; NCT03260491; NCT04610528; NCT02980341	
EGFR	Amplified of overexpressed in many tumor types	ABT-414	MMAF (MTi)	Non-cleavable (thioether)	NCT02343406, NCT02573324	Enrollment halted since 2019 due to lack of survival benefit
		MRG003	MMAE (MTi)	Protease cleavable (vc)	NCT05188209; NCT04868344, NCT05126719; NCT04868162; NCT04838964; NCT04838548	
		M1231	Hemiasterlin (MTi)	Protease cleavable (vc)	NCT05188209; NCT04868344, NCT05126719; NCT04868162; NCT04838964; NCT04838548	Anti-MUC1/EGFR bispecific ADC
		BB-1705	Eribulin (MTi)	Non-cleavable (isopeptide)	NCT05263869	Residue-specific conjugation technology (RESPECT®)
		AZD9592	n.d. (TOPi)	Cleavable (n.d.)	NCT05647122	Anti-C-Met/EGFR bispecific ADC

(continued)

Table 14.3 (continued)

Target	Description	ADC	Drug MOA	Linker	Example clinical trials	Comments
B7-H3	Immune checkpoint protein; expression in various solid tumors is associated with advanced diseases	DS-7300a	DXd (MTi)	Protease cleavable (GGFG)	NCT04145622	
		MHB088	n.d. (TOPi)	Cleavable (n.d.)	NCT05652855	
C-Met	Transmembrane receptor tyrosine kinase; activated upon binding of hepatocyte growth factor and is dysregulated in various cancer types	ABBV-399 (telisotuzumab vedotin)	MMAE (MTi)	Protease cleavable (vc)	NCT04928846; NCT04662580	
		MYTX-011	MMAE (MTi)	Protease cleavable (vc)	NCT05652868	pH-sensitive antibody utilizing FateControl™ technology
		RC108	MMAE (MTi)	n.d.	NCT05628857	
		BYON3521	Duocarmycin (DNA alkylation)	Protease cleavable (vc)	NCT05323045	ByoZine® and ByonShieLD® conjugation technology
Claudin 18.2	Tight junction protein in normal tissues which is overexpressed in various types of cancer	SOT102	PNU-159682 (TOPi)	Non-cleavable (peptide)	NCT0552586	
		RC118-ADC	MMAE (MTi)	Cleavable (n.d.)	NCT05205850	

Abbreviations: *B7-H3* B7 Homolog 3 Protein, *DXd* deruxtecan, *EGFR* epidermal growth factor receptor, *HER* human epidermal growth factor receptor, *MMAF* Monomethyl auristatin F, *MMAE* Monomethyl auristatin F, *MTi* microtubule inhibitor, *n.d.* not disclosed, *RTK* receptor tyrosine kinase, *TOPi* topoisomerase inhibitor, *vc* Valine-Citrulline protease-sensitive linker

Moving forward, the main hurdles in molecular targets are on answering key questions regarding the required tumor antigen density and heterogeneity, and the acceptable levels of expression in normal tissues and organs. Many ADC studies have attempted to correlate the therapeutic effect with antigen density and show many inconsistencies in the relationship, with results including a linear relationship, lack of correlation, and even an inverse relationship. For example, a study on an anti-CD79b ADC (polatuzumab vedotin, Polivy®) showed that, in cell lines with CD79b expression levels above a minimum threshold, there was a nonlinear correlation of expression levels with in vitro activity (Dornan et al. 2009). On the other hand, studies on the HER2 receptor expression and intracellular ADC exposure have shown a linear relationship in some studies (Maecker 2023) and even an inverse relationship in others (Ram 2014). Taken together, target expression levels may not be the main determinant for ADC efficacy and appears to depend on the ADC target.

For the issue of off-target expression in healthy tissues, important considerations include whether the expression is in vital organs or regenerative tissues, and also the accessibility to the ADC. For instance, expression in bone marrow may represent a manageable toxicity, since B cells and myeloid cells can regenerate in most cases. Another good example is prostate-specific membrane antigen (PMSA), which is expressed on both prostate cancer cells and healthy prostate tissue. Prostatic toxicity is not a serious health concern though, and most prostate cancer patients would have their prostate removed. However, PMSA is also expressed on the apical membrane in kidney and small intestine. While toxicity in these organs would be problematic, the apical membrane of the cells is not easily accessible to ADCs, which are constrained to the basolateral side (Horoszewicz et al. 1987; Israeli et al. 1993). Finally, the tissue factor receptor, the target for tisotumab vedotin (Tivdak®), is widely expressed in fibroblasts, pericytes, and smooth muscle cells but is majorly inaccessible under normal conditions due to localization in sequestered intracellular pools (Mandal 2006). Further understanding of the relationship between expression/specificity of molecular targets and efficacy/side-effect profiles will aid in the future design and development of ADCs.

14.1.2 Antibody Considerations

The antibody component is the important determinant of the targeting efficiency and pharmacokinetics (PK) of the ADC (Fig. 14.1). The main factors related to the antibody component of ADCs are the size, trafficking, intrinsic antibody effector functions, and Fc-receptor binding. Although some differences exist due to factors related to drug conjugation, ADCs typically have low clearance and long half-lives similar to the parent antibody. The long half-life is imparted by the Fc receptor-recycling pathway following binding of the Fc-region of the antibody (Janeway et al. 2001). As will be discussed later, the half-life and clearance of ADCs can differ slightly than the parent antibody component due to the drug-antibody ratio (DAR) and the additional metabolic pathway of drug cleavage.

One of the main considerations for the antibody design is its subclass (i.e., IgG1, IgG2, IgG3, or IgG4), which can affect its intrinsic effector function due to differences in the Fc region. IgG1 and IgG3 subclasses are typically more active in both antibody-dependent cell-mediated cytotoxicity (ADCC) and complement-dependent cytotoxicity (CDC), while IgG4 tends to lack effector functions (Janeway et al. 2001). Engagement of these secondary immune functions is considered beneficial for antitumor activity but may have the unintended effect of reducing tumor localization. IgG1, the most common subclass in ADCs, is used for 9 of the 11 approved ADCs and is the subclass for all the ADCs approved since 2017 (Fig. 14.5a). As expected, the ADCs with the IgG1 subclass have shown ADCC activity (Junttila et al. 2011; Oflazoglu et al. 2007), while ADCs with an IgG4 variant have no ADCC or CDC activity (Bross et al. 2001).

In addition to the subclass of the Ab, the backbone is another important selection factor. The development of hybridoma technology has been crucial to the field of antibody therapeutics, including ADCs. However, administration of murine antibodies to patients may lead to the development of human anti-murine antibodies (HAMAs), causing infusion-related reactions and/or diminishing bioavailability of the ADC (Tjandra 1990). This issue can be mitigated by the use of chimeric

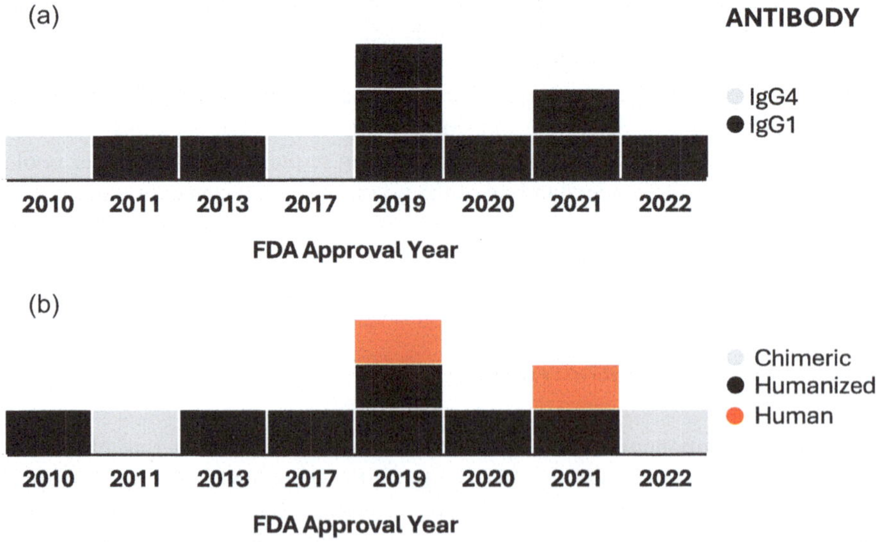

Fig. 14.5 Comparison of types of antibody used in approved ADCs. The currently approved ADCs utilize (**a**) IgG1 (Adcetris®, Kadcyla®, Padcev®, Polivy®, Trodelvy®, Elahere®, Enhertu®, Tivdak®, and Zynlonta®) or IgG4 subtypes (Mylotarg® and Besponsa®), and (**b**) consist of chimeric (Adcetris® and Elahere®), humanized (Besponsa®, Enhertu®, Kadcyla®, Mylotarg®, Polivy®, Trodelvy®, and Zynlonta®), or human (Padcev® and Tivdak®) antibodies

antibodies, used in two out of the 11 FDA-approved ADCs (Adcetris® and Elahere®, Fig. 14.5b), where the constant regions are replaced with the equivalent regions from a human antibody resulting in around 25–30% murine sequences. Antibody humanization, where everything except the complementarity-determining regions is replaced with human antibody sequences, further reduces the murine content to less than 10%. The majority of the FDA-approved ADCs utilize humanized antibodies, including both IgG4 subclass antibodies (Mylotarg® and Besponsa®) along with five IgG1 subclass antibodies (Enhertu®, Kadcyla®, Polivy®, Trodelvy®, and Zynlonta®). The most recent state of the art is fully human antibodies which account for two of the more recently approved ADCs (Padcev® and Tivdak®). The humanized and the fully human antibodies carry a lower risk of inducing an immune response in humans compared to the chimeric antibodies (Hwang and Foote 2005). However, it is important to note that even human antibodies can be immunogenic (Harding 2010).

With regard to future improvements on the antibody component of the ADC, one approach is utilization of smaller antibody fragments (Wu and Senter 2005). These fragments, including "diabodies", "minibodies", and "nanobodies", conceivably offer an advantage of higher tumor penetration due to their smaller size (Wu and Senter 2005) but have not yet been successful in the clinic. One perceivable drawback is a reduction in ADC half-life due to its smaller size. A second approach to improve antibody function, engineering strategies to allow for site-specific modification, will be discussed in the next section.

14.1.3 Chemical Conjugation Strategies

The site of conjugation and DAR both play important roles in the success of the ADC since they can affect the PK/pharmacodynamics (PD), binding affinity and aggregation state. The majority of the ADCs in the clinics utilizes nonspecific conjugation chemistry and attaches the cytotoxin to the ε-amino group of lysine or sulfhydryl group of cysteine (Udova 2024; Lyon et al. 2012). In the case of lysine modification, the drug molecules are distributed across many lysine residues in the antibody with a range of DARs. Modification of cysteine residues involves reduction of disulfide bonds in the hinge region of the antibody and results in an even number DAR. Kadcyla® (3.5 DM1 molecules per antibody) and Adcetris® (four monomethyl auristatin E [MMAE] molecules per antibody) are produced via the lysine modification and cysteine coupling routes, respectively. Of note, one of the HER2-targeting ADCs, trastuzumab deruxtecan (Enhertu®), utilizes a high DAR of 8 in order to treat low-expression HER2. Although conjugation of more drug molecules to the antibody component would result in delivery of a larger payload, it often results in higher aggregation of the final product, as well as poor PK and low-binding affinity compared to the parent antibody. With conjugation, the half-life of the ADC tends to be shorter (several days) than the unmodified antibody (several weeks) (Chari 2008; Hamblett et al. 2004; Senter and Sievers 2012). Therefore, the average DAR is usually maintained at 3:4 in order to preserve the long half-life and high antibody-binding efficiency.

One of the other major drawbacks in the current chemical conjugation strategies is the generation of a heterogeneous final product, with multiple DARs at various sites. It has been calculated that conjugation through modification of lysine residues can distribute the drug molecules to approximately 40 different sites, resulting in $>10^6$ different ADC species (Wang et al. 2005). Conjugation via cysteine residues results in less heterogeneity but can still yield >100 ADC species (Hamblett et al. 2004). Although none have yet reached the clinics, several different companies are focusing on site-specific modification methods to overcome this issue, resulting in the generation of a homogeneously loaded ADC. The major technologies include the use of engineered cysteine residues, incorporation of unnatural amino acids, and enzymatic conjugation as described below.

14.1.3.1 Cysteine-Specific Conjugation

One of the strategies to enable site-specific modification involves the introduction of cysteine residues into the antibody sequence by single-point mutation. This technology developed by Genentech, termed ThioMAb, first utilizes phage-display technology to select a suitable conjugation site. Next, cysteine residues (typically $n = 2$) are introduced to the antibody amino acid sequence by single-point mutation. The drug is then conjugated to the engineered cysteines via a maleimide linker (Junutula et al. 2008; Shen et al. 2012). The ThioMAb technology has also been recently combined with XTEN polypeptide technology to incorporate a series of cysteine residues to be used as conjugation sites. Using this combined technology, homogeneous ADCs with high DAR of up to 18 have been generated (Zacharias 2022).

Another type of cysteine-specific conjugation utilizes REsidue-SPEcific Conjugation Technology (RESPECT®). Here, the unique interchain disulfide bond in a rabbit antibody light chain is exploited, where one of the two cysteine residues is humanized to generate a free cysteine available for conjugation (Albone et al. 2017). This technology is utilized in an anti-EGFR ADC, BB-1705, currently being tested in clinical trials.

14.1.3.2 Incorporation of Unnatural Amino Acids

Two different companies, Sutro Biopharma and Ambrx, are using non-natural amino acid-based approaches to site-specifically modified ADCs. In this approach, unique functional groups not present in natural amino acids can be incorporated and specifically modified using bio-orthogonal chemistries. These techniques overcome limitations of maleimide and disulfide exchange chemistries seen with cysteine modification and do not require further enzymatic processing steps. In Ambrx technology, a *p*-acetylphenylalanine group (pAcPhe) is incorporated, which introduces a keto group that can be selectively coupled to an alkoxy-amine derivatized drug to form a stable oxime bond linkage. Sutro Biopharma technology involves the introduction of a *p*-azidomethyl-*L*-phenylalanine (pAMF) group into the antibody, which can be reacted with the drug molecule via strain-promoted azide-alkyne cycloaddition (SPAAC) click-chemistry (Zimmerman et al. 2014).

14.1.3.3 Enzymatic Conjugation

Innate Pharma is utilizing the enzymatic posttranslational modification to enable precise site-specific conjugation. This technology requires either the removal of *N*-linked glycans from the antibody or the use of mutant aglycosylated variants generated by site-directed mutagenesis. Subsequently, cadaverine-derived drug substrates are conjugated to the antibody using bacterial transglutaminase (BTGase), resulting in the site-specific conjugation (Behrens and Liu 2014).

In addition to these approaches, many other site-specific technologies are being actively developed, including conjugation to native glycan residues (GlycoConnect™ by SynAffix), targeting the cysteine-rich hinge region (ThioBridge by PolyTherics and Stapled ADC technology by Igenica) and the "medtope-enabled" technology by Medtope Biosciences.

14.1.4 Linkers

Linker stability in the ADC must be sufficient enough in systemic circulation to reduce overall toxicity at off-target sites and maintain a prolonged half-life, but it must allow for the release of sufficient free drug at the target tumor site to be efficacious. The linkers utilized for ADCs are classified as either cleavable or non-cleavable (Table 14.1). Cleavable linkers take advantage of the different extracellular versus intracellular properties to release the conjugated drug inside the cells. For non-cleavable linkers, the antibody portion of the ADC must be degraded in order to release the cytotoxic drug molecule.

Of the FDA-approved ADCs, all but one (Kadcyla®) utilize cleavable linkers, which include acid-cleavable (Mylotarg®, Besponsa® and Trodelvy®), protease-cleavable (Adcetris®, Polivy®, Padcev®, Enhertu®, Zynlonta® and Tivdak®), and glutathione-cleavable (Elahere®) linkers (Table 14.1, Fig. 14.5). In addition, most of the new ADCs being tested in the clinic also use cleavable linkers (Udofa 2024). The most common type of cleavable linkers used is the protease-sensitive type (Table 14.1, Fig. 14.6). These linkers take advantage of the specific localization of lysosomal enzymes (e.g., cathepsin or plasmin) present at elevated levels in certain tumor tissues (Dubowchik and Firestone 1998). These proteases are also usually only active in the moderately acidic environment of the lysosome, offering an additional layer of specificity. Most ADCs with protease-sensitive linkages utilize a Seattle Genetics (Seagen)–developed maleimido-containing valine-citrulline

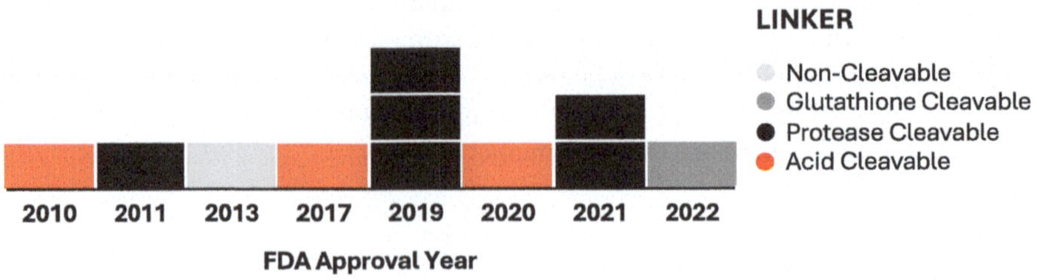

Fig. 14.6 Comparison of types of linkers used in approved ADCs. The currently approved ADCs utilize non-cleavable (thioester) linkers (Kadcyla®) or cleavable linkers, including glutathione-cleavable (Elahere®), protease-cleavable (Adcetris®, Polivy®, Padcev®, Enhertu®, Zynlonta® and Tivdak®), and acid-cleavable (Mylotarg®, Besponsa® and Trodelvy®) linkers

(val-cit or vc) linker, cleavable by cathepsin B. Other protease-sensitive linkers include the valine-alanine (val-ala or va) linker, also cleavable by cathepsin B, and the more recently developed glycine-glycine-phenylalanine-glycine (GGFG) tetra-peptide linker (as in Enhertu®), which is cathepsin B- and L-cleavable. In Zynlonta®, the val-ala linker is combined with a polyethylene glycol (PEG) spacer. This linker was designed to evade multidrug resistance (MDR), which has been linked with poor clinical responses in ADCs (Walter et al. 2003; Linenberger 2005). The results showed that the PEG-linked conjugates were more potent in MDR1-expressing cells, presumably due to the increased hydrophilicity of the PEG-linked DM1, which keeps the drug inside the cell (Kovtun et al. 2010).

The second most common cleavable linker utilized in the approved ADCs is the acid-sensitive linker (Table 14.1, Fig. 14.6). Following RME, the ADCs are exposed to a mild to moderate acidic environment in the endosomes and lysosomes (Fig. 14.2). Taking advantage of this condition, several ADCs have been made utilizing pH-sensitive linkers, such as 4-(4-acetylphenoxy) butanoic acid (AcBut, Mylotarg® and Besponsa®) or CL2A (Trodelvy®), to conjugate the drug to the antibody via a hydrazone or carbonate bond, respectively.

The most recently approved ADC, Elahere®, utilizes a disulfide linker which exploits the higher concentration of glutathione inside cells relative to the blood stream. These linkers are relatively stable in circulation and can be further stabilized by introducing steric hindrance by flanking disulfide bonds with methyl groups (Sapra et al. 2011). Of note, this linker is combined with targeting of the folate receptor, which has been shown to have limited lysosomal exposure but is sorted to late endosomes that contain glutathione. Therefore, the use of a glutathione-sensitive linker, as opposed to a lysosomal protease-sensitive linker, may be optimal for this drug.

The ADCs generated with non-cleavable linkers mainly utilize a thioether bond to link the drug to the antibody. ADCs with non-cleavable linkers are found to be better tolerated, presumably due to a higher stability in systemic circulation. One main consideration of utilizing a non-cleavable linker is whether or not the cytotoxic drug is still active despite being chemically modified. For example, Kadcyla®, linked via a non-cleavable thioether linkage, releases the free drug as a lysine-MCC-DM1 analog, which has similar potency to DM1 (LoRusso et al. 2011).

Despite the numerous advances in linker technology, there is no universal formula in their application to ADCs. Most studies have shown that the linker function is highly dependent on the main components of the ADC (the drug and antibody), and on the target site. Therefore, each antibody must be optimized on an individual basis to determine the most appropriate linker. Future improvements in linker technology focus on optimizing stability in circulation, using more hydrophilic linkers to improve pharmacokinetic properties and using branched linkers to improve the drug load.

14.1.5 *Cytotoxic Payloads*

There are four main classes of drugs used in ADCs, antimitotic drugs that disrupt microtubule assembly and play an important role in mitosis (microtubule inhibitors or MTis, e.g., maytansine analogs DM1 and DM4, auristatin analogs MMAE and monomethyl auristatin F [MMAF]), drugs that bind to the minor groove of DNA causing DNA double-strand cleavage (DNA cleavage, e.g., calicheamicin analogs), drugs that are DNA minor groove alkylators that crosslink DNA strands together (DNA alkylators, e.g., pyrrolobenzodiazepine dimers), and drugs that prevent DNA topoisomerases from unwinding DNA strands, creating DNA breaks leading to cell death (topoisomerase inhibitors or TOPis, e.g., deruxtecan and SN-38). The microtubule inhibitors are most commonly used in approved drugs (Table 14.1, Fig. 14.7), and also have an additional layer of specificity due to their preferential destruction of highly proliferating cells and to the increased sensitivity of malignant cells to mitotic catastrophe. However, one of the main drawbacks of the microtubule inhibitors is that they are much less effective against static cancer cells (Florian 2016).

More recent classes of drugs include DNA damaging agents such as topoisomerase inhibitors (Trodelvy® and Enhertu®) and the DNA alkylators (Zynlonta®). These molecules destroy DNA structures and can target the whole cell cycle. Topoisomerase inhibitors have recently been investigated clinically as a promising class of

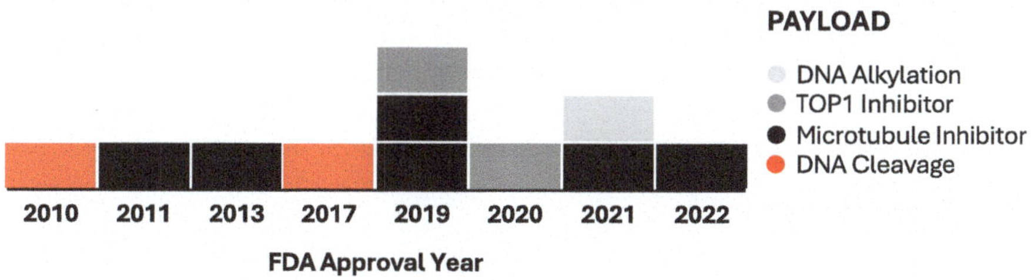

Fig. 14.7 Comparison of types of cytotoxic payload used in approved ADCs. The currently approved ADCs utilize cytotoxic payloads with mechanisms including DNA alkylation (Zynlonta®), topoisomerase 1 (TOP1) inhibition (Trodelvy® and Enhertu®), microtubule inhibition (Adcetris®, Elahere®, Kadcyla®, Padcev®, Polivy®, and Tivdak®), or DNA cleavage (Mylotarg® and Besponsa®)

anticancer agents. However, these drugs have many inherent obstacles including chemical instability and short plasma half-lives (DeMan 2018). The use of ADCs to target the delivery of topoisomerase inhibitors has helped overcome these limitations in maximizing their potential to treat cancer. Compared to other ADC payloads, TOPis have moderate potency compared to other ADC payloads (Nakada 2016). In order to overcome this limitation, TOPi-ADCs, including Trodelvy® (SN-38) and Enhertu® (DXd), utilize higher DARs of 7:8 to carry more cytotoxic payloads to the target cells. The DNA alkylators, such as pyrrolobenzodiazepine (PBD) dimers, are naturally occurring antitumor antibiotics (e.g. anthramycin, sibiromycin) produced by *Streptomyces* that initiate cell damage through DNA cross-linking, leading to cell death (Hartley and Hochhauser 2012; Smellie et al. 2003). Some PBD analogs are more potent than currently used drugs in ADCs and are also less sensitive to the P-glycoprotein (Pgp)-mediated drug resistance (Aird et al. 2008; Guichard et al. 2005). Pgp-mediated drug resistance has been linked with poor clinical responses in ADCs (Walter et al. 2003; Linenberger 2005); therefore, overcoming this MDR mechanism is important. Although ADCs are targeted toward the tumor site, it is estimated that only 1–2% of the administered dose will accumulate in cancer cells (Teicher and Chari 2011), indicating the requirement for high drug potency. Therefore, all classes of ADC drugs have a much higher potency (100- to 1000-fold) compared to conventional chemotherapy agents such as doxorubicin and paclitaxanes (Pietersz et al. 1994; Tolcher et al. 1999).

Although there is a great deal of variability in the antigen/antibody and linker components of ADCs, the types of drugs used are somewhat limited. Developing new drugs suitable for ADCs is a difficult task. In order to work with the ADC, the drug needs to have a high cytotoxicity, or the DAR needs to be increased to compensate for lower potency, the target of the drug needs to be inside the cell, its size should be small to reduce the risk of immunogenicity and issues with aqueous solubility, it must be stable in circulation, and it should have an available linkage site in order to conjugate it to the antibody. Therefore, there is still plenty of room to grow in the identification of new, suitable drugs for ADCs. Avenues of development in the payload include prodrug-based payloads to mitigate off-tumor toxicity, creation of hydrophilic cytotoxic payloads to retain potency with better pharmacokinetic properties, improved conjugation strategies to allow for higher DARs without diminishing physical stability or targeting efficiency of the antibody, and utilization of bifunctional payloads to induce greater antitumor activity.

14.2 Conclusion

After several decades of trials and errors, ADCs as the magic bullets for cancer treatment have become a reality, forming the newest class of cancer therapeutics. With eleven FDA-approved drugs performing well in the clinic and more than 500 in various stages of clinical investigations, the future of ADC drug discovery and development look extremely bright. The paradigm of cancer treatment might shift to antigen identification and targeting from the current one based on the cancer origin of organ. This will facilitate the individualization or personalization of cancer therapy with greater efficacy and reduced side effects. Optimization of the payload, the targeting antibody, and the linker technology will generate the next generation of ADCs, with greater drug delivery efficiency, desirable adverse reaction profile and reduced chance of drug resistance and cancer metastasis.

References

Abdool A, Yeh CH, Kantarjian H, O'Brien S, Bruey JM, Giles F, Albitar M (2010) Circulating CD33 and its clinical value in acute leukemia. Exp Hematol 38(6):462–471

Aird RE, Thomson M, Macpherson JS, Thurston DE, Jodrell DI, Guichard SM (2008) ABCB1 genetic polymorphism influences the pharmacology of the new pyrrolobenzodiazepine derivative SJG-136. Pharmacogenomics J 8(4):289–296. https://doi.org/10.1038/sj.tpj.6500465

Albone EF, Spidel JL, Cheng X, Park YC, Jacob S, Milinichik AZ, Vaessen B, Butler J, Kline JB, Grasso L (2017) Generation of therapeutic immunoconjugates via Residue-Specific Conjugation Technology (RESPECT) utilizing a native cysteine in the light chain framework of Oryctolagus cuniculus. Cancer Biol Ther 18(5):347–357. https://doi.org/10.1080/15384047.2017.1312232

Behrens CR, Liu B (2014) Methods for site-specific drug conjugation to antibodies. MAbs 6:46. https://doi.org/10.4161/mabs.26632

Bretz T (2024) Antibody drug conjugates market poised for unprecedented growth through 2032. https://wiredrelease.com/press-release/antibody-drug-conjugates-market-poised-for-unprecedented-growth-through-2032-market-us-2024-05-20

Bross PF, Beitz J, Chen G, Chen XH, Duffy E, Kieffer L, Roy S, Sridhara R, Rahman A, Williams G, Pazdur R (2001) Approval summary: gemtuzumab ozogamicin in relapsed acute myeloid leukemia. Clin Cancer Res 7(6):1490–1496

Caron P, Schwartz MA, Sung M, Queen C, Finn R, Graham M, Divgi C, Larson S, Scheinberg DA (1994) Murine and humanized constructs of monoclonal antibody M195 (anti-CD33) for the therapy of acute myelogenous leukemia. Cancer 73(7):1049–1056

Carter PJ, Senter PD (2008) Antibody-drug conjugates for cancer therapy. Cancer J 14(3):154–169. https://doi.org/10.1097/PPO.0b013e318172d704

Chari RV (2008) Targeted cancer therapy: conferring specificity to cytotoxic drugs. Acc Chem Res 41(1):98–107. https://doi.org/10.1021/ar700108g

Cheng X (2024) A comprehensive review of HER2 in cancer biology and therapeutics. Genes 15(7):903

De Man FM, Goey AKL, van Schaik RHN, Mathijssen RHJ, Bins S (2018) Individualization of irinotecan treatment: a review of pharmacokinetics, pharmacodynamics, and pharmacogenetics. Clin Pharmacol 57:1229–1254

Dinndorf PA, Andrews RG, Benjamin D, Ridgway D, Wolff L, Bernstein ID (1986) Expression of normal myeloid-associated antigens by acute leukemia cells. Blood 67(4):1048–1053

Dornan D, Bennett F, Chen Y, Dennis M, Eaton D, Elkins K, French D, Go MA, Jack A, Junutula JR, Koeppen H, Lau J, McBride J, Rawstron A, Shi X, Yu N, Yu SF, Yue P, Zheng B, Ebens A, Polson AG (2009) Therapeutic potential of an anti-CD79b antibody-drug conjugate, anti-CD79b-vc-MMAE, for the treatment of non-Hodgkin lymphoma. Blood 114(13):2721–2729. https://doi.org/10.1182/blood-2009-02-205500

Dubowchik GM, Firestone RA (1998) Cathepsin B-sensitive dipeptide prodrugs. 1. A model study of structural requirements for efficient release of doxorubicin. Bioorg Med Chem Lett 8 (23):3341-3346

Esapa B, Jiang J, Cheung A, Chenoweth A, Thurston DE, Karagiannis SN (2023, 1845) Target antigen attributes and their contributions to clinically approved antibody-drug conjugates (ADCs) in haematopoietic and solid cancers. Cancers 16(6)

Florian S, Mitchison TJ (2016) Anti-Microtubule drugs. Methods Mol Biol 1413:403–421

Guichard SM, Macpherson JS, Thurston DE, Jodrell DI (2005) Influence of P-glycoprotein expression on in vitro cytotoxicity and in vivo antitumour activity of the novel pyrrolobenzodiazepine dimer SJG-136. Eur J Cancer 41(12):1811–1818. https://doi.org/10.1016/j.ejca.2005.04.026

Hamblett KJ, Senter PD, Chace DF, Sun MM, Lenox J, Cerveny CG, Kissler KM, Bernhardt SX, Kopcha AK, Zabinski RF, Meyer DL, Francisco JA (2004) Effects of drug loading on the antitumor activity of a monoclonal antibody drug conjugate. Clin Cancer Res 10(20):7063–7070. https://doi.org/10.1158/1078-0432.CCR-04-0789

Hammood M, Craig AW, Leyton JV (2021) Impact of endocytosis mechanisms for the receptors targeted by the currently approved antibody-drug conjugates (ADCs)—a necessity for future ADC research and development. Pharmaceuticals 14(7):674

Harding FA, Stickler MM, Razo J, DuBridge RB (2010) MoAbs 2(3):256–265

Hartley JA, Hochhauser D (2012) Small molecule drugs—optimizing DNA damaging agent-based therapeutics. Curr Opin Pharmacol 12(4):398–402. https://doi.org/10.1016/j.coph.2012.03.008

Horoszewicz JS, Kawinski E, Murphy GP (1987) Monoclonal antibodies to a new antigenic marker in epithelial prostatic cells and serum of prostatic cancer patients. Anticancer Res 7(5B):927–935

Hwang WY, Foote J (2005) Immunogenicity of engineered antibodies. Methods 36:3–10

Israeli RS, Powell CT, Fair WR, Heston WD (1993) Molecular cloning of a complementary DNA encoding a prostate-specific membrane antigen. Cancer Res 53(2):227–230

Janeway C, Travers P, Walport M, Shlomchik M (2001) Immunobiology. The immune system in health and disease. Garland Science, New York. (add page numbers?)

Junttila TT, Li G, Parsons K, Phillips GL, Sliwkowski MX (2011) Trastuzumab-DM1 (T-DM1) retains all the mechanisms of action of trastuzumab and efficiently inhibits growth of lapatinib insensitive breast cancer. Breast Cancer Res Treat 128(2):347–356. https://doi.org/10.1007/s10549-010-1090-x

Junutula JR, Raab H, Clark S, Bhakta S, Leipold DD, Weir S, Chen Y, Simpson M, Tsai SP, Dennis MS, Lu Y, Meng YG, Ng C, Yang J, Lee CC, Duenas E, Gorrell J, Katta V, Kim A, McDorman K, Flagella K, Venook R, Ross S, Spencer SD, Lee Wong W, Lowman HB, Vandlen R, Sliwkowski MX, Scheller RH, Polakis P, Mallet W (2008) Site-specific conjugation of a cytotoxic drug to an antibody improves the therapeutic index. Nat Biotechnol 26(8):925–932. https://doi.org/10.1038/nbt.1480

Kovtun YV, Audette CA, Mayo MF, Jones GE, Doherty H, Maloney EK, Erickson HK, Sun X, Wilhelm S, Ab O, Lai KC, Widdison WC, Kellogg B, Johnson H, Pinkas J, Lutz RJ, Singh R, Goldmacher VS, Chari RV (2010) Antibody-maytansinoid conjugates designed to bypass multidrug resistance. Cancer Res 70(6):2528–2537. https://doi.org/10.1158/0008-5472.CAN-09-3546

Linenberger ML (2005) CD33-directed therapy with gemtuzumab ozogamicin in acute myeloid leukemia: progress in understanding cytotoxicity and potential mechanisms of drug resistance. Leukemia 19(2):176–182. https://doi.org/10.1038/sj.leu.2403598

LoRusso PM, Weiss D, Guardino E, Girish S, Sliwkowski MX (2011) Trastuzumab emtansine: a unique antibody-drug conjugate in development for human epidermal growth factor receptor 2-positive cancer. Clin Cancer Res 17(20):6437–6447. https://doi.org/10.1158/1078-0432.CCR-11-0762

Lyon RP, Meyer DL, Setter JR, Senter PD (2012) Conjugation of anticancer drugs through endogenous monoclonal antibody cysteine residues. Methods Enzymol 502:123–138. https://doi.org/10.1016/B978-0-12-416039-2.00006-9

Maecker H, Jonnalagadda V, Bhakta S, Jammalamadaka V, Junutula J (2023) Exploration of the antibody–drug conjugate clinical landscape. mAbs 15(1). https://doi.org/10.1080/19420862.2023.2229101

Mandal SK, Pendurthi UR, Rao LVM (2006) Cellular localization and trafficking of tissue factor. Blood 107(12):4746–4753. https://doi.org/10.1182/blood-2005-11-4674

Nakada T, Masuda T, Naita H, Yoshida M, Ashida S, Morita K, Miyazaki H, Kasuya Y, Ogitani Y, Yamaguchi J, Abe Y, Honda T (2016) Novel antibody drug conjugates containing exatechan derivative-based cytotoxic payloads. Bioorganic and Medicinal Chemistry Letters 26:1542–1545

Oflazoglu E, Stone IJ, Gordon KA, Grewal IS, van Rooijen N, Law CL, Gerber HP (2007) Macrophages contribute to the antitumor activity of the anti-CD30 antibody SGN-30. Blood 110(13):4370–4372. https://doi.org/10.1182/blood-2007-06-097014

Peiper SC, Ashmun RA, Look AT (1988) Molecular cloning, expression, and chromosomal localization of a human gene encoding the CD33 myeloid differentiation antigen. Blood 72(1):314–321

Pietersz GA, Rowland A, Smyth MJ, McKenzie IF (1994) Chemoimmunoconjugates for the treatment of cancer. Adv Immunol 56:301–387

Ram S, Kim D, Ober RJ, Ward ES (2014) The level of HER2 expression is a predictor of antibody-HER2 trafficking behavior in cancer cells. mAbs 6(5):1211–1219 https://doi.org/10.4161/mabs.29865

Sapra P, Hooper AT, O'Donnell CJ, Gerber HP (2011) Investigational antibody drug conjugates for solid tumors. Expert Opin Investig Drugs 20(8):1131–1149. https://doi.org/10.1517/13543784.2011.582866

Senter PD, Sievers EL (2012) The discovery and development of brentuximab vedotin for use in relapsed Hodgkin lymphoma and systemic anaplastic large cell lymphoma. Nat Biotechnol 30(7):631–637. https://doi.org/10.1038/nbt.2289

Shen BQ, Xu K, Liu L, Raab H, Bhakta S, Kenrick M, Parsons-Reponte KL, Tien J, Yu SF, Mai E, Li D, Tibbitts J, Baudys J, Saad OM, Scales SJ, McDonald PJ, Hass PE, Eigenbrot C, Nguyen T, Solis WA, Fuji RN, Flagella KM, Patel D, Spencer SD, Khawli LA, Ebens A, Wong WL, Vandlen R, Kaur S, Sliwkowski MX, Scheller RH, Polakis P, Junutula JR (2012) Conjugation site modulates the in vivo stability and therapeutic activity of antibody-drug conjugates. Nat Biotechnol 30(2):184–189. https://doi.org/10.1038/nbt.2108

Smellie M, Bose DS, Thompson AS, Jenkins TC, Hartley JA, Thurston DE (2003) Sequence-selective recognition of duplex DNA through covalent interstrand cross-linking: kinetic and molecular modeling studies with pyrrolobenzodiazepine dimers. Biochemistry 42(27):8232–8239. https://doi.org/10.1021/bi034313t

Teicher BA, Chari RV (2011) Antibody conjugate therapeutics: challenges and potential. Clin Cancer Res 17(20):6389–6397. https://doi.org/10.1158/1078-0432.CCR-11-1417

Tjandra JJ, Ramadi L, McKenzie IF (1990) Development of human anti-murine antibody (HAMA) response in patients. Immune Cell Biol 68:367–376

Tolcher AW, Sugarman S, Gelmon KA, Cohen R, Saleh M, Isaacs C, Young L, Healey D, Onetto N, Slichenmyer W (1999) Randomized phase II study of BR96-doxorubicin conjugate in patients with metastatic breast cancer. J Clin Oncol 17(2):478–484

Udofa E, Sankholkar D, Mitragotri S, Zhao Z (2024) Antibody drug conjugates in the clinic. Bioeng Transl Med 9:e10677. https://doi.org/10.1002/btm2.10677

Walter RB, Raden BW, Hong TC, Flowers DA, Bernstein ID, Linenberger ML (2003) Multidrug resistance protein attenuates gemtuzumab ozogamicin-induced cytotoxicity in acute myeloid leukemia cells. Blood 102(4):1466–1473. https://doi.org/10.1182/blood-2003-02-0396

Wang L, Amphlett G, Blattler WA, Lambert JM, Zhang W (2005) Structural characterization of the maytansinoid-monoclonal antibody immunoconjugate, huN901-DM1, by mass spectrometry. Protein Sci 14(9):2436–2446. https://doi.org/10.1110/ps.051478705

Wu AM, Senter PD (2005) Arming antibodies: prospects and challenges for immunoconjugates. Nat Biotechnol 23(9):1137–1146. https://doi.org/10.1038/nbt1141

Zacharias N, Podust V, Kajihara K, Leipold D, Rosario G, Thayer D, Dong D, Paluch M, Fischer D, Zheng K, Lei C et al (2022) A homogeneous high-DAR antibody–drug conjugate platform combining THIOMAB antibodies and XTEN polypeptides. Chem Sci 13:3147
Zimmerman ES, Heibeck TH, Gill A, Li X, Murray CJ, Madlansacay MR, Tran C, Uter NT, Yin G, Rivers PJ, Yam AY, Wang WD, Steiner AR, Bajad SU, Penta K, Yang W, Hallam TJ, Thanos CD, Sato AK (2014) Production of site-specific antibody-drug conjugates using optimized non-natural amino acids in a cell-free expression system. Bioconjug Chem 25(2):351–361. https://doi.org/10.1021/bc400490z

Index

J. L. Zaro, J. Wang (eds.), *Antibody-Drug Conjugates*, AAPS Advances in the Pharmaceutical Sciences Series 17, https://doi.org/10.1007/978-3-031-83005-1

The manufacturer's authorised representative in the EU is Springer Nature Customer Service Centre GmbH, Europaplatz 3, 69115 Heidelberg, Germany. If you have any concerns regarding our products, please contact ProductSafety@springernature.com

Printed and bound by CPI Group (UK) Ltd, Croydon, CR0 4YY
01/10/2026
02225928-0005